BEHAVIORAL MODELING AND SIMULATION

FROM INDIVIDUALS TO SOCIETIES

Committee on Organizational Modeling: From Individuals to Societies

Greg L. Zacharias, Jean MacMillan, and Susan B. Van Hemel, *Editors*

Board on Behavioral, Cognitive, and Sensory Sciences
Division of Behavioral and Social Sciences and Education

NATIONAL RESEARCH COUNCIL
OF THE NATIONAL ACADEMIES

THE NATIONAL ACADEMIES PRESS
Washington, D.C.
www.nap.edu

THE NATIONAL ACADEMIES PRESS 500 Fifth Street, N.W. Washington, DC 20001

NOTICE: The project that is the subject of this report was approved by the Governing Board of the National Research Council, whose members are drawn from the councils of the National Academy of Sciences, the National Academy of Engineering, and the Institute of Medicine. The members of the committee responsible for the report were chosen for their special competences and with regard for appropriate balance.

The study was supported by Award No. FA8650-04-2-6542 between the National Academy of Sciences and the U.S. Department of the Air Force. Any opinions, findings, conclusions, or recommendations expressed in this publication are those of the author(s) and do not necessarily reflect the view of the organizations or agencies that provided support for this project.

Library of Congress Cataloging-in-Publication Data

Behavioral modeling and simulation : from individuals to societies / Committee on Organizational Modeling:From Individuals to Societies ; Greg L. Zacharias, Jean MacMillan, and Susan B. Van Hemel, editors ; Board on Behavior, Cognitive, and Sensory Sciences, Division of Behavioral and Social Sciences and Education.
    p. cm.
  Includes bibliographical references.
  ISBN 978-0-309-11862-0 (pbk.) — ISBN 978-0-309-11863-7 (pdf) 1. Psychology, Military. 2. Sociology, Military. 3. Human behavior—Simulation methods. 4. Organizational behavior—Simulation methods. I. Zacharias, Greg. II. MacMillan, Jean. III. Van Hemel, Susan B. IV. National Research Council (U.S.). Committee on Organizational Modeling: From Individuals to Societies.
  U22.3.B44 2008
  355.001'9—dc22
                           2008019733

Additional copies of this report are available from the National Academies Press, 500 Fifth Street, N.W., Lockbox 285, Washington, DC 20055; (800) 624-6242 or (202) 334-3313 (in the Washington metropolitan area); Internet, http://www.nap.edu.

Suggested citation: National Research Council. (2008). *Behavioral Modeling and Simulation: From Individuals to Societies*. Committee on Organizational Modeling: From Individuals to Societies, Greg L. Zacharias, Jean MacMillan, and Susan Van Hemel, editors. Board on Behavioral, Cognitive, and Sensory Sciences, Division of Behavioral and Social Sciences and Education. Washington, DC: The National Academies Press.

THE NATIONAL ACADEMIES
Advisers to the Nation on Science, Engineering, and Medicine

The **National Academy of Sciences** is a private, nonprofit, self-perpetuating society of distinguished scholars engaged in scientific and engineering research, dedicated to the furtherance of science and technology and to their use for the general welfare. Upon the authority of the charter granted to it by the Congress in 1863, the Academy has a mandate that requires it to advise the federal government on scientific and technical matters. Dr. Ralph J. Cicerone is president of the National Academy of Sciences.

The **National Academy of Engineering** was established in 1964, under the charter of the National Academy of Sciences, as a parallel organization of outstanding engineers. It is autonomous in its administration and in the selection of its members, sharing with the National Academy of Sciences the responsibility for advising the federal government. The National Academy of Engineering also sponsors engineering programs aimed at meeting national needs, encourages education and research, and recognizes the superior achievements of engineers. Dr. Charles M. Vest is president of the National Academy of Engineering.

The **Institute of Medicine** was established in 1970 by the National Academy of Sciences to secure the services of eminent members of appropriate professions in the examination of policy matters pertaining to the health of the public. The Institute acts under the responsibility given to the National Academy of Sciences by its congressional charter to be an adviser to the federal government and, upon its own initiative, to identify issues of medical care, research, and education. Dr. Harvey V. Fineberg is president of the Institute of Medicine.

The **National Research Council** was organized by the National Academy of Sciences in 1916 to associate the broad community of science and technology with the Academy's purposes of furthering knowledge and advising the federal government. Functioning in accordance with general policies determined by the Academy, the Council has become the principal operating agency of both the National Academy of Sciences and the National Academy of Engineering in providing services to the government, the public, and the scientific and engineering communities. The Council is administered jointly by both Academies and the Institute of Medicine. Dr. Ralph J. Cicerone and Dr. Charles M. Vest are chair and vice chair, respectively, of the National Research Council.

www.national-academies.org

Acknowledgments

This report is the result of over 3 years of effort by a committee of 13 experts. The study was performed at the request of the United States Air Force. The committee gathered and reviewed literature on human behavior modeling efforts, listened to briefings and presentations by military modelers and model users, and, using this information and its combined expertise, has attempted to provide the Air Force with its best advice on planning for future organizational modeling research.

Members of the study committee, volunteers selected from several academic and professional practice specialties, found the project an interesting and stimulating opportunity for interdisciplinary collaboration. They cooperated in work groups, learned each other's technical languages, and exemplified in their work the collegial qualities that are among the National Academies' unique strengths. We are grateful to them for their hard work, expertise, and good humor.

On behalf of the committee, we would like to express our appreciation to the many other people who contributed to this project. Janet Miller at the Air Force Research Laboratory served as project monitor and provided guidance as needed. Michael Young at the Air Force Research Laboratory and John Tangney at the Air Force Office of Scientific Research (now at the Office of Naval Research) were also helpful in supporting our work. John Allen, Ted Fichtl, Alex Kott, Alex Levis, Robert Popp, and William Rouse served as unpaid consultants and provided briefings that helped the committee understand the role of organizational modeling in military applications and the needs that such modeling could fill.

At the National Research Council (NRC), Susan Van Hemel, study director for the project, Christine Hartel, director of the Center for Studies of Behavior and Development, and Anne Mavor, director of the Committee on Human Factors, provided support for the project. Two senior program assistants, Jessica Martinez and Kristin Martin, provided administrative and logistic support over the course of the study. Kristen A. Butler served as research assistant and did extensive manuscript preparation work. The executive office reports staff of the Division of Behavioral and Social Sciences and Education, especially Christine McShane and Yvonne Wise, provided valuable help with editing and production of the report. Kirsten Sampson-Snyder managed the report review process.

This report has been reviewed in draft form by individuals chosen for their diverse perspectives and technical expertise, in accordance with procedures approved by the Report Review Committee of the NRC. The purpose of this independent review is to provide candid and critical comments that will assist the institution in making the published report as sound as possible and to ensure that the report meets institutional standards for objectivity, evidence, and responsiveness to the study charge. The review comments and draft manuscript remain confidential to protect the integrity of the deliberative process.

We thank the following individuals for their participation in the review of this report: Robert Axelrod, Gerald R. Ford School of Public Policy, University of Michigan; Stephen J. DeCanio, University of California, Santa Barbara Washington Program, Washington, DC; Larry Hirschhorn, Center for Applied Research, Inc., Philadelphia, PA; Daniel R. Ilgen, Psychology and Management, Michigan State University; Marco A. Janssen, School of Human Evolution and Social Change, School of Computing and Informatics Center for the Study of Institutional Diversity, Arizona State University; Michael Prietula, Information Systems and Operations Management, Emory University; and Amy Pritchett, School of Aerospace Engineering, H. Milton Stewart School of Industrial and Systems Engineering, Georgia Institute of Technology.

Although the reviewers listed above have provided many constructive comments and suggestions, they were not asked to endorse the conclusions or recommendations, nor did they see the final draft of the report before its release. The review of this report was overseen by R. Duncan Luce, Institute for Mathematical Behavioral Science, University of California, Irvine, as review coordinator. Appointed by the National Research Council, he was responsible for making sure that an independent examination of this report was carried out in accordance with institutional procedures and that all reviewers' comments were considered carefully. Responsibility for the final content of this report, however, rests entirely with the authoring committee and the institution.

Greg L. Zacharias and Jean MacMillan,
Cochairs
Committee on Organizational Modeling:
From Individuals to Societies

Contents

Executive Summary

The Air Force and the other military services are increasingly interested in using models of the behavior of humans, as individuals and in groups of various kinds and sizes, to support the development of doctrine, strategies, and tactics for dealing with state and nonstate adversaries, for use in analysis of the current political and military situation, for planning future operations, for training and mission rehearsal, and even for the acquisition of new systems. In this report we refer to this broad class of models as individual, organizational, and societal (IOS) models. There are many lines of research on such models, which span several disciplines, have different goals, and often use different terminologies.

The National Research Council was asked by the U.S. Air Force to review relevant IOS modeling research programs in the various research communities, evaluate the strengths and weaknesses of the programs and their methodologies, determine which have the greatest potential for military use, and provide guidance for the design of a research program to effectively foster the development of IOS models useful to the military. The formal statement of task for the study includes the following specific items:

- Review the state of the art of the subset of the social sciences perceived as having the greatest payoff in terms of informing future computational model developments.
- Review the state of the art in societal[1] modeling applications serving

---

[1] In this study, the committee broadened the scope to include individual and organizational models as well because of the inseparability of all three, given the intended usage.

1

the U.S. Department of Defense (DoD) and related agencies, with special emphasis given to computational modeling and simulation-based approaches.

- Review the state of the art in the three computational modeling communities outside DoD (cognitive science and individual behavioral modeling, network analysis and multiagent organizational modeling, and multiresolution modeling and simulation) and identify strengths and shortcomings in each.

- Identify how gaps in societal behavioral modeling applications serving DoD and related agencies might be filled by conceptual models in the social sciences; computational modeling approaches now under way in the social science community; and closer linkages between the cognitive science community, the network/organizational modeling community, and the multiresolution modeling and simulation community.

- Develop a research and development roadmap to fill current application gaps, for the near, mid-, and far term.

Today's military missions have shifted away from force-on-force warfare—fighting nation-states using conventional weapons—toward combating insurgents and terrorist networks in a battlespace in which the attitudes and behaviors of civilian noncombatants may be the primary effects of military actions. These new missions call for agile, indigenously sensitive forces capable of switching quickly and effectively from conventional combat to humanitarian assistance and able to defuse tense situations without, if possible, the use of force. IOS models are greatly needed for planning, supporting, and training for these forces and for evaluating the technology with which they fight. Models of human behavior in social units—teams, organizations, cultural and ethnic groups, and societies—are needed to understand, predict, and influence the behavior of these social units.

For example, models could be used to predict the effects of actions intended to disrupt terrorist networks, to predict the response of insurgents and the local population to the presence of friendly forces in a given area, or to predict the effects of alternative diplomatic, military, and economic courses of action on the attitudes and behaviors of the population in a region of interest. Models could also be used in training and mission rehearsal to create simulation environments in which military units could, for example, experience the effects of their actions on the (simulated) behavior of a crowd that might either disperse or turn hostile. Models could also be used to evaluate the likely results of proposed changes intended to make military command and control organizational structures more agile and adaptive, and to assess the effects of introducing new technology capabilities on the performance of these organizations.

CONCLUSIONS

We use a framework of modeling pitfalls, lessons learned, and future needs to characterize our major conclusions in a way that will be most useful to the sponsors in the design of future research programs. The problems or pitfalls identified by the committee are organized in terms of five major categories:

1. *Modeling strategy—matching the problem to the real world:* Difficulties in this area are created either by inattention to the real world being modeled or by unrealistic expectations about how much of the world can be modeled and how close a match between model and world is feasible.
2. *Verification, validation, and accreditation:* These important functions often are made more difficult by expectations that verification, validation, and accreditation (VV&A)—as it has been defined for the validation of models of physical systems—can be usefully applied to IOS models.
3. *Modeling tactics—designing the internal structure of a model:* Problems are sometimes generated by unwarranted assumptions about the nature of the social, organizational, cultural, and individual behavior domains, and sometimes by a failure to deliberately and thoughtfully match the scope of the model to the scope of the phenomena to be modeled.
4. *Differences between modeling physical phenomena and human behavior—dealing with uncertainty and adaptation:* Problems arise from unrealistic expectations of how much uncertainty reduction is plausible in modeling human and organizational behavior, as well as from poor choices in handling the changing nature of human structures and processes.
5. *Combining components and federating models:* Problems arise from the way in which linkages within and across levels of analysis change the nature of system operation. They occur when creating multilevel models and when linking together more specialized models of behavior into a federation of models.

To summarize, IOS modeling is a complex, emerging science with roots in many different disciplines. Its advancement requires that researchers maintain awareness of each other's work and build on each other's results, yet the multidisciplinary nature of IOS modeling has created a fragmented field. For the field to advance, researchers need better frameworks and forums in which to compare, discuss, and evaluate their results. The field currently features a multitude of complex models created using different data and different theories to address different problems, making comparative analysis nearly impossible. Common datasets and challenge problems are needed in order to learn which modeling approaches and sets of variables are most useful for specific types of problems.

It seems clear there is no single right model and probably will never be. The committee thinks that a federated modeling approach, in which different models at different levels are linked together and component submodels can be swapped in and out, is promising for attacking complex IOS modeling problems. Considerable research needs to be done to make this federated vision a reality, however. Standards, architectures, methods, and tools are needed to lower the barriers for developing, linking, and validating federated models.

Different modeling purposes require different types of models. In the committee's judgment, the purpose of the model should drive the appropriate variables to be included in the model. To do this successfully requires a clear specification of model purpose and criteria for usefulness for that purpose, which in turn requires that model developers work closely with the eventual users of the model.

The committee also recommends validation for action, in which the purpose of the model drives its validation criteria. IOS models cannot be validated "in general"—they must be validated for a specific use. A cross-disciplinary community of interest needs to establish and promulgate accepted standards for validation of IOS models. Triangulation methods that combine expert judgment, qualitative results and theoretical work, and quantitative results should be further refined and more widely used. Common challenge problems and datasets are needed to facilitate docking of models for comparative purposes.

Finally, models of human beings and their individual and collective behaviors necessarily include a large amount of inherent uncertainty. This uncertainty is not a flaw of the model and cannot be designed out of the model. Human behavior is dynamic and adaptive over time, and it is impossible at the moment (and into the foreseeable future) to make reliably exact predictions about it. Researchers need to develop ways to estimate the probability of plausible outcomes and express those estimates in ways that are clear and meaningful to model users, who can then judge whether the results meet their needs. It is important also to avoid raising expectations about the capabilities of IOS models beyond what can realistically be expected.

RECOMMENDATIONS

Recommendations for an IOS modeling research and development program fall into three broad categories: (1) large-scale, integrated cross-disciplinary research programs, focused around representative challenge problems and common datasets; (2) research in six independent areas that will advance the capabilities to address these integrated problems; and (3) multidisciplinary conferences, workshops, and other information exchange forums, with attendees to include not only model developers but also government program managers and military decision makers.

Integrated Cross-Disciplinary Research Programs

We suggest the funding of multiple large-scale, multiyear research programs that focus on comparing and, if appropriate, integrating models from different disciplines, different perspectives, and different levels of detail. The goal would be to create a level playing field on which the capabilities of different approaches could be compared and the strengths of each assessed. The ultimate goal is to move IOS modeling science forward through the process of comparison, docking, and integration.

It is essential for all participants in each program to focus on the same well-defined challenge problem instantiated in a common testbed and to use a common dataset. At the heart of each program would be a representative problem that is critical for military operations, defined in detail. We have chosen five representative problems as a starting point for choosing the problems to be addressed.

The research teams for these efforts should be multidisciplinary, and the program team should also include military users with operational experience in the domain for which the models are to be developed. These users will be ultimate judges of whether model results are useful and will provide advice on how the results can best be presented. The use of a common challenge problem and a common testbed will facilitate docking of the different models for purposes of comparison. The development of challenge problems should be a major focus early in the development of research programs.

These integrated programs will encourage mutual education between modelers and operational users. Results should be presented at workshops for program participants and other interested parties and at public conferences as well as published in the open literature.

Independent Research Thrusts

In support of the integrated programs we recommend, we have identified six independent areas in which research is needed. Progress in each of these areas could increase the ability to develop the integrated modeling capabilities that are needed to address military problems. In each area, we suggest the funding of multiple research teams from multiple perspectives, with periodic workshops for researchers to exchange results. We also suggest that operational users as well as government program managers participate in these workshops.

Thrust 1: Theory Development

Models should be conceptually correct and grounded in the underlying fundamentals of what is known about individual human and group social behavior. However, current theory in this area does not answer all of the questions needed to structure models that address relevant issues. Basic research is needed for theory development, especially for the low-level social behaviors that are the building blocks for larger scale social behavioral patterns. This theory development work must involve multiple disciplines and perspectives with periodic workshops to exchange results.

Theory development challenge problems should be defined to guide the work, but these can be nonmilitary and need not involve the level of military detail necessary for the integrated problems discussed above. A series of workshops should be conducted with researchers to identify key theory gaps.

Academic institutions are key players for theory development, but they need information, incentives, and funding to address these theoretical issues. There is a need to educate researchers in military domains, establish conferences and journals in which their results can be presented, provide postdoctoral support, and provide funding that allows researchers to spend time learning about military domains in depth.

Thrust 2: Uncertainty, Dynamic Adaptability, and Rational Behavior

Models must deal with the inherent uncertainty and the dynamic adaptation that characterizes human behavior. Models must also be capable of modeling both rational and nonrational behavior.

Basic research is needed in each of these areas. Issues include

- How should models capture the "uncertainty-in-the-small" associated with individuals and small groups? How can model structures and parameters capture this variability, and how much of this variability must be included for the purposes of the model?
- How should models capture the "uncertainty-in-the-large" associated with populations and variations in population distributions? How much variability must be included for the purposes of the model?
- How can models capture adaptation and learning over time and as the results of actions by others? For example, people have multiple

overlapping identities and allegiances. How can these be captured in a model, and how can one estimate the effects of actions and events on the primacy of these multiple allegiances as they affect decisions and actions?

- What are the factors that contribute to rational, adaptive behavior, and what factors induce behavior that appears irrational? Models of both rational and nonrational behavior must capture all the key factors—cognitive, affective, cultural, and contextual—that motivate and shape behavior of specific individuals in specific situations.

Better techniques are needed for understanding the implications of diversity and variability for model-based sensitivity analysis. Better automated technology is needed to put the model through its paces to explore the parameter space effectively and produce robust results.

Thrust 3: Data Collection Methods

The difficulty of obtaining data is an ongoing challenge for IOS modeling. Research is needed to develop better data collection processes through field studies, experiments, and potentially massively multiplayer online games (MMOGs).

Although a variety of ethnographic data collection techniques are currently in use, they need to be better tailored to the needs of IOS models. For field data collection, it is necessary to bring modelers and data collectors together to develop data ontologies, joint specifications, and data collection methodologies and tools that are specifically tuned to IOS models.

MMOGs are a potential untapped resource for collecting social and behavioral data on a large scale. We recommend the creation of a MMOG facility and the funding of basic research to determine if MMOGs can be used to test, verify, and validate IOS models. We recommend that funding be put into developing the science of MMOGs. We note that funding MMOGs is a risky endeavor, but we think that the potential benefits outweigh the risks.

Thrust 4: Federated Models

It is a fundamental conclusion of the committee that no single modeling approach can provide all the capabilities needed by DoD. We recommend a federated approach in which modeling components are created to be interoperable across levels of aggregation and detail. For example, a federated model might include a detailed representation of a few key individuals, linked to group-level models of different cultural groups and terrorist organizations, linked to geographic sector–level models of the level of unrest in a city. This approach is flexible and extensible, allowing the addition or subtraction of models at different levels of detail as needed for the problem to be addressed.

Combining model components to create federated models in the sense being recommended requires deep semantic interoperability (i.e., theoretical consistency) and presents difficult challenges. To create semantic interoperability, it is necessary to recognize that the links among components are themselves elements of the model. Research is needed on:

- How to ensure that the models being federated embrace compatible assumptions regarding concept abstractions, entity resolution, time scale resolution (tempo), uncertainty, adaptability, docking standards, input-output, semantics, etc.
- How the components of the federated model should be encapsulated and which elements must be exposed to other components.
- How specific classes of models should be linked (e.g., cognitive models to social network models).
- How to ensure dynamic extensibility.

In addressing these issues, IOS modelers should maintain awareness of research and development in model federation in the larger modeling and simulation community.

Thrust 5: Validation and Usefulness

Current VV&A concepts and practices were developed for the physical sciences, and we argue that different approaches are needed for IOS models. Specifically, we recommend that a "validation for action" approach be used that assesses the usefulness of a model for the specific purposes for which it was developed. It is thus very important that the purpose(s) and criteria for judging success be clearly stated a priori for all models. We recommend organizing national workshops to agree on appropriate processes for VV&A of IOS models and to outline a roadmap for developing improved processes and standards. On the basis of the results of this workshop, we recommend that a DoD-wide authority develop and disseminate VV&A processes and standards for IOS models. Basing model validation on the usefulness of the model for specific problems requires that model purposes be clearly stated by model users and clearly understood by model developers. We suggest that, as part of developing a VV&A standard for IOS models, clear guidelines be developed for specifying model purpose.

Thrust 6: Tools and Infrastructure for Model Building

It is important to reduce the barrier to entry for developing models, modeling tools, frameworks, and testbeds. Scientists should be able to build and validate models without the large overhead currently associated with many DoD modeling and simulation investments. It should be possible to easily tailor existing models for specific purposes.

Sharing of IOS modeling knowledge across disciplines, as facilitated by the conferences and workshops recommended below, will support this goal. Work is also needed in developing an infrastructure for IOS modelers, including a national network of possible collaborators, common databases for model development and testing, and frameworks and toolkits for rapid model development.

The limited data that exist for IOS models are often not accessible to model developers. We recommend national web-accessible data repositories that are open to researchers who seek to inform and test models. For militarily relevant domains in which some data are classified, we recommend an investment in automated tools to sanitize the data.

We also recommend the development and maintenance of an online web-based catalog of general approaches, models, simulations, and tools. The notion is to develop something along the lines of the Defense Modeling and Simulation Office's Modeling and Simulation Resource Repository or the clearinghouse at Carnegie Mellon's CASOS site (http://www.casos. cs.cmu.edu). To be effective, the envisioned site needs careful consideration in terms of organization, content, currency, and usability. This cannot be a one-time effort. It needs significant startup funding and continued support over its lifetime.

Multidisciplinary Conferences and Workshops

A number of the issues and problems identified by the panel were the results of the failure of different disciplines to exchange information, or they resulted from misunderstandings among government funders of model development efforts, military users of models, and model developers. Because of the diversity of this group, there is no natural forum for them to exchange information. We recommend the organization of special-purpose workshops around the integrated research programs recommended above, as well as workshops for the independent research thrusts described above.

IOS modelers need to be educated on:

- The nature of the military decisions for which models are relevant.
- Desired model functionality.

- The most useful form for presenting model results.
- The value of work performed by others outside their discipline.
- Feasible and appropriate VV&A approaches for IOS models.

Operational users and managers need to be educated on:

- The value of multidisciplinary approaches and the need for review of models from multiple perspectives.
- The inherent uncertainty associated with model predictions.
- The value of models for sensitivity and trade-off analysis (versus the one right answer).
- The design of virtual experiments to assess results over a range of conditions.
- Reasonable definitions of validation for IOS models, feasible approaches for VV&A testing, and why these approaches differ from those used for physics-based models.

The recommended workshops should involve model developers, operational military users of the models, and government personnel making funding decisions regarding model development.

Roadmap for Future Research and Development

The committee recommends a use-driven research program to extend the state of the art in IOS modeling, focused around a series of challenge problems—clear specifications of the uses to which the model is to be put, defined to be relevant to military needs, and expanded over time as progress is made in modeling approaches, tools, and technologies. The purpose of the model, as captured in the challenge problems, drives the theory to be applied, the data to be used, and the model development. Model development is made easier by modeling tools and infrastructure and relies on federation standards to ensure the interoperability of model components. Once the model is developed it is validated by asking the question: Is the model useful for its intended purpose?

The recommended program proceeds in a cyclical fashion. Based on the answers to the question "Is the model useful?" new models may need to be developed, new theory and new data (and new types of data) may be needed, and new interoperability standards, tools, and infrastructure may be required. Depending on the results, the problem itself may need to be redefined, clarified, or expanded. These challenge problems, combined with periodic workshops and conferences to compare and exchange results, serve as a unifying force and a common ground for the fragmented field of IOS modeling, providing a foundation on which scientific progress can be made.

Part I

BACKGROUND AND NEED FOR ORGANIZATIONAL MODELS

1

Introduction

In 1951, Isaac Asimov published a science fiction novel, *Foundation*, that imagined a future world in which a maverick genius, Hari Seldon, invented a new science—psychohistory—that was capable of mathematically predicting "the reactions of human conglomerates to fixed social and economic stimuli" (Asimov, 1951, p. 19). In Asimov's novel, Seldon's psychohistory equations are used to predict the collapse of a galactic empire, allowing a band of scientists (the Foundation) to act to preserve human knowledge and greatly shorten the period of chaos that follows the galactic collapse.

Asimov's vision has inspired generations of scientists. Today scientists find themselves at the edge of what he imagined—working on computational mathematical models of aggregate human behavior that allow them to understand, assess, and, to a very limited extent, predict "the reactions of human conglomerates." This report assesses how close they have come to that vision and what still remains to be done.

The study was requested by the Human Effectiveness Division of the U.S. Air Force Research Laboratory, with additional funding from the Air Force Office of Scientific Research. The Air Force and the other military services are increasingly interested in using models of the behavior of humans, as individuals and in groups of various kinds and sizes to support the development of doctrine, strategies, and tactics for dealing with state and nonstate adversaries, in support of military planning and operations, acquisition programs, and as training and simulation tools. In this report, we are calling them individual, organizational, and societal (IOS) models. There are many lines of research on such models, in academia, industry, and the military, and it would be difficult for a military program office staff to become thoroughly familiar with all of them or to evaluate the potential of each research program for use by the military. The modeling efforts span several disciplines, have different goals, and often use different terminologies.

The Air Force therefore asked the National Research Council (NRC) to review several relevant IOS modeling research programs, evaluate the strengths and weaknesses of the programs and their methodologies, determine which have the greatest potential for general military use (i.e., not just Air Force specific), and provide the Air Force with guidance for the design of a research program to effectively foster the development of IOS models useful for the military. One of the great strengths of the NRC is its ability to convene committees of experts from a broad range of disciplines and facilitate their cooperative work on the study of a cross-disciplinary topic like this one.

STUDY TASK AND OBJECTIVES

The formal statement of task from the cooperative agreement between the NRC and the Air Force for this study is as follows:

- Review the state of the art of the subset of the social sciences perceived as having the greatest payoff in terms of informing future computational model developments. These will include
    - o   key conceptual models in the areas of anthropology, sociology, social psychology, political science, organizational theory, and similar social sciences specialties
    - o   efforts in developing computational models, "artificial life" simulations, and the like being undertaken by these communities
- Review the state of the art in societal[1] modeling applications serving the Department of Defense (DoD) and related agencies, with special emphasis given to computational modeling and simulation-based approaches
- Review the state of the art in the three computational modeling communities outside DoD, and identify strengths and shortcomings in each:
    - o   cognitive science and individual behavioral modeling
    - o   network analysis and multiagent organizational modeling
    - o   multiresolution modeling and simulation

---

[1]In this study, the committee broadened the scope to include individual and organizational models as well, because of the inseparability of all three, given the intended usage. Additional discussion appears in the Concepts and Definitions section below, as well as in chapters following.

- Identify how gaps in societal behavioral modeling applications serving DoD and related agencies might be filled by:
  o conceptual models in the social sciences
  o computational modeling approaches now under way in the social science community
  o closer linkages—via shared research, common development frameworks, interlinked computational models and the like—between the cognitive science community, the network/organizational modeling community, and the multiresolution modeling and simulation community
- Develop a research and development roadmap to fill current application gaps, for the near, mid-, and far term[2]

NATIONAL ACADEMIES' RESPONSE

The NRC, an operating arm of the National Academies, responded by appointing a committee of 13 experts, drawn from the social sciences and from several human behavior modeling communities and disciplines, following the procedures mandated for all NRC committee appointments. These procedures are designed to ensure that committee members are chosen for their expertise, independence, and diversity and that the committee's membership is balanced and without conflicts of interest. The appointments were finalized after the discussion of sources of potential bias and conflict of interest at the committee's first meeting in April 2005. Brief biographies of the committee members appear in Appendix D.

THE COMMITTEE'S APPROACH

The committee developed its approach to the task at the first meeting. We discussed each member's expertise and identified information needs in several domains, including the military's needs and uses for IOS modeling, research now under way under military contracts (and often not available in the open literature), and the current state of the art of modeling efforts in the social science and computational modeling communities listed in the task statement. We developed plans for obtaining and analyzing the needed information and for organizing the report. The committee also discussed the scope of its task and determined what would and would not be attempted.

---

[2]In our recommendations, we distinguish actions to be taken in the first year, years 2–4, and beyond. These may be interpreted to correspond to near-, mid-, and far-term horizons.

Defining the Project Scope

To achieve the objectives of the study, the committee needed to review the state of the art in several modeling disciplines and communities of practice. We decided that it would be neither feasible nor useful, for the purposes of this study, to produce an exhaustive literature review. Rather we decided to summarize relevant knowledge in each of the modeling areas, and to organize our summary review of each area using a template of significant features developed by the committee. The template focused on the applicability of each type of modeling approach to the DoD's IOS modeling needs.

Gathering Data

The committee used a variety of data-gathering methods, mainly over the period of the first three meetings. We reviewed pertinent literature, scholarly and applied, including publicly available military documents, such as the Quadrennial Defense Review (U.S. Department of Defense, 2006), with each committee member concentrating on his or her area of expertise. We invited the sponsor and other military experts to brief us on the particulars of DoD's needs for IOS models and on the expectations of potential model users, and we invited managers of DoD modeling research programs to tell us about their programs. We appointed three military operations experts with some knowledge of IOS modeling as consultants to the committee, enlisting their help in developing representative scenarios of situations in which models might be used by DoD, as one way of understanding the need for IOS models.

Data Analysis and Review

In our later meetings, the committee discussed the information we had found, developed a framework for presenting our findings and conclusions, and developed recommendations for the study sponsor. The report structure is straightforward: we discuss DoD's need for modeling and the current knowledge and capabilities (state of the art) in the modeling community. We then highlight the important gaps between the state of the art and the identified needs and discuss ways to bridge the gap in a research program.

Concepts and Definitions

Because the field of IOS modeling is spread among several disciplines and domains, the same terms are often used with different meanings by different authors. We felt it necessary to agree on common definitions for some important terms and then to use the terms in only the defined senses.

First, what should we call this area of modeling that spans the range from individual actors who are members of a small group to whole nations, societies, or ethnic groups? We are using the term "individual/ organizational/societal" modeling to convey this broad meaning, to cover the modeling communities that are concerned with the full span of human behavior, including individuals, teams, small groups, large groups (including different cultures and ethnic/religious groups), societies, nations, and national coalitions. We are of course not viewing this as merely a one-dimensional progression in the size of the "human conglomerate," but rather as a rich tapestry of many dimensions that are complexly interlinked via relationships that are only now being recognized, let alone understood. Through the course of this report, we hope to point out some of the key relationships, as well as the considerable distance there is to go in terms of understanding the fundamental interdependencies and interactions that exist, in a manner that supports meaningful and useful models.

Second, what should we call these different levels of "knowledge representation" that start with empirically based observations of human activity[3] and end with computational instantiations (specifically, computer-based simulations) of human behavior, often referred to simply as "models"? It is certainly beyond the scope of the committee to develop an ontology of human behavior representation, but we think that it is appropriate to attempt to identify at least four levels that proliferate in the modeling community. Going from the general to the specific, they are

- *Theory:* This is an explanation of how something works, in this case how one of the human conglomerates behaves for a given set of traits (or culture) in a given situation or environment. Theories may be global (e.g., at the individual level, a "unified theory of cognition"), or they may be local (e.g., the decay of working memory). Theories may be formal or informal, mathematical or verbal, well formed and founded, ill formed and unfounded, and everything in between.
- *Architecture:* This is a more specific statement of a theory, one that places a structure under it, and attempts to either: (a) break down the theory into smaller and perhaps more readily understood components or subtheories (e.g., at the individual level again, a "cognitive architecture") or (b) link the theory with collateral theories to explain behavior at a larger scale or in more complex environments (e.g., an architecture to link cultural influences on social networks). "Good" architectures attempt to maintain as much generality as

---

[3]Although not always: some might argue that one starts with internalized theoretical constructs that shape what one observes, rather than the other way around.

possible (i.e., are parameter-free, domain independent, etc.), so
as to be able to accommodate the broadest set of behaviors and
situations.

- *Model:* This is a yet more specific representation of the human
  conglomerate, one that can be directly derived from a corresponding theory, or, more indirectly, instantiated from an associated
  architecture, in which the specific instantiation takes into account,
  for the entity being modeled, that entity's inherent characteristics
  (e.g., personality traits, religious beliefs, social connectivity, etc.),
  the associated domain-specific knowledge base (e.g., knowing the
  local village streets), and the specific situation and environment of
  interest (e.g., crowd formation in a village).[4] Like theories, models
  can be global or local, and well founded or not.

- *Simulation:* This is a still yet more specific representation of the
  human conglomerate, this time instantiated in executable software. Simulations can be developed directly from theories (e.g., by
  coding up, say, a mathematical equation embodying a particular
  theory), from architectures (by developing a simulation within the
  software/system development environment associated with a particular architecture, if available), or from models instantiated for a
  specific situation (giving rise to the term "computational models").[5]
  The power of a simulation is several-fold: simulated "data" can be
  compared with empirically collected data for model validation
  purposes; simulations can be used to explore the range of potential
  outcomes; and simulations can be used to drive new theory development and empirical data collection efforts, via the generation of
  new hypotheses based on simulation-based "experiments."[6]

Again, we emphasize that this is not intended to be a definitive ontology of behavior modeling and simulation, but merely an attempt to clarify
terms somewhat, terms that are often used interchangeably in the literature
(including, occasionally, this report).

---

[4] Although models can be directly instantiated from theories, there is a trend toward increasing use of "intermediate" architectures, driven both by the practical benefits gained by the
model developers in being able to instantiate well-grounded models quickly for specific situations and by the lessons learned gained by the architecture developers with each new model
instantiation.

[5] Again, the trend in the well-established modeling and simulation community is to dis-
courage "direct coding" from theory to simulation and instead move through the levels outlined here, because of the advantages gained from established architectures and model-specific
databases (which may be reused), although clearly the development overhead is higher.

[6] In addition to the other uses identified in Chapter 2.

CAUTIONS FOR IOS MODELING

In our discussions with military personnel, and in interactions outside the committee deliberations, the committee became aware that many people may have unrealistic expectations of what a model or simulation of human behavior is able to do. No model is ever likely to be able to predict exactly what an individual or group will do, except in a situation so constrained, with alternatives so well understood, that a model is not needed. Human behavior, individually and in groups, is governed by so many variables, including many that are not likely to be susceptible to capture in a model, that the best any model will do is to narrow the range of plausible behavioral outcomes of a defined situation. For example, a model may be able to forecast the most likely range of outcomes of a potential course of action. It may be able to direct attention to situational variables that are known to be important but may have been overlooked in a particular engagement. A well-designed model may draw a decision maker's attention to possible unintended consequences ("second-order effects") of a planned course of action. But it will *not* be able to make point predictions, such as "If we take Action A, the adversary will attack at Point B early tomorrow morning with three simultaneous improvised explosive devices (IEDs)." So we speak of models *forecasting* a range of outcomes, rather than making precise *predictions*. Certainly models that can produce such forecasts are a worthwhile objective. They can serve many useful purposes, from supporting training, to serving as tactical decision aids, to examining possible outcomes of alternative strategies or policies.

Some of the known difficulties of developing and implementing models are discussed later in the report, but a few may bear mention at this point. The most desirable data to put into a model that would provide the most accurate forecasts often will not be available: the data may not be accessible, may not be in a usable form, or may not be verifiably accurate, timely, or complete. In fact, it is common knowledge that adversaries will often attempt to provide false data (disinformation) if they think it will be believed and used. So the development of a model, in itself, is only a small part of the work that must be done to use it, and there is never a guarantee that good information will be available to implement the model when it is needed. These issues must be taken into account in the design of models.

The work of developing models of adversary behavior is never complete, because any worthy adversary, once it realizes that its modus operandi is known and defenses are being used against it, will make changes in its organization, operations, etc., designed to invalidate the model. So we have the ever-changing methods used by insurgencies to attack friendly forces[7] or innocent citizens in the Middle East: car bombs, IEDs, suicide bombers, each adapted or supplanted by a different method as soon as effective countermeasures are devised. This means that a modeling effort must include ongoing maintenance and updating functions if it is to remain useful.

Another challenge is that some of the research on modeling for military purposes must necessarily be conducted at high security levels, in secure environments. It is likely that much of the fundamental research for the design of modeling methods and tools can be done in open venues by researchers with low or no security clearances, but any work that includes specific and current field information on individuals or groups, specifics on friendly or adversary force capabilities, or detailed operational plans must of necessity be highly classified to prevent the adversary anticipation, adaptation, and/or exploitation discussed above.

Finally, it is important to recall that the predecessor report by the NRC in this area (National Research Council, 1998, p. 8) noted that "the modeling of cognition and action by individuals and groups is quite possibly the most difficult task humans have yet undertaken. Developments in this area are still in their infancy." This situation has not changed significantly in the mere 10 years since the publication of that report. But the world has, and, as a result, it has become ever more clear that human behavioral modeling at all levels is critical to DoD specifically and to the nation more generally.

ORGANIZATION OF THE REPORT

The report is organized into three parts. Part I provides background information and explains the need for organizational models. Chapter 1 gives the background of the study and the committee's approach to the work. Chapter 2 discusses evolving missions of the military and the applicability of IOS modeling to those missions. It includes an introduction to a set of military scenarios that are used throughout the report as exemplars of situations that could benefit from the use of modeling.

Part II contains extensive descriptions of the major modeling methodologies and model types the committee reviewed. Models take many forms, ranging from loose conceptual models to precise mathematical models (Lave and March, 1975). They include agent-based models, cognitive models, expert systems, dynamical systems, and input-output models. The diverse expertise of the committee members contributed greatly to the complete-

---

[7]The term "friendly forces" is used to refer to forces that are either formally or informally allied with the United States and that support its objectives. It may thus refer to the armed forces of allied nations or to forces representing nonstate organizations or factions.

ness of this review but also made it challenging to agree on an organizing framework for presenting the review results. Refined through multiple iterations, the organizing framework that we developed represents a significant product of the study, as discussed in the introduction to Part II.

Chapter 3 presents conceptual and cultural (verbal) models. The subsequent model descriptions are then organized according to the level of granularity of the models. We have differentiated "macro" models that describe organizations and their behaviors on a large scale (Chapter 4); "micro" models dealing on a level as detailed as the individual actors within groups or organizations (Chapter 5); and "meso" or intermediate models somewhere between these two, as well as integrated, multilevel models (both in Chapter 6). We discuss games separately in Chapter 7 because although they incorporate formal models, they do not easily fit into any of the other categories.

For each methodology, we describe the method and its current state of development, often with some history of the field. We review the applicability of the methodology to the modeling requirements identified in Chapter 2, its major limitations, issues of data, verification and validation, and needs for continued research and development.

 The discussions of models and methodologies are not exhaustive. We have attempted to provide an overview of a broad range of model types and modeling methods, although the committee members, chosen for their range of modeling expertise, naturally discussed in greatest depth the areas with which they are most familiar.

In Chapter 8 we discuss some generic issues, such as integration across levels of models, modeling frameworks and tools, model verification and validation, and data sources and quality. Chapter 9 summarizes the state of the art of IOS modeling as presented in Part II and its utility for the applications discussed in Part I.

In Part III we identify the gaps between the current modeling capabilities and the military's modeling needs, and, in Chapter 10, we discuss common problems or pitfalls that may impede the development and application of models or reduce their utility. In Chapter 11 we present recommendations for a research roadmap, a program of use-inspired IOS modeling research and development designed to reduce the gaps and develop the needed capabilities.

The report ends with four appendixes. Appendix A provides a list of acronyms and abbreviations used in the report, spelled out, with some information on their meanings. Appendix B contains detailed military scenarios that served as exemplars for considering how models could be used for military purposes. Appendix C provides detailed material relevant to the discussion in Chapter 8 of DIME/PMESII modeling paradigms. Appendix D provides biographical sketches of committee members and staff.

REFERENCES

Asimov, I. (1951). *Foundation.* New York: Random House.

Lave, J., and March, J.G. (1975). *An introduction to models in the social sciences.* New York: Harper & Row.

National Research Council. (1998). *Modeling human and organizational behavior: Application to military simulations.* R.W. Pew and A.S. Mavor (Eds.). Panel on Modeling Human Behavior and Command Decision Making: Representations for Military Simulations. Commission on Behavioral and Social Sciences and Education. Washington, DC: National Academy Press.

U.S. Department of Defense. (2006). *Quadrennial defense review report.* Washington, DC: Author.

2

Military Missions and
How IOS Models Can Help

Computational modeling and simulation (M&S) technology have long been useful military tools, although these models have focused primarily on physical effects, such as the predicted capabilities of sensors or weapons systems. Today, the changing nature of military missions is driving the need for new types of computational models that focus on human behavior, specifically on human behavior in social units, such as organizations and societies.

The military has traditionally made use of computational modeling in three broad areas of activity:

1. *Analysis and forecasting for planning.* Models are used for the fusion of fragmented and incomplete information about enemy activities and capabilities. For example, models of enemy equipment can be used to interpret fragmentary data on the performance of that equipment (e.g., what capabilities in the equipment could have resulted in the observed performance). Forecasting models are used to develop courses of action (COAs) based on the desired outcomes and their estimated likelihood of achieving those outcomes. At a simple level, for example, models are used to forecast the effectiveness of different types of weapons against different kinds of targets.

2. *Simulation for training and rehearsal.* Models are used in simulations that create training and rehearsal environments. For example, pilots practice complex and dangerous combat maneuvers in simulators before encountering them in exercises or combat, and tank

commanders practice ground combat missions before an actual engagement. In both situations, considerable effort goes into modeling the environment (e.g., aerodynamics and terrain), simulating the dynamics of the friendly and enemy sensors and weapons systems, and providing the critical performance feedback to trainees needed for skill improvement and "learning to criterion."

3. *Design and evaluation for acquisition.* When a system is designed, built, and acquired, models are used throughout the process to predict performance and make design decisions based on cost-benefit trade-offs. For example, detailed physical and electronic models can be used to predict the additional range of a sensor accruing from a proposed enhancement (and increased cost), to support a cost-benefit trade-off.

In this chapter we argue that the successful performance of all three of these activities in today's military environment requires not only the traditional set of physically based models and simulations now used, but also computational models of human behavior, particularly computational models of human behavior in social units. We begin by describing today's changing military missions in order to explain why—in the current environment—analysis, planning, training, and acquisition require models of human behavior at many levels: at the individual level, at the team or organizational level, and at the societal level. We then give specific examples of how these individual, organizational, and societal (IOS) models could be used by the military. Finally, we briefly review current military IOS modeling efforts and summarize the major challenges involved in meeting current needs. Subsequent chapters provide a broader review of state-of-the-art IOS behavioral modeling approaches, assess the extent to which those approaches have the potential to meet military needs, identify major shortfalls and gaps, and recommend a plan of action to address them.

MILITARY MISSIONS NOW AND INTO THE FUTURE

This section reviews the changing nature of today's military missions to explain why effective forecasting, training, and acquisition require computational IOS models.

Overarching Strategy and Operational Enablers

The changing nature of current and future military missions is made quite explicit in the Department of Defense's (DoD) *Quadrennial Defense Review* (U.S. Department of Defense, 2006). Coming out of a long tradition of "attrition-based" conventional warfare and backed ultimately by nuclear-based mutual assured destruction (MAD), DoD is now undergoing a shift of tectonic proportions to operationalize the National Defense Strategy of "fighting the long war" and has identified five critical operational enablers:

1. *Defeating multinational multiethnic terrorist networks* that "seek to break the will of nations that have joined the fight alongside the United States by attacking their populations" and "use intimidation, propaganda and indiscriminate violence in an attempt to subjugate the Muslim world under a radical theocratic tyranny" (U.S. Department of Defense, 2006, p. 13).

2. *Defending the homeland in depth* against both terrorist networks and hostile states with weapons of mass destruction (WMD) capabilities. Globalization enables "the spread of extremist ideologies" and "the movement of terrorists" and "empowers small groups and individuals" with the result that "nation-states no longer have a monopoly over the catastrophic use of violence" (U.S. Department of Defense, 2006, p. 36).

3. *Shaping the choices of countries at strategic crossroads* to protect the "future strategic position and freedom of action of the United States, its allies and partners" by shaping the choices of "major and emerging powers . . . in ways that foster cooperation and mutual security interests" (U.S. Department of Defense, 2006, p. 39). In addition to the Middle Eastern region, countries of particular concern are India, China, and Russia.

4. *Preventing the acquisition or use of WMD* by hostile states (e.g., Iran) or nonstate actors (e.g., Osama bin Laden). "Based on the demonstrated ease with which uncooperative states and non-state actors can conceal WMD programs and related activities, [we] must expect further intelligence gaps and surprises" (U.S. Department of Defense, 2006, p. 45).

5. *Refining DoD's force planning construct for wartime* to move gradually from a two-front conventional campaign capability to more loosely defined "distributed, long-duration operations, including unconventional warfare, foreign internal defense, counterterrorism, counterinsurgency, and stabilization and reconstruction operations" (U.S. Department of Defense, 2006, p. 36).

This is a remarkable shift in emphasis since the terrorist attacks in the United States on September 11, 2001, and may very well be a turning point away from more than 50 years of conventional force planning (backed by MAD) and the start of a much more agile and indigenously sensitive force. The United States is no longer fighting nation-states using conventional weapons but instead is fighting a very different kind of organization—

terrorist networks—in a battlespace in which effects may be defined by the attitudes and behaviors of civilian noncombatants rather than by bombs on targets.[1] In order to analyze, plan, train, and acquire effective technology for this new battlespace, models are needed to help people understand and interpret fragmentary information about terrorist activities and understand the likely effects of U.S. actions on the attitudes and behaviors of diverse multicultural civilian populations. People need to understand the forces that drive individuals to join terrorist organizations, how these organizations function, and how they organize action. People need to understand the factors that contribute to the stability of neighborhoods and regions and how military actions as well as political, diplomatic, and economic actions contribute to that stability. People need to understand complex shifting cultural allegiances and how U.S. actions affect those allegiances. Models of sensor and weapons systems are not adequate tools for fighting this long war. The nation's defense planners need IOS models that capture the richness of individual, team, organizational, societal, and cultural influences that can help to address the key dimensions of the new battlespace.

Dimensions of the New Battlespace

In this section we examine some of the drivers of the changing DoD mission to gain insight into what this shift in mission means for IOS modeling requirements.

The Impact of Urbanization

One of the key drivers in this shift has been the growing recognition that fundamental world demographics are changing: "The world's urban population reached 2.9 billion in 2000 and is expected to rise to 5 billion by 2030. Whereas 30 per cent of the world population lived in urban areas in 1950, the proportion of urban dwellers rose to 47 per cent by 2000 and is projected to attain 60 per cent by 2030. . . . At current rates of change, the number of urban dwellers will equal the number of rural dwellers in the world in 2007" (United Nations, 2002, Part I, p. 5). The military implica-

---

[1] A note of caution is appropriate here. Although it is true that at the time of this writing the United States is not engaged in a conventional war, that is not to say that it will not be engaged in one at some point in the future. Thus, there is always the danger that the nation will be "preparing for the last war" (e.g., today's Afghanistan and Iraq campaigns) via a wholesale shift in focus to nonconventional strategies, tactics, and weapons systems. DoD recognizes this, as noted in the fifth "operational enabler" cited above (U.S. Department of Defense, 2006, p. 36), identifying the desire to "move *gradually* [emphasis added] from a two-front conventional campaign capability. . . ." Clearly, the operative issue is how long this transition takes and to what extent it transforms the services' force structure.

tions of this fact are explored in depth in two recent RAND studies (Glenn, 2000; Vick et al., 2002). Key issues and implications that emerge from these studies and others include:

- Most, if not all, of the future conflicts the nation will face will have an urban component, based both on historic precedent and on the fact that the adversaries are no match for U.S. forces in "open field" engagements.
- It will no longer be sufficient to avoid urban and surrounding built-up areas during military operations, as has so long been U.S. doctrine. According to a 2002 Joint Chiefs of Staff report, "urban areas are the natural battleground for terrorists: the effects of terrorist acts are greater and more noticeable and the terrorist groups more difficult to locate and identify" (Joint Chiefs of Staff, 2002, p. III-27). From a "hearts and minds" standpoint, there is also a clear political advantage of having a close connection with the noncombatant urban population.
- Urban operations are extremely difficult, with the operational environment characterized by high densities and tempos, inherent complexity, and constraints. The battle tempo can be extremely high, forcing rapid assessments, decisions, and actions. Collateral damage issues covering critical infrastructure losses, damage to symbolic edifices, and noncombatant loss of life are critical.

Urban operations are also complicated by the fact that mission objectives can vary dramatically in both time and space, running from all-out conflict to infrastructure rehabilitation. This spatiotemporal nonuniformity has been referred to as the "three-block war" by the former commandant of the Marine Corps, General Charles C. Krulak: "In one moment in time, our service members will be feeding and clothing displaced refugees, providing humanitarian assistance. In the next moment, they will be holding two warring tribes apart—conducting peacekeeping operations—and, finally, they will be fighting a highly lethal mid-intensity battle—all on the same day . . . all within three city blocks. It will be what we call the 'three block war'" (Krulak, 1997, p. 139).

In these stability and support operations (SASO) stages, it becomes increasingly important to interact with and not alienate the local population, get their support to identify social networks of adversaries (and potential allies), and anticipate first- and second-order effects (i.e., unintended consequences) of actions that are within the scope of the unit's capabilities (i.e., executing a search-and-destroy mission) but that may be highly counterproductive in the long run. It also follows that as the mission becomes dictated less by military objectives than by social and political objectives, there is a need to ensure greater interaction with other organizations outside the local unit's normal sphere of interest. Not only does this imply a greater reliance on joint operations (coordinating the sister services), and increasingly a reliance on coalition (non-U.S.) partners, but it also implies greater interagency coordination, both national (e.g., the State Department, the intelligence agencies, the organs of public diplomacy, U.S.-based nongovernmental organizations or NGOs), international (e.g., sister intelligence services, non-U.S. NGOs), and private-sector economic interests. As a consequence, in order to address and achieve the peacemaking objectives in the new theaters of war, planners must somehow consider and assess the aggregated complex interactions of entire social systems, both regional in behaviors and global in influence, at resolutions of fidelity neither needed nor attempted in prior military history.[2]

The objectives and technologies of peacemaking in this environment are very different from those of conventional warfare, most notably, a substantially increased emphasis on peacekeeping, disaster relief, and nation-state building (see, for example, the Urban Sunrise study of the Air Force Research Laboratory, 2004). The urban operational environment serves to transform what was once viewed as a strictly military (and tactically difficult) engagement into something that is now considerably more holistic and focuses primarily on social, organizational, and cultural factors involving key individuals, nonmilitary groups, local crowds, and indigenous populations, all within a rich tapestry of a complex local infrastructure overlaid by local, national, and transnational economic markets, organizational and social structures, traditions, cultures, and religious beliefs.

The Growing Importance of Pre- and Postconflict Operations

The changing nature of military missions is putting increasing focus on operations that occur before and after periods of overt conflict. These pre- and postconflict operations may persist much longer than the conflict itself, as is all too well illustrated by the current situation in Iraq.

In the doctrine for Joint Urban Operations (JUO) (Joint Chiefs of Staff, 2002) five phases are recognized—understand, shape, engage, consolidate, and transition (USECT, emphasis added):

---

[2]While the military is the branch of the U.S. government having primary responsibility for projecting U.S. power overseas, it may be a classic case of "mission creep" for the military to be taking a leading role in economic development, political reconstruction, diplomacy, disaster relief, and intercultural communication. But this is exactly what is happening in today's conflicts, with young servicemen serving effectively as "mayors" of Iraqi villages, see http://www.washingtonpost.com/wp-dyn/content/article/2007/01/11/AR2007011101576.html. And this is likely to remain the case until other U.S. agencies or NGOs can take the lead, or the United States successfully transitions these functions back to the local population.

1. *Understand:* "The JFC [joint forces commander] evaluates the
    urban battlespace, including the urban triad [the physical terrain,
    the urban infrastructure, and the population] and the threat, to
    determine the implications for military operations. This evalua-
    tion extends from complex terrain considerations to the even more
    *complex impact of the sheer number of actors operating in an
    urban battlespace.* On one hand there may be adversary military
    troops, criminal gangs, vigilantes, and paramilitary factions oper-
    ating among the noncombatant population. On the other hand,
    especially in MOOTW [military operations other than war], the
    situation may be further complicated by the presence of nonmilitary
    government departments and agencies, to include intelligence, law
    enforcement, and other specialized entities" (Joint Chiefs of Staff,
    2002, Chapter II, pp. 8-9).
2. *Shape:* "Shaping includes all actions that the JFC takes to seize the
    initiative and set the conditions for decisive operations to begin.
    The JFC shapes the battlespace to best suit operational objectives
    by *exerting appropriate influence on adversary forces, friendly
    forces, the information environment,* and particularly the elements
    of the urban triad. Methods of shaping may include . . . the phased
    deployment and employment of joint forces. Rather than deploying
    combat forces initially, the JFC may, in many cases, need to *deploy
    noncombat forces early, such as civil affairs (CA), public affairs
    (PA), medical support, and psychological operations (PSYOP) units.*
    . . . Critical to shaping operations is the isolation of the urban area
    to support the campaign" along physical, *informational, and moral
    dimensions* (Joint Chiefs of Staff, 2002, Chapter II, p. 11).
3. *Engage:* "To engage, the JFC brings the full dimensional capabilities
    of the force to bear in order to accomplish operational objectives.
    Engagement can range from full combat in war to FHA [foreign
    humanitarian assistance] and logistic support for disaster relief
    operations. It consists of those actions taken by the JFC against
    a hostile force, a political situation, or a natural or humanitarian
    predicament that will most directly accomplish the mission. In all
    cases, the speed and precision with which the JFC engages will
    largely determine any degree of success. . . . [S]uccessful engage-
    ment requires . . . the seizure, disruption, control, or destruction of
    the adversary's critical factors," which include their "capabilities,
    requirements, and vulnerabilities" and may include

    o   "tangible components of the infrastructure such as power
        grids, communications centers, transportation hubs, or basic
        services."

o   *"intangible socio-economic or political factors* such as financial
centers and capabilities, particular *demographic groups and
sites, and cultural sensitivities."*

In addition, "both offensive and defensive JUOs will probably entail
heavy use of IO [information operations] and CMO [civil military
operations]" (Joint Chiefs of Staff, 2002, Chapter II, p. 12).

4. *Consolidate:* "In war and MOOTW, the focus of consolidation
is not just on protecting what has been gained, but also retaining
the initiative to disorganize the adversary in depth. . . . Consolidation may place heavy emphasis on logistic support and CMO. The
nature of the urban triad ensures that the JFC will have to contend
with issues concerning physical damage, *noncombatants,* and infrastructure as part of consolidation. *CMO and PSYOP units may
continue to be especially critical* in this aspect, as well as engineering efforts ranging from destruction to repairs to new construction.
Equally important are the expected issues of infrastructure collapse
and the tasks of FHA and disaster relief" (Joint Chiefs of Staff,
2002, Chapter II, pp. 12-13).

5. *Transition:* "In general, the end state of JUOs is the termination
of operations after strategic and operational objectives have been
achieved. This may include the transfer of routine responsibilities
over the urban area from military to *civilian authorities,* another
military force, or *regional or international organizations.* . . . In
JUOs, transition may occur in one part of an urban area while
engagement still is going on in another [three-block war]" (Joint
Chiefs of Staff, 2002, Chapter II, p. 13).

Note the overall emphasis on the social and organizational interactions
of a diverse set of actors, including noncombatants, noncombat forces,
and local and multinational civilian agencies. There is also a focus on the
effects of informational, socioeconomic, and political factors on attitudes
and behaviors in the urban battlespace.

Changes in the Nature and Scale of Intervention Operations

Urbanization and the broader view of military USECT interventions
yield a dramatic expansion of considerations of scale, in both spatial and
temporal dimensions, as well as an expansion in the nature and types of
intervention to be considered.

In the *spatial dimension,* urban operations demand a much finer view of
the battlespace: it is no longer sufficient to consider high-level aggregates of
large units and large geographic areas of responsibility, such as one might do

in planning conventional operations at the division level and above. Instead, the urban domain demands a block-by-block (if not building-by-building) geographic focus, at squad-level units consisting of only a few individual soldiers. At the other end of the spectrum, the broad considerations of USECT phasing of an engagement call for understanding wide-ranging geopolitical factors, including the nation-states involved, and the associated ethnic, cultural, religious, and economic factors in the region. These are typically not small or geographically focused but may in fact encompass huge spatial overlay regions of the potential battlespace (e.g., the Middle East). As a consequence, there are simultaneous demands to have a very fine spatial focus (at, say, the building level) while simultaneously being highly sensitive to the very large regional characteristics of the battlespace.

In the *temporal dimension*, a similar situation exists. The fine-scale urban focus, with its short "interaction distances," typified by an improvised explosive device (IED) or a rocket-propelled grenade, demand a very fine-grained temporal view of events for assessment, planning, and execution. Planning horizons are short, and urban operations demand a high temporal resolution of activities if operations are to succeed.[3] The time available to plan operations is likewise compressed, and planning windows are compressed, often down to minutes. At the other end of the spectrum, USECT phases can take months or years to accomplish and are often characterized by considerably slower temporal dynamics and windows, in both the planning and the execution of activities. Thus, as in the situation with the spatial dimension, there is a simultaneous stretching of the temporal dimension from both ends, from very quickly occurring events at a high temporal resolution (e.g., building clearing), to activities that evolve at a considerably slower pace, demanding low temporal resolution but long time horizons (e.g., nation building).

A key issue for modeling IOS behavior is the spatiotemporal "coverage" that must be accommodated in models. One can clearly no longer expect that a high-level aggregate model of, say, an armored division covering miles of open plain will be up to the challenge of anticipating the outcome of a fast-paced short-range small-unit urban engagement. Nor will the small-unit model be any indicator of overall outcome in the big picture of the overall military engagement. And neither is up to the challenge of anticipating outcomes in the larger USECT tableau, with its many other dimensions beyond the application of military force.

Growth of the spatiotemporal scale is also accompanied by an expansion of *intervention options* available in urban operations over the several USECT phases. This is a natural consequence of the additional dimensions

---

[3]This is perhaps best illustrated with the detailed step-by-step choreography that goes into the planning of a simple room clearing by a four-man squad.

and structures that make up the urban environment and its indigenous population, as illustrated in the deliberately simplified three-layer structure of Figure 2-1. Shown here is the conventional *physical* structure (and infrastructure) that is the focus of traditional military campaigns, on which is superimposed an *information* structure associated with elements of the underlying physical entities, in turn superimposed by a *cognitive* structure characterized by individual and group perceptions, beliefs, intentions, plans, and actions (Air Force Research Laboratory, 2004).

The focus for planning military operations is increasingly on understanding and forecasting[4] "nonkinetic" effects. Kinetic effects are associated with the use of "kinetic weapons"—conventional bullets and bombs. Nonkinetic weapons and defenses are associated primarily with IO, which include the triad of electronic warfare, computer network operations (both defensive and offensive), and influence operations, which include PSYOPS, military deception, and operations security (OPSEC). Nonkinetic options also include the use of nonlethal weapons at the individual or crowd level (e.g., high-powered microwaves) and at the population level (e.g., disabling or destroying one or more components of, say, an urban infrastructure).

In this expanded battlespace, planning and executing effects-based operations (McCrabb, 2001) require analysis of the potential effects that a given set of diplomatic, information, military, and economic (DIME) actions will have across the full range of the political, military, economic, social, information, and infrastructure (PMESII) context. To be useful for analysis and planning, behavioral models must capture not only the separate effects of each action in each of these areas but also the *interactions* of these factors.

HOW IOS BEHAVIORAL MODELS CAN HELP THE MILITARY

The changing nature of DoD's mission has greatly increased the need for IOS models that capture the cognitive, organizational, societal, and cultural factors that are critical in the urban battlespace. IOS models are needed across the full spectrum of operations, particularly during urban

---

[4]We introduce the term "forecasting" here, in place of *predicting*, to reemphasize the difficult problem of anticipating individual or organizational behavior (see Chapter 1), in comparison to that of anticipating the consequences of well-understood physical or engineering laws, the latter operating under conditions in which there is neither agency nor feedback involved (e.g., when you swing a hammer, the hammer does not deliberately try to avoid the nail in order to dissuade you from further swinging, so that your dynamic model of the muscle-hammer system is reasonably "predictive"). The term "forecasting" is also loaded with weather analogies, serving to remind us of how weather "point predictions" (in time and space) are almost always wrong and how "bounding envelope forecasts" are much more likely to capture the future trajectory of the weather, especially as the spatial and temporal resolution grows more coarse (i.e., with larger geographic areas covering "climate zones" and longer time windows covering "seasonal variations"). See also the extensive discussion of forecasting in Chapter 8.

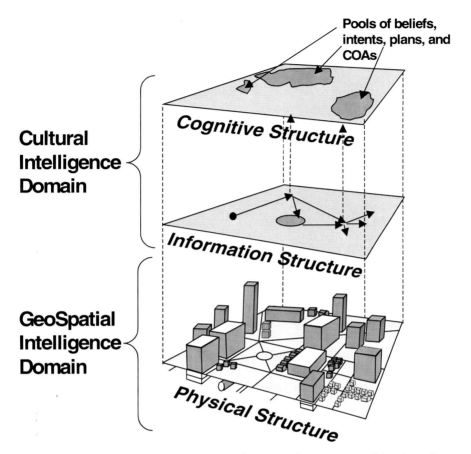

FIGURE 2-1 Heterogeneous structures that must be represented in the urban environment.
SOURCE: Air Force Research Laboratory (2004, p. 10).

operations, as indicated by the number one recommendation of the recent Joint Urban Operations Workshop: "Employ high-resolution modeling, simulations, and other decision support tools that incorporate friendly, enemy, and neutral forces, plus the urban population in order to conduct rehearsals, assess courses of action, and make better decisions faster than the enemy in an urban operation" (Mahoney, 2005).

This section reviews how IOS models can contribute to today's missions in the three broad areas: (1) analysis and forecasting for planning, (2) training and rehearsal, and (3) design and evaluation for acquisition. Another view of such applications is found in Axelrod (2004).

Potential Use of IOS Models for Analysis, Forecasting, and Planning

In general military operations, COA development and planning has been traditionally a completely manual operation, with a heavy reliance on staff experience and seat-of-the-pants "mental models" of the adversary and its likely response to potential military activities. Consequently, it is often the case that only a few COAs are generated, evaluated, and planned for—often with only minimal computer-based support.

Figure 2-2 illustrates the essential closed-loop nature of military planning and operations[5] and indicates where models could be of use. A walk around the loop begins with the external world or battlespace shown at the bottom of the figure. Many actors populate the space, including blue (friendly) forces, red (adversary) forces, and a range of others depending on the particular environment (e.g., whether it is urban or not). Some of the blue assets include sensors and data collection systems that pick up incomplete and uncertain information about the battlespace and, via associated communications assets, transmit it to a variety of data processing facilities and data storage centers. Some support relatively short-term data needs (Intelligence [INTEL] data) for current operations, while others may support long-term development of background data and knowledge bases.

The INTEL data support "inner loop" situation assessment—that is, short-term assessment of the state of the battlespace—to estimate the current situation in the face of collected information that is incomplete, noisy, and stale, which may also be compromised by reporting errors, communications failures, and deliberate disinformation on the part of the adversary. This is clearly a complex estimation process. Given this estimate of the current situation, decisions can be made and orders/requests can be modified or generated, triggering a set of general action requirements defining how to use a range of blue assets (data collection systems, weapons platforms, etc.), thus closing the loop.

Also shown in the figure is the use of background data and long-term knowledge bases to support "outer loop" situation forecasting[6] of the future evolution of the battlespace, based on the inner loop's current assessment of the situation and any available behavioral, cultural, or historical knowledge pertinent to the conflict, geography, and population. This is clearly another complex process fraught with uncertainty, both because of the attempt to forecast into the future based on knowledge of the current situation and the reliance on uncertain information stored in the knowledge

---

[5]This is essentially a more detailed version of the OODA (observe, orient, decide, act) loop of Colonel John Boyd, USAF (Ret.). For more information about John Boyd and his writings, see Defense and the National Interest (2007).

[6]The term "prediction" is shown in the figure to be consistent with the original USAF study from which the figure is adapted.

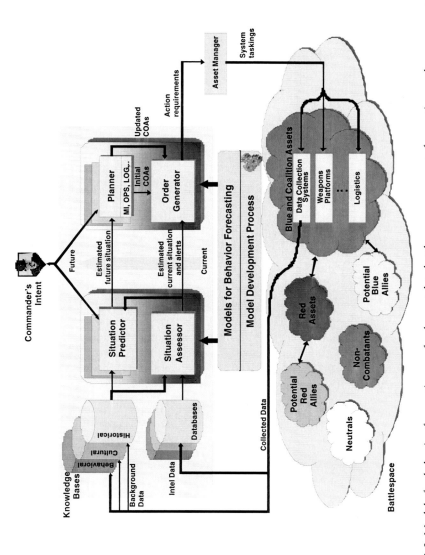

FIGURE 2-2 Models for behavior forecasting are fundamental to battlespace assessment, forecasting, and management.
SOURCE: Adapted from U.S. Air Force Scientific Advisory Board Study (2002a).

bases and generated by the inner loop situation assessment activity. The objective is to generate an estimate of the future situation (or envelope of future situations) at different time scales and geographic resolution, so as to be able to plan accordingly, across a range of time horizons, areas of responsibility, and military functional specialties (INTEL, OPS, logistics, etc.). The net results of this process are COAs and plans generated at all echelons, to support the inner loop action generation activities, as indicated in the figure.

Whether focusing on the inner loop or outer loop activities, it should be clear that current estimates and future forecasts must naturally rely on IOS behavior models of some sort. They may be implicit seat-of-the-pants mental models held by the personnel performing the intelligence, planning, and operational functions, or they may be explicit and simplified computational models (possibly instantiated at vastly different time scales or spatial resolution), but they all implicitly attempt to forecast behavior by using a model as an "extrapolation engine" operating on the current assessed state of the situation, using the best available information and knowledge collected from the battlespace, and knowing what future blue asset activities are likely to be.[7]

IOS behavior models, their associated simulations, and model-derived tools are needed to track, identify, and target critical individuals and resources and to assess the relative ability of various courses of action to influence adversary behavior and to win the hearts and minds of the indigenous population. Whether the issue is mapping the human terrain (Kipp, Grau, Prinslow, and Smith, 2006; Schaffer, 2005),[8] or understanding the atmospherics, evaluating the impact of interventions to promote or inhibit state failure, forecasting hot spots of activities in urban settings, or providing more cultural and cognitive situation awareness, IOS behavior models and their derivatives (simulations and tools) are clearly needed.

Models for Understanding, Forecasting, Shaping, and Responding to Adversary Behavior

Reliable anticipation and forecasting of individual human and collective organizational behavior on the part of the adversary is the highest goal of all military commanders. This view is embraced by the Army and the Marine Corps, in their call for doctrine and tools that enable "predic-

---

[7]This last component supporting the forecasting process assumes that blue assets behave according to plan and is predicated on the notion that "the best way to predict the future is to create it" (The Drucker School, Claremont Graduate University, 2008). See http://www.cgu.edu/pages/4181.asp.

[8]See http://www.army.mil/professionalwriting/volumes/volume4/december_2006/12_06_2.html.

tive analysis" (Kasales, 2002), especially as potential future engagements become more asymmetric and urban, less influenced by traditional formalisms of conventional military doctrine, and more determined by the contextual influences of society: political and military organizations, ethnic groups, national cultures, and transnational religious organizations (Brown, 1997; Lwin, 1997; Staten, 1998). A similar view is held by a host of military and other groups:

- The Air Force Scientific Advisory Board's (SAB's) Predictive Battle-space Awareness Study (U.S. Air Force Scientific Advisory Board, 2002a, 2002b)
- A more recent SAB study on the need for behavior modeling in urban operations (U.S. Air Force Scientific Advisory Board, 2005)
- Ongoing science and technology efforts being conducted by the Defense Advanced Research Projects Agency (DARPA) (e.g., the RAID, Integrated Battle Command, and Integrated Crisis Early Warning System programs)
- The Office of Naval Research's Affordable Human Behavior Modeling program (see http://www.onr.navy.mil/sci_tech/personnel/342/training_afford.asp)
- The USAF Commander's Predictive Environment program (see http://www.wintersim.org/abstracts06/Mil.htm)
- JFCOM's Urban Resolve program (see http://www.jfcom.mil/about/experiments/uresolve.htm and the like)
- Conferences and workshops focused on the problem (e.g., the 2003 DMSO-sponsored conference on organizational simulation, see https://www.dmso.mil/public/)
- The 2003 Army Research Institute-sponsored Workshop on Cognition and Multi-Agent Interaction, the Air Force Office of Scientific Research (AFOSR) Workshop on Culture and Adversary Modeling, 2005, 2006
- The Annual International Conference on Complex Systems (see http://necsi.org/events/conferences.html)

Before and during military operations, IOS models can serve as decision aids and as guides for data collection. Models can, have been, and should be developed to support tactical, operational, and strategic missions. Key uses for IOS models include model-based INTEL fusion and situation assessment, forecasting (projecting), planning (COA development and assessment), mission rehearsal, execution monitoring, and postexecution assessment. IOS models can be used to gather information ("see"); assess or evaluate the current state ("identify"); explain, understand, and forecast behavior ("think"); shape, manage, and disrupt oneself or the enemy ("do"); and aid in decision making or strategizing ("reflect"). IOS models hold the promise of aiding the war-fighter by providing a better toolkit for knowing the enemy.

From the tactical to the strategic levels, there is a need to forecast adversarial reasoning. IOS models can be used to provide guidance on the space of actions that the adversary might take and why, thereby reducing surprise. Moreover, these models can suggest what actions are the most probable and provide insight into the general order of actions. Such models, however, do not and should not be expected to provide guidance on exactly what action will be taken when (as we discuss later). The deployment of IOS models needs to be accompanied by training in their appropriate use and in the interpretation of the results generated.

Models for Understanding, Forecasting, and Shaping Societal Behavior

Increasing military involvement in military operations, peacemaking, and peacekeeping is creating a need for the military to understand, forecast, shape, and respond to the larger context of societal norms, expectations, perceptions, and behavior. Examples abound as to this need:

- Understanding the local society, its history, and its current over-lapping networks increases the likelihood that one might be able to identify those who would harbor terrorists or turn to terrorism.
- Understanding the local culture and its homogeneity, or lack thereof, is necessary for planning effective PSYOPS campaigns, and assessing the impact afterward.
- Shore leave has repercussions for the local population, including increasing the monetary inflow that can stabilize some local businesses while leading lead to an increase in corruption and a change in the power base.

IOS models, in general, hold the promise of enabling the identification of geographic locations where and periods when threats are likely to emerge as a function of current events and action (or inaction) by U.S. and coalition forces. Two notable DARPA-sponsored programs have focused on the identification of potential global hot spots and the forecasting of their likely evolution over time: the Pre-Conflict Anticipation and Shaping (PCAS) program (described later in this chapter), directed at forecasting the likelihood of a nation-state collapse (Popp et al., 2006), and the follow-on Integrated Crisis Early Warning System, which has as its goal "the development of state-of-the-art computational modeling capabilities that can monitor, assess, and forecast, in near-real time, a variety of phenomena associated with country instability" (see http://www.darpa.mil/ipto/solicitations/open/07-10_PIP.pdf).

Multiagent models and system dynamic simulations that take social and cultural factors into account could be used to assess the likely consequences of the COAs executed by the U.S. military and coalition forces on state stability. They could be used to assess the potential impact of a multipronged initiative with DIME, and to assess the consequences in PMESII. This is exactly the focus of an ongoing DARPA-sponsored program in Integrated Battle Command (see http://www.darpa.mil/sto/solicitations/IBC/).

IOS models could also be used for identifying what COAs will have the least negative or most positive effects on civilians and neutrals. In all cases, the value of such models and simulations could be enhanced if information on the underlying social and organizational networks and available resources were taken into account and if the models were combined with effects-based operations models. Models could also be used to determine the effects of U.S. information operations activities designed to influence attitudes and behaviors of individuals in different cultures. In this case, the effectiveness of these information activities is likely to be enhanced by linking social network models with psychological profile information, cultural models, and psychosocial models.

Models for Understanding Enemy Command and Control Structures

Understanding enemy command and control structures through IOS behavioral models enables the identification of vulnerabilities and strengths *before* planning friendly activities. This was pointed out in a previous study on human behavioral modeling (National Research Council, 1998) but has only recently been acted on because of the post-9/11 focus on counterterrorism and the rapid development and dissemination of supporting tools.

In particular, support tools, such as text and data mining facilities, are beginning to be used for extracting information from open sources (e.g., news articles, websites, etc.) to identify events and structures that enable the detection and recognition of terrorist and insurgent networks and organizational structures. Much of this work is classified, but significant insight can be gained from parallel efforts ongoing in the commercial world (see, for example, the growing conference on "text analytics" at http://www. textanalyticsnews.com/usa/program.shtml).

However, the strength of these tools could be considerably increased if they were combined with social and dynamic network modeling techniques, to enable a model-based approach to text and data mining and information fusion. Such tools could then be used in an "alert" mode to identify what data should be collected, as well as in an "evaluate" mode to suggest the impact of various courses of action. Given an understanding of the enemy's command and control structure, targets could be identified for disrupting the enemy, and various courses of action developed to achieve those goals. COAs that are intended to demoralize, disrupt, or inhibit action or recruitment on the part of the enemy could then be evaluated if behavioral models incorporated realistic affective, social, and cultural influences. These models could be made even more effective if they were placed in data-farming environments so that the space of COAs could be more effectively mapped.

Models for Training and Mission Rehearsal

IOS behavior models have the potential for providing significant benefit to U.S. and coalition forces in training (of general skill sets) and mission rehearsal (for mission-specific expertise), based on years of experience of simulation-based training in more conventional areas (e.g., flight training, tank tactics training, etc.), the critical dependence of learning on performance feedback and "after action review" insight, and, perhaps most importantly, the opportunity to learn from errors that might be devastatingly fatal in the real world. Key uses of models for training include model-based simulation of virtual actors (including simulated entities, such as teammates, adversaries, and noncombatants), games to provide immersive experiences, and models to preassess potential new training tools. Model-based training simulations and systems can provide training that:

- supports a number of activities, such as teaching individuals how to be more culturally aware, training teams how to coordinate and fight as a unit, training commanders how to evaluate the organizational health of their battalion;
- enables live, large-scale war-gaming with truly dynamic enemies— these training systems can be constructive or done using virtual reality or gaming systems;
- takes into account social, cultural, or organizational factors and can be used for realistic training of individuals, teams, or organizations; and
- crosses operational activities and enables joint or coalition training.

Training at all levels, up to and including higher headquarters staff, is vital to ensuring successful joint and coalition operations. New demands are being placed on training and rehearsal systems that increase the need for modeling to support training. Effective training and rehearsal systems immerse the trainee in realistic scenarios, provide information about roles and responsibilities, enable the development of technical skills, and provide experience working in joint or coalition task forces, facing new, dynamic, and culturally distinct enemies. The problem is providing such an immersive training environment in less time, for less money, for more personnel, using models and simulations that can be rapidly adapted to changing missions and new adversaries.

The changing nature of military missions and the increased emphasis on peacekeeping, disaster relief, and nation-state building means that training systems are needed for more than conventional training in weapons usage, war-fighting tactics, and basic survival. Indeed, there is now a need for training in cultural awareness, crowd behavior, negotiation, management, and city planning. Training systems need to be more flexible to capture the changing nature of the enemy from nation-states to insurgents, from one monolithic actor to federations of loosely aligned tribes, and from large-scale weapons to improvised explosive devices. Finally, new training technologies are required due to changes in military staffing, such as fewer, more computer-savvy recruits, increased use of reserves, and just-in-time training.

Massively multiplayer online games (MMOGs) can, in principle, provide such an immersive environment. The value of such systems for mission rehearsal will be increased if the cultural and social models embedded are more socially realistic than those in current games. Dynamic network models hold the promise of providing a dynamic adversary for war-gaming, and the value of such systems will be increased if links can be made between the network models and models of action, planning, and goal attainment.

Training can be provided for teams and larger units by populating scenarios with socially and culturally realistic artificial actors as team members, but there is a need for simulation infrastructure to rapidly develop socially realistic and culturally differentiated artificial actors.

Training for future scenarios can be provided by using system dynamic and multiagent computational models that allow the user to look ahead and do what-if analysis of alternate scenarios, such as the impact of tsunamis or hurricanes on various regions of the world or the impact of avian flu on military personnel. The value of these systems will be increased if they move beyond military and economic factors to consider social, political, diplomatic, and information factors.

Expert systems and cultural models can be used to increase cultural awareness and train military personnel in crowd control behavior. The value of these systems will be increased if they can be rapidly populated with data as new adversaries arise.

In general, IOS behavioral models can be used effectively for improved training, but more realistic models of actors, groups, and nation-states are needed. A key aspect of the current training and rehearsal process is that, during training, military personnel are provided access to people (or their model-based surrogates) with whom they will be working in the field. For example, at joint war games, Air Force, Army, Navy, and Marine personnel meet, plan, and execute together. This increases their transactive knowledge of who knows what and who can do what, which in turn improves group performance. If IOS models can help support this function, war games can be reduced in size, conducted more frequently, and tuned to specific individuals and organizations needing specific training or rehearsal.

Models for Military Systems Development, Evaluation, and Acquisition

DoD is in the midst of two revolutionary changes (Frost, 1998): a revolution in military affairs (RMA) and a revolution in business affairs (RBA). The RMA involves the military requirements and concepts envisioned in light of the threat environment and advances in technology (Joint Chiefs of Staff, 2000). The RBA addresses how to leverage technology and commercial business processes to the how-to-buy problem. The use of M&S has been recognized as a key facilitator to addressing military training and education and the acquisition of military systems (Office of the Secretary of Defense, 1996). For example, the Air Force's Modeling and Simulation Strategic Plan (Johnson, 2004) spells out a number of points of focus to meeting the challenges of RMA and RBA by providing a persistent synthetic battlespace infrastructure to support the exploration, design, development, analysis, and testing of new war-fighting systems and concepts (as well as more conventional military training and mission rehearsal activities).

Inherent in the necessary M&S infrastructure to support system acquisition is the requirement to provide realistic representations of battlespace entities (blue, red, and neutral), natural and cultural features (terrain, locale), and physics-based effects (sensor processing, missile flyout, etc.). Such an infrastructure must provide a framework to support the rapid integration of these synthetic representations across all simulation levels (i.e., campaign, mission, engagement, or engineering). In the Air Force, this M&S capability has direct relevance to important acquisition programs, such as the Joint Strike Fighter, the Multi-sensor Command and Control Constellation, the Joint Distributed Engineering Plant, and the Distributed Mission Operations/Mission Rehearsal initiative. The other services have their own acquisition programs in which M&S capabilities are directly relevant to effective acquisition. As a consequence, a substantive effort is needed to develop the requisite M&S components to address the evolving threat environment.

IOS behavior models that could support automated means to generate and adapt red strategies/tactics in line with asymmetric warfare can be a key enabling component for meeting the objectives needed to support acquisition. Key uses here are to preevaluate the value of new technology in a variety of scenarios that are both physically and culturally accurate, assess the need for particular skills in soldiers, and assess how generational changes in soldiers may lead them to need or utilize technologies differently from their predecessors. Realistic IOS models could also be put to good use while systems are under development, especially as more acquisition programs adopt the "spiral development" paradigm, in which the requirements for each new spiral are determined not only by the evaluation results of the past spiral, but also by the changing battlespace requirements generated by an adaptive and resourceful adversary. (The counter-IED development work is an excellent example of rapidly changing tactics and countertactics; see http://www.nationaldefensemagazine.org/issues/2006/jan/adaptive_foe.htm.)

Models for Enabling Command and Control Weapons Systems

In the current network-centric operating environment (Alberts and Hayes, 2003), command and control (C2) organizations and the information infrastructure that supports them are becoming increasingly important. One could say that C2 organizations, their information architectures, and their underlying communication infrastructures have in fact become the new weapons. This was underlined in 2004 when General Jumper, then Air Force chief of staff, officially designated the Air and Space Operations Center as a weapons system. And it has been formally recognized with the newly formed Cyber Command in the Air Force. As stated by the organization's first commander, General Elder: "The Air Force now recognizes that cyberspace ops is a potential center of gravity for the United States and, much like air and space superiority, cyberspace superiority is a prerequisite for effective operations in all warfighting domains" (Wait, 2007).

The Army and the Navy similarly recognize the leverage obtainable from effective C2 systems, especially given the Army's commitment to the networked Future Combat System program[9] and the Navy's effective invention of the term "network-centric warfare" through Admiral Cebrowski's leadership,[10] but it is fair to say that all three services have tended to focus on the hardware and software infrastructure (communications pipes, fusion algorithms, decision aids, visualization techniques, etc.), with less emphasis on the human and organizational component of effective C2.

What is becoming increasingly clear, especially in light of the current conflicts in Afghanistan and Iraq, is that there is a critical need for the rapid design and redesign of military units, including the architecture of their C2 systems, to meet changes in missions and to respond to innovations in enemy activities. Related to this is the need to be able to identify vulnerabilities in current C2 structures. IOS models have significant potential for assessing, designing, and evaluating the impact of new technologies or new C2 procedures on potential vulnerabilities, strengths, shared situ-

---

[9]See http://www.army.mil/fcs/.
[10]See http://www.oft.osd.mil/biographies/cebrowski_with_pic.cfm.

ation awareness, work distribution, and adaptability to enhance friendly operational effectiveness, while defending against enemy actions, across a full spectrum of cultures, nations, and nonnation-state actors. Using such models has the potential for moving the military beyond logistics planning to organizational planning, facilitating improved recruitment strategies, and enabling just-in-time team design.

Military personnel often remark that, as soon as they get to the field, the plan goes out the window. One way of building in flexibility (or resilience) is to support the design and redesign of the various units to take into account changes in mission, changes in technology, attrition, rotation, or incorporation of joint and coalition forces. Many times such design needs to be made on the fly as the situation changes. In this case, the commander is faced with the problem of identifying experts quickly and incorporating them in a "tiger team." A related problem is assessment of the unit's organizational health, its vulnerabilities, its shared situation awareness, and its overall war-fighting effectiveness.

A traditional approach to organizational design has been to identify structures that are optimized to meet some organizational criteria. This approach is insufficiently flexible in many cases, as the military needs to operate in a responsive and adaptive mode. Criteria for designing adaptive units are under investigation, and emerging behavior modeling efforts are beginning to afford new possibilities for organizational design (Levchuk, Yu, Levchuk, and Pattipati, 2006, 2004; Pattipati, Meirina, Pete, Levchuk, and Kleinman, 2002; Neal Reilly, 2006; Levchuk, Levchuk, Meirina, Pattipati, and Kleinman, 2004; Levchuk, Levchuk, Luo, Pattipati, and Kleinman, 2002a, 2002b; Entin, 1999).

IOS models, simulations, and assessment tools could be used to pre-evaluate the impact of new technology on the unit, identifying potential ways in which the unit's structure should change in response to this insertion. Dynamic network models linked to various databases with streaming information on personnel could enable real-time assessment of shared situation awareness and organizational health. Text mining tools and shared mental model assessment tools could be used to improve information flow and rapidly process incoming data. IOS models of unit needs could be used to form a "smart" command center that could be used to push information to people only when they need it. Design tools and smart command center tools are well within the reach of current technology. The key problems are those of scalability, handling streaming data, and linkage of noninvasively collected data to dynamic network metrics of organizational health and text-mining evaluations of information flows.

Representative Model-Addressable Problems in a Scenario Context

We now illustrate how different behavioral models might be used to address specific questions raised by the commander and his staff during the course of operations, how models might be used to train for particular skill sets and missions, and how models might be used to help specify an "optimal" organizational design for a given mission. The intent is not to be exhaustive but merely illustrative, with the goal of motivating a closer look at the use of detailed behavioral models in a range of military activities.

To orient our analysis and review of IOS modeling approaches, we developed scenario elements that are derivative of the one detailed in TRADOC PAM 525-3-90 O&O 22 JUL 2002 (U.S. Army, 2002), which describes a mission in the trans-Caucasus region. Three vignettes are developed to provide a construct for the purpose of addressing the potential of behavioral models supporting operations of a brigade combat team as part of a joint campaign. These vignettes center around

- tactical operations in entry operation (entry),
- operational maneuver by air, combined arms operation for urban warfare (transition), and
- secure portion of a major urban area (JUO).

Details of the scenarios and vignettes are given in Appendix B. In conjunction with an Army subject matter expert, we have specified representative model-addressable questions for portions of three vignettes. We think that these vignettes and the associated model-addressable questions only begin to scratch the surface in terms of providing suitable challenge problems to stimulate the modeling community and to provide a common reference frame for discussing alternative approaches to the same problem. In fact, we propose, in Chapter 11, that an initial effort in a large-scale, multiyear research program—focusing on comparing and integrating different disciplines, perspectives, and levels of detail—be dedicated to the definition of a number of well-defined and highly focused challenge problems that can serve as a common basis for comparing and contrasting different approaches. If the vignettes and questions presented here can serve as a launching point, some effort might be saved in the long term, but the primary purpose of presenting these in this study has been to focus the committee on relevant military problems and to provide the reader with some sense of the broad range of challenges that exist in the military domain.

Box 2-1 shows the resulting representative high-level model-addressable questions. Given these representative problems and issues, a number of more specific questions were generated, to illustrate the kinds of specific questions that might be asked during the unfolding of the vignettes.

BOX 2-1
Representative Model—Addressable Problems and
Issues of Interest to the Commander

("We" in this box refers to the commander and his forces.)

Analysis and Forecasting for Planning

Disrupt terrorist networks. Fuse uncertain and partial information from multiple sources to identify the dynamic network structure of a terrorist organization. How can we best disrupt those networks?

- Tribal leader Muhkta is on the fence about whether or not to support the intervention. Which is likely to be the most effective way of gaining his support—overt recognition, overt financial reward, covert financial reward, covert protection of family, or a combination of methods?
- We need to disable/disrupt the clan of followers of Sheik Mustafa while our troops are moving toward the city. If we ensure he is disconnected from his clan during this phase of the operations, is it likely to degrade the clan's decision making as related to their willingness to conduct offensive military operations?
- In order to reduce IED attacks, are the terrorist networks with their support base in our target city more vulnerable to selective attacks on their leadership or interruption of their recruitment programs?
- Abdul X is the leader of a terrorist network. Mohamed is on the network council and more radical than Abdul X. If Abdul X is killed, how likely is it that Mohamed will become the leader of the network?

Forecast adversary response to COAs. In an urban operation, forecast the likely response of local insurgents to friendly force movements, basing, and logistics. Identify likely counters to proposed COAs and identify early harbingers of those counters.

- What will impact the local economy the least: denial of transportation fuels or denial of electricity?
- The JTF can plan on placing its logistics support base either within the bounds of the city or in the adjacent countryside. Which population in the area, urban or rural, will be less hostile to the presence of the logistics base?
- To establish crowd control early in the urban environment, is controlling an area, like the civilian neighborhood, or a point of special interest, like a mosque, more likely to mitigate crowd behavior?
- In neighborhoods not committed to radicalism, what is the most influential means to insert forces: in combat vehicles or on foot?
- JTF wants to use disinformation to partially protect our intentions of moving from forward operating base (FOB) to the city. Is the most effective point of insertion of the disinformation the few public media outlets or the informal rumor mill/tribal network?

BOX 2-1 Continued

Societal forecasting. Forecast the effects of alternative diplomatic, infrastructure, military, economic courses of action on attitudes and behaviors of residents in a region of interest. Assess the likelihood of state failure and identify actions that will lead to escalation of violence.

- Troops give a lot of meals ready to eat (MREs) to locals. Considering the items in MREs and the local culture, will MREs be a better giveaway than basic grains and cooking oil?
- Entry phase combat will be kept at the lowest level possible. Given local conditions and the impacts of the blockade, which will the locals respond better to initially: engineers/civil works or medical response teams?
- Considering the effect of the blockade, which will have the psychological effect most supportive of our mission end state: overwhelming force or "helping hand" intervention?
- Which approach will least offend locals as we travel from the initial entry area to the city: keeping civilian vehicles in a separate convoy or infusing them into tactical convoys?
- Can we forecast the response by the local religious leaders to the presence of female soldiers on the streets of the city?
- A specific Mosque is known to be the headquarters of a particular militia. Joint forces will destroy the mosque in order to deny access by the militia. Which will produce the least negative impact in the neighborhood: announcing our intentions to destroy the mosque or destroying it unannounced?
- How do attitudes differ between the tribal regions of the country and the urban area we are targeting?
- What is the formal communication dynamic between the host national government (HNG) and the population? What is the informal communication dynamic? (How do people get information on a day-to-day basis—coffeehouses, religious structures, etc.?) How great is the delta between formal and informal communication dynamics?
- What are the expectations of the population about the government's ability to provide services?
- Is the HNG a government on the road to collapse?
- Are there indicators of popular support for the alternative power structure? Are they reflected in local media and among the local intelligentsia?

Training and Rehearsal

Crowd control training. Create an immersive virtual training environment in which soldiers can learn to take appropriate action based on the correct interpretation of the behavior of small groups of citizens and understand the triggering mechanisms for violent responses by the crowd.

continued

BOX 2-1 Continued

- To effectively control crowds we need to know where the leaders are. In this setting, are crowd leaders more likely to be leading from the front, urging from the rear, or not on site? Given the answer, should we use information operations or force to control the crowds?
- Given the nature of the small villages along the route from our FOB to the city, is it likely there will be crowds along the route, are they likely to be friendly or hostile, and in either case will stopping to interact with them be likely to alter their feelings?

Design and Evaluation for Acquisition

Organizational design: force composition and command and control architecture. The Army is moving toward modular forces focused on joint and expeditionary capabilities. These units of action will be rapidly reconfigured and equipped for specific mission requirements. The Navy is fielding expeditionary strike groups that include marine expeditionary units capable of amphibious operations attached to Navy ships. The Navy and the Marines follow different doctrine and are in the process of defining flexible supporting and supported relationships that allow them to function effectively as a combined fighting unit.

- Develop a recommended force composition (systems, equipment, units and personnel) for a humanitarian assistance mission.
- What command and control architecture will be most effective for this mission?
- What are the appropriate organizational coordination points for most effectively working with NGOs during the humanitarian assistance mission?
- Is the force composition structure recently used for a humanitarian assistance mission appropriate for a disaster relief operation that requires immediate deployment?
- Are new roles needed to take advantage of the information-rich network-centric environment? For example, would an information commander/coordinator role result in more effective mission performance?

OVERVIEW OF CURRENT DOD IOS MODELING EFFORTS

In this section we briefly review major IOS behavioral modeling efforts under way to address military questions such as those described above, pointing out some of the major challenges that confront these efforts.

The DMSO Master Plan for Modeling and Simulation

In 1995, DoD published a master plan for M&S, in an attempt to unify efforts across all services, identify needed areas of development (gaps), and minimize duplication of efforts (overlaps). The plan was "the Department of Defense's first step in directing, organizing, and concentrating its M&S capabilities and efforts on resolving commonly shared problems" (U.S. Department of Defense, 1995, p. i). The Defense Modeling and Simulation Office (DMSO) was given six major objectives under this plan, including "provide authoritative representations of human behavior" (U.S. Department of Defense, 1995). The DoD M&S master plan also specified a set of more detailed subobjectives for achieving these goals, as well as a detailed timetable for initiating and concluding some of these activities. The lofty goals and aggressive timelines of the DMSO master plan have not been achieved, 10 years after they were first promulgated.[11] A quick review of DMSO's Modeling and Simulation Resource Repository (MSRR) at http://www.msrr.dmso.mil/ would appear to demonstrate this. The MSRR system is maintained by the Modeling and Simulation Information Analysis Center (MSIAC). It includes five nodes representing the three services (Army, Navy, and Air Force), the DoD system, and the Defense Intelligence Agency and provides "retrieval of metadata descriptions of modeling and simulation resources" (Defense Modeling and Simulation Office, 2007), including models, simulations, frameworks/toolkits, background reference material, and the like. The following bullets summarize the results of a recent (December 2006) search of the three service nodes:

- The Army node (see http://www.msrr.army.mil/) indexes 926 models, simulations, and simulators. Of these, fewer than 20 relate to individual human cognition, behavior, or performance. Of those, four focus on human visual performance (e.g., VISEO), three on human-in-the-loop (HIL) simulators, one on anthropometry (Jack), and the remaining few on four distinct behavior models: IMPRINT, IUSS, MATREX, and OneSAF (more on these later). Of the same 926 modeling resources, only three relate to group or organizational modeling: C3GRID (built on MATREX), a crowd model based on diffusion kinetics (RDEBBSM), and a software tool for building an organizational model (C3TRACE). Searching for models associated with the keywords "culture/cultural," "economic," "ethnic,"

---

[11]It is beyond the scope of this study to attempt to do a forensic analysis of DMSO performance in this area. A number of factors may have contributed: a problem scope that was simply "too big" for the funding and personnel resources available to DMSO; a science and technology portfolio decision that emphasized simulation engineering issues over basic science and technology; a political/economic environment that pitted DMSO against the entrenched M&S agencies in the services (Army, Navy, Air Force) and other agencies; etc. But it certainly would be worth revisiting the office's past history, should recommendations be made to rejuvenate the office or to create a new one with similar responsibilities.

"political," "religion/religious," and "social" yielded no hits on the database.[12]

- The Air Force node (see http://afmsrr.afams.af.mil/) indexes 54 models, 39 simulations, and 26 simulators. Of these, less than a dozen relate to individual human cognition, behavior, or performance, and of these, two refer to HIL simulators, one to anthropometry (INTERMEDIATE), one to decision aiding (for target prioritization), and the remaining few on generic frameworks (DIAS, FLAMES, ICET) or distinct behavior models (CART/ IMPRINT, JSAF, OMAR, and STELLA). Searching for models associated with the keywords "political" and "social" called up the DIAS generic framework and the IO suite for command and control warfare, both developed by the Air Force Agency for Modeling and Simulation (AFAMS), but neither explicitly representing human behavior. Models associated with economic features were focused on acquisition, and no models were associated with the terms "culture/cultural," "ethnic," or "religion/religious."

- The Navy node (see http://nmso.navy.mil/) indexes 832 models and simulations. Of these, fewer than a dozen relate to individual human cognition, behavior, or performance, with most focusing on HIL simulations or human visual performance. Only one distinct (cognitive) behavioral model is called out: the Air Defense Commander simulation (full name Autonomous Agent-Based Simulation of an AEGIS Cruiser Combat Information Center Performing Battle Group Air-Defense Commander Operations), which models small-team performance in C2 (Navy Modeling and Simulation Office, 2004). Searching for models associated with the keywords "culture/cultural," "economic," "ethnic," "organization/organizational," "political," "religion/religious," or "social" yielded no hits on the database.[13]

One might be led to conclude on the basis of these results that the M&S community is not active in developing models of individual and group behavior. This is not the case. Rather, MSIAC is simply not keeping pace with the explosive development and application of behavioral models that

---

[12]The term "economic" did identify two tools not related to human economic behaviors, and the "social" search term did identify the SPECTRUM facility at the National Simulation Center at Ft. Leavenworth, which claims to "use a subject matter expert developed database to describe the political, economic, and social characteristics of the region being simulated" for use in HIL war-gaming simulations. The SPECTRUM description was last updated in 1998.

[13]The "organization/organizational" keyword did identify several organization-level trainers used by the Navy and dependent on HIL operation but not organization-level behavior models.

started in the mid-1990s and continues to grow today, both inside DoD and in the behavioral research and computational modeling community. And no one else is keeping pace with the M&S effort either, even within DoD. There simply is no comprehensive archive or summary of all human behavioral models developed for or applied by DoD, although several organizations have specialized "snapshots" with associated information on their state of technical readiness, limitations, availability, etc. Clearly, an across-DoD survey, maintained in a regularly updated fashion, would be particularly valuable, especially if it went beyond a simple verbal description and attempted to describe each M&S resource in a common ontology or framework, so that comparisons could be made across models and simulations.

Selected Current DoD Behavioral Modeling Efforts

A complete survey of the state of IOS model development and application (both inside and outside DoD but having potential for use in DoD applications) goes beyond the charge of this committee. We can, however, give a brief overview of some of the more visible efforts[14] on the basis of surveys conducted outside MSIAC and on the basis of the committee's knowledge of the domain. It is appropriate to note that our focus here is *not* on the traditional M&S tools used by the military operations research and training communities (e.g., AASPEM, CASTFOREM, CBS, CCTT, CSSTSS, EAAGLES, EADSIM, EAGLE, JANUS, JCATS, JCM, JWARS, MTWS, TACBRAWLER, TACSIM, WARSIM2000), which focus on the physical aspects of the battlefield and the associated sensor/weapon/C2 systems. Instead, it is on the complex behaviors generated by the individuals, teams, and organizations of people populating the battlespace.[15] In fact, few IOS models are being used on a daily basis by war-fighters, military planners, or military trainers. Most existing models accredited[16] for use by military personnel are large-scale models of physical systems that do not take social, cultural, organizational, or affective factors into account. Key exceptions are identified in Appendix Table 2-A1 at the end of this chapter, which tabulates some of the major current efforts in this latter area of

---

[14]Many are classified or simply buried in organizational stovepipes, leading to significant overlap or duplicative activity.

[15]A brief overview of these military simulations is given in National Research Council (1998, pp. 33–50).

[16]Verification, validation, and accreditation is a well-defined DoD process; an overview of this overall "certification" process is given in the section entitled "Military Approaches to Verification and Validation" in Chapter 8. In simple terms, accreditation occurs when the accrediting agency (the owner of the simulation) places its stamp of approval on the validation results.

militarily relevant IOS behavioral modeling. In the paragraphs that follow, we describe only a few of these activities to provide a general sense of the overall effort—since a full review is beyond the scope of this study.

OneSAF Family of Models and Simulations

The OneSAF family of simulations includes One Semi-Automated Forces (OneSAF); OneSAF Objective System (OOS); OneSAF Testbed (OTB); Joint Semi-Automated Forces (JSAF); and ModSAF (Modular Semi-Automated Forces) and provides some capabilities for modeling human behaviors that vary by culture. Underlying this is the OneSAF Test Bed, which is a model derived from ModSAF (Parsons and Wittman, 2004). OneSAF has behavior representations that are effectively implemented in code and facilities that enable the user to rapidly develop instantiated models of new groups, communities, etc., considering a set of social, economic, and political factors. Although OneSAF is more flexible and provides better culturally sensitive modeling than was previously possible, it still has limitations. One is that OneSAF is still under development but is nearing government acceptance testing for the initial operating capability. A second limitation is that it is not clear at this time how alternative models could be linked to or federated with OneSAF. Finally, the structure by which cultural variables are included in OneSAF may limit the type of cultural factors that can be included.

Task Network Models and Tools

Task network models describe actors' behaviors in terms of interdependent tasks to be accomplished in order to achieve an overall goal. These models have their foundations in the Navy's PERT[17] chart development in the early 1950s and owe their popularity to the ease of constructing them and the clear visualization they afford in terms of task interdependencies and task completion progress. MicroSAINT[18] popularized their use in the 1970s in modeling human performance in tasks via task networks by (1) adding simple human performance parameters to each block in the network (the likelihood of correct task completion, time to complete, etc.); (2) making graphical task network construction easy to do by the nonspecialist; and (3) providing a discrete-event standalone simulation environment for exercising the model over time.

Many task network models have been developed for simulating military tasks, and the basic MicroSAINT language has been extended by develop-

---

[17]PERT stands for Program Evaluation Review Technique, a methodology closely related to the Critical Path Method used to identify bottlenecks in overall task progress.

[18]See http://www.adeptscience.co.uk/products/mathsim/microsaint/.

ment supported by ARL under the IMPRINT program, as well as by subsequent extensions by the Air Force Research Laboratory (AFRL) under the CART program, to support embedding into other simulation environments. Several derivatives have been developed by the behavior modeling community, including C3HPM, which builds on IMPRINT, and C3TRACE and HOS, which build directly on MicroSAINT. More sophisticated researchers, particularly from the ACT-R community, have made efforts to integrate MicroSAINT models with more traditional cognitive architectures.

Cognitive and Cognitive-Affective Architectures and Models

A wide variety of cognitive and cognitive-affective architectures and models are represented in Appendix Table 2-A1. Although their history may not be as long as the task network models (going back perhaps 25 years to the pioneering work of Anderson, 1983), they are remarkably diverse in their underlying structures, their associated computational implementations and development tools, and their applications, both military and nonmilitary. This includes the "pure cognitive" architectures/models, which tend to be standalone and used within their own communities (ACT-R, CLARION, COGNET, EPIC, OMAR and D-OMAR, SAMPLE, and Soar); the "hybrid cognitive" architectures/models, which bridge the gap between communities by combining models (EPIC-ACT-R, IMPRINT-ACT-R, Soar-EPIC, and others); and the cognitive-affective architectures, which extend the pure cognitive into the affective domain (MAMID, MINDS, and PMFServe).

One major commonality among all of the architectures/models is that they were developed—initially at least—with the goal of modeling the *individual* human faced with dealing with some sort of cognitive task. That focus on the individual has been maintained while extensions have been made in many different directions (perception, motor control, affect, memory, multitasking, among others). It is only recently that significant effort has begun to be devoted to dealing with modeling groups of individuals, from small teams to large organizations. As described in the agent-based modeling (ABM) section of Chapter 6, one of the primary barriers to representing the behaviors larger groups of individuals using cognitive and cognitive-affective models is the computational constraints: These models tend to be very fine-grained, and running a large number of them on a single host quickly brings the simulation to a grinding halt. However, this is expected to be less of a problem as the hardware's computational speed increases, and better use can also be made of parallelism across multiple platforms. But a more fundamental problem exists: the lack of social knowledge in most of these representations. Cognitive modelers are keenly aware of the need to incorporate mental models of the environment they are interacting with, but they seem to be less so inclined regarding the mental models of the other agents they are interacting with, perhaps because of the infinite regress involved. This is clearly a needed direction for further research if this category of ABM is to succeed in modeling larger collections of cognitive and cognitive-affective agents.

Multiagent Systems

ABM environments and multiagent systems trade off the complexity of individual cognitive-affective agents for an increase in the sheer number of agents and a concomitant increase in the complexities in interagent interactions. These are described in more detail in Chapter 6, but it is worth commenting briefly on the three multiagent models highlighted in Appendix Table 2-A1 and how they have been extended and applied to DoD questions of interest. Construct is a multiagent network simulation framework that supports the modeling and analysis of *dynamic* agent networks that evolve over time as a function of agent-to-agent interactions, and it clearly has direct applicability to the growth of terrorist networks. CORES is a multiagent environment that supports the inclusion of DIME/PMESII factors in the agent interactions, to support understanding of broader contextual factors in agent and network behaviors. BioWar combines multiagent models of social networks, disease models, and population demographics into a single integrated model of the impact of a biological warfare attack on a city. Additional multiagent models and frameworks developed at Carnegie Mellon University's Center for Computational Analysis of Social and Organizational Systems include DyNet, NetWatch, OrgSim, and VISTA, and the reader is referred there for further information (see http://www.casos.cs.cmu.edu/).

For truly large-scale multiagent model development efforts, a number of communities are developing domain-free MAS frameworks and toolkits. These include SWARM, developed in the Center for the Study of Complex Systems (see http://www.cscs.umich.edu) at the University of Michigan; the Java-based REPAST agent simulation environment (North et al., 2005; Tatara et al., 2006); and MASON, another Java-based multiagent simulation environment, developed at George Mason University (see http://cs.gmu.edu/~eclab/projects/mason/). At the time of this writing, it is unclear what, if any, inroads have been made into the DoD M&S community.

Massively Multiplayer Online Gaming

America's Army is an MMOG developed by and for the Army (Zyda, Mayberry, McCree, and Davis, 2005). The game was designed as a recruiting (Belanich, Sibley, and Orvis, 2004) and training (Farrell, Klimack, and Jacquet, 2003) tool to paint a realistic portrait of combat in the U.S. Army. The game falls into a first person shooter (FPS) game genre, and all the game features are based on the real world. However, it goes well beyond being an FPS game (Nieborg, 2004), since social and cultural factors are increasingly being embedded in both the scenarios and the attributes of the roles that the players can take on. Additional information on America's Army is provided in Chapter 7.

DIME/PMESII Models

A number of behavior modeling efforts aimed at understanding large-scale behaviors—at the societal and nation-state levels—are under way to explore the effects that DIME actions will have across the full range of the PMESII context. These include DARPA's IBC, PCAS, and ICEWS programs, the Air Force's SROM effort, and JFCOM's SEAS program.

The Integrated Battle Command (IBC) program (Allen, 2004) emphasizes linked and networked behavior models that can support military planning and decision making for dealing with asymmetric threats embedded in an urban environment. The approach clearly recognizes the importance of obtaining and maintaining a clear understanding of the complex sociopolitical context. In terms of planning and executing effects-based operations (McCrabb, 2001), this translates into the analysis of the potential effects that a given set of DIME actions will have across the full range of PMESII variables. The key to successfully executing such encompassing analyses lies in the development of the embedded behavior models representing the full range of PMESII variables and how they can be individually and collectively affected by specific DIME actions.

A conceptual representation[19] of the model "space" is shown in Figure 2-3, in which the dimensions are the DIME dimensions, the PMESII dimensions, and the modeling paradigms themselves, this last shown as modeling "families." As noted in the program description (Allen, 2004; see http://www.afcea.org/events/pastevents/documents/AFCEAIICPanel.ppt):

> Each model in the family may represent its portion of the domain in a manner and level of fidelity quite different from other models. . . . The Modeling Paradigms include techniques such as: concept maps, social network models, influence diagrams, differential equations, causal models, Bayesian networks, Petri nets, event-based simulation, and agent based

---

[19]Clearly, this is not intended to represent modeling "reality" in any sense but is merely an attempt to illustrate (1) the concept of different modeling paradigms/families covering different portions of the DIME/PMESII modeling space; (2) the potential for their interacting, e.g., outputs of one driving the inputs of another; and (3) the possibility of uncovering "unintended consequences" through these interactions. But it must be recognized that, fundamentally, the figure is merely an illustration of the concept of multiple models interacting at multiple levels and nothing more.

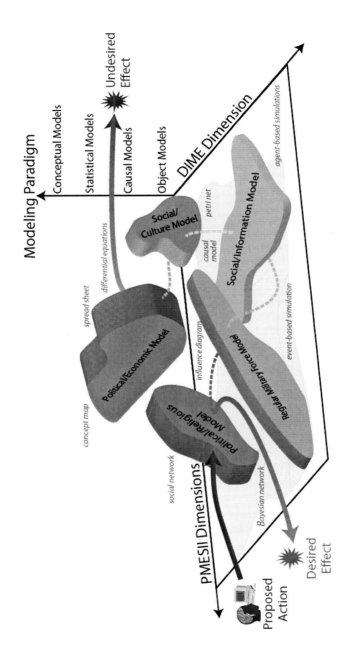

FIGURE 2-3 IBC modeling space.
SOURCE: Adapted from Allen (2004).

simulation. The need for a variety of modeling paradigms also stems from the fact that the different domains of knowledge do not lend themselves to being represented by one common paradigm such as an influence network. Also, human subject matter experts have preferences in the use of different paradigms and different paradigms fit different styles of thought.[20]

The figure also illustrates how different models in different families interact via their interconnections (inputs, outputs, and state interactions). An analyst may investigate the impact of a DIME action and a model may forecast a primary PMESII outcome, but that effect may also "stimulate another model that predicts an effect that stimulates another model and in a cascade manner, the family of models, in a symbiotic manner, may predict another effect. Such cascading can produce astonishing results because, while a human may grasp and master a single model, it is unlikely that a human can predict the complex interactions between models!" (Allen, 2004; see http://www.afcea.org/events/pastevents/documents/AFCEAIICPanel.ppt).

As noted earlier, DARPA has sponsored two other programs focused on the identification of potential global hot spots and the forecasting of their likely evolution over time: the PCAS program, directed at forecasting the likelihood of a nation-state collapse (Popp et al., 2006), and the follow-on Integrated Crisis Early Warning System, which has as its goal "the development of state-of-the-art computational modeling capabilities that can monitor, assess, and forecast, in near-real time, a variety of phenomena associated with country instability." The latter program is in its early stages of development (see http://www.darpa.mil/ipto/solicitations/open/07-10_PIP.pdf).

The Air Force's Stabilization and Reconstruction Operations Model (SROM) (Robbins, Deckro, and Wiley, 2005) analyzes the organizational hierarchy, dependencies, interdependencies, exogenous drivers, strengths, and weaknesses of a country's PMESII systems using a complex set of interdependent system dynamics representations. SROM models a country system in a lumped-parameter fashion as a national model (NM), which is then defined in terms of its n regional submodels that interact with each other and the NM. Each regional submodel contains six functional submodels: the demographics submodel, the insurgent and coalition military submodel, critical infrastructure, law enforcement, indigenous security

---

[20]The two assertions made here are based on the program manager's long experience in the M&S world and generally match what the modeling community has long known, namely, that (1) different domains often call for different modeling paradigms (e.g., modeling a social network is probably better represented by network modeling methods, than, say, by an argumentation framework) and (2) different domain experts have different preferences for representing their knowledge to others (e.g., some may be more expressive with a declarative expert system approach while others may be more facile with a graphically based Bayesian network formalism).

institutions, and public opinion. The utility of SROM has been demonstrated using Operation Iraqi Freedom as a case study.

Simulation Frameworks and Tools

In addition to these domain-focused modeling efforts, there are many efforts devoted to the development of general-purpose frameworks that make the modeler's job easier. As noted in Table 2-A1, these include the C2 modeling framework C3GRID, the team/organizational modeling framework DDD, the generic M&S frameworks FLAMES, ICET, and MATREX, the social network analysis tool ORA, and the collaboration decision aid framework SIAM. Many others exist in or are in development inside DoD, as well as outside in the academic and commercial worlds. Of particular note are the multiagent development and simulation environments commented on earlier (e.g., MASON, REPAST, and SWARM).

Other Efforts

In addition to these large efforts, there are hundreds of development efforts either recently concluded or just under way, varying dramatically in scale and focus, to produce representative and useful IOS models for the military. At one end is the spectacularly unsuccessful and terminated Joint Simulation Systems (JSIMS) effort, which attempted to be all things to all people, serving as DoD's general M&S environment. At the other is the IUSS/IWARS M&S program, which is successfully focusing on small-team behavior at the squad level. In between are efforts like the now concluded MIDAS effort of the National Aeronautics and Space Administration to build an "end-to-end" model of the human operator (of rotorcraft), and DARPA's RAID program, aimed at forecasting adversary behavior in the urban environment. However, there is no general inventory of what models exist and at what level of technical readiness. As a result there is duplication of efforts and too little effort at making these existing models interoperable. Furthermore, there is a trend for well-educated military personnel with some computational training to develop small, special-purpose IOS models that meet specific needs. A little programming training, however, does not make a good modeler, especially when that modeler is unaware of the importance of cognitive, affective, organizational, social, and cultural factors.

Major Challenges for Development of IOS Models for Military Applications

The current status of IOS modeling in DoD is the result of the funding profile for M&S in the last 10-15 years. Beginning in 1995, DMSO began centralizing funding and development efforts in M&S. The High Level Architecture (HLA) and the JSIMS systems are archetypes of efforts funded and managed by DMSO during that decade, with funding to other systems at the service level cut to focus on these centralized efforts. In the end, after spending some $1.8 billion in development funds, JSIMS was canceled and the services all had to try to recover from the loss of JSIMS plus the loss of development time, effort, and funding on service-specific M&S efforts.

During this same decade, however, there was an increase in the funding of models at the basic and applied research levels. This led to the development of a large number of models that are particularly relevant to the military and may even be used at commands, but that are generally not yet accredited. Examples of tools that evolved in this period are SIAM, ORA, and SEAS. In addition, work in this period gave rise to the model-driven experimentation paradigm (MacMillan, Diedrich, Entin, and Serfaty, 2005).

Interoperability Challenges

While some utility has been derived from HLA, its requirement that everything be statically defined ahead of time and its reliance on interoperability at the source code change level mean that the interoperability of defense simulations and their ability to change as threats rapidly change are greatly diminished. Had a bit of time been spent in the mid-1990s to design a dynamically extensible, semantically interoperable simulation infrastructure, defense M&S interoperability would now be more advanced. Furthermore, such an effort would have paved the way for incorporation of some of the IOS models now emerging.

Another difficulty with the centralized approach is that it assumes that modeling needs can be predefined. It is apparent that, as the military mission changes, M&S needs change, and new models are often needed immediately. Hence, an alternative distributed paradigm is needed that enables rapid access to new models and enables the military to make use of the increasing number of models that exist even when they were not developed expressly for military purposes. A possible alternative paradigm has a plug-and-play distributed infrastructure with data distributed across sets of servers with appropriate access controls; multiple models and simulations for different purposes with appropriate access control; and documentation, intelligent tools for aiding the user in determining which tools can be used with which data, and web enablement. In this way, any developer could place a model in the distributed system.

Data Collection and Validation Challenges

As noted in the Urban Sunrise report (Air Force Research Laboratory, 2004), most models to be used in real-world settings need to be tied to data. For example, models of insurgents often need as a basis data on the insurgency, such as the number of insurgents, modus operandi, sources of support, means of interaction, weapons, and location of activities. Data collection, however, is often done piecemeal by relying on subject matter experts to go out and collect data *after* a need for those data has been demonstrated, or opportunistically, as when a soldier, adversary, or civilian provides unsolicited intelligence (e.g., when an insurgent group posts a video of an IED attack on the web).

This means that new ways of thinking about validation are needed, and it means that the models need to operate with uncertain and incomplete data, but the science of model creation and validation with incomplete and uncertain data does not exist. The nature of the data that are, or can be, collected is often not consistent with the data requirements of the existing models. For example, a model may require data on who actually interacts with whom when all that is available is who is known to have participated in what events.

Applications-focused tools do not exist, such as expert systems for identifying for the user what models in their arsenal can be used given their data. The data are streaming, and time and location information is critical for data to inform action. For example, knowing the location and time of IED attacks is critical to identifying courses of action to protect U.S. soldiers from future attacks. However, in many cases, databases do not contain the time and location data. Moreover, even when the data exist, many types of models cannot make use of that information.

Because important data are classified, many models are developed in a vacuum, without access to the real data. Representative and unclassified data would be highly valuable and would get a wider range of model developers involved. However, the disadvantage is that the models are often tested and validated using proxy data that are conceptually different from, and may not even have the same data fields as, the classified data. This can result in erroneous assumptions of model validity at the classified level and in erroneous assumptions by the modelers about what needs to be modeled.

In addition, there are across-the-board needs for better modeling infrastructure, methods to link models to streaming data, and improved model visualization systems. Finally, there is a need for socially intelligent tools for collecting and interpreting intelligence information, particularly on insurgents and terrorists. IOS model-based fusion and data collection management techniques are needed. IOS models for identifying potential missing or erroneous data should also be developed.

A key to successful IOS models in this area is the development of measures and procedures that are robust with respect to scale and missing data. For example, IOS models of the adversary may need to be able to scale to 10^7 (ten million) actors. Even when there are massive amounts of missing and erroneous data, IOS measures need to be robust, appropriate, and meaningful. Adversarial models need to be sensitive to cultural factors, particularly to alternative goals, preferences for actions, and gender roles.

CONCLUSION

Current military missions and today's operating environment have created a pervasive need for models that can capture and forecast the behavior of humans acting in social units, ranging from small groups and teams to neighborhoods, cultural and ethnic groups, and entire societies. IOS models are needed to understand adversary and nonadversary behavior and to forecast the effects of alternative courses of action on that behavior. Today's broader missions focus not just on COAs for conventional combat with well-identified adversaries, but also on COAs for influencing the attitudes and behaviors of noncombatants at levels of detail ranging from block-by-block urban operations to the stability of nation-states. The COAs to be analyzed include not just military actions but the broader DIME/PMESII dimensions that may influence behavior. IOS models are also needed for training and rehearsal, to create realistic environments in which the military may test planned COAs and learn new skills associated with cultural awareness, joint and coalition operations, and stability and support operations. IOS models are valuable for design, evaluation, and acquisition as well. They can support the evaluation of potential contributions of new technologies to effective operations as well as the design of command and control organizational architectures that are effective for rapidly changing missions and new environments.

Efforts are under way to meet the military's needs for IOS models, but they are fragmented and uncoordinated, with no central direction, little information sharing, and no mechanisms to guard against duplication of effort in multiple locations. All of the current efforts face challenges for interoperability, with models developed from different perspectives unable to communicate in any meaningful way. Models also face data collection and validation challenges, with data collection efforts often piecemeal and unrelated to modeling requirements, and validation strategies frequently absent altogether.

The chapters in Part II review the state of the art in IOS modeling to evaluate the extent to which current approaches can meet military requirements as outlined above. On the basis of that review, we analyze where broad gaps exist and recommend a plan of action to fill those gaps.

Appendix TABLE 2-A1 Selected IOS Models

Acronym	Acronym Expansion	Description
ACT-R	Adaptive Control of Thought	A cognitive architecture in which a neural network activation system controls the activation of a rule-based production system model for simulating and understanding detailed human cognition. ACT-R continues to evolve to perform the full range of human perceptual, cognitive, and motor tasks, supported largely only in the academic community. Has been hybridized with other models, notably IMPRINT and Soar.
ADC	Air Defense Commander	Models the operations of an AEGIS Cruiser Combat Information Center (CIC) team performing air defense duties in a battle group using multiagent system (MAS) technology implemented in the Java programming language.
America's Army		Massively multiplayer online game (MMOG), starting as a first-person shooter game and now evolving to more complex environments and tasks and used as a recruiting tool.
BioWar		Combines computational models of social networks, communication media, disease models, demographically accurate agent models, wind dispersion models, and a diagnostic error model into a single integrated model of the impact of a biological warfare attack on a city. BioWar moves beyond existing epidemiological models by accounting for the heterogeneity of social networks and the geographical distribution of people when forecasting disease outbreaks.
C3GRID	Command, Control, Communication Grid Model	Parametric C4ISR modeling capacity for network-centric warfare. Provides the capability to simulate the common operating picture management for a given force structure at the platform level.
C3HPM	C3 Human Performance Model	Provides high-resolution modeling of individual human operators in terms of task performance and human decision processes in the execution of combat tactics, techniques, and procedures. Built on top of IMPRINT and operates in the MATREX simulation environment.

Category	Sponsor/ Research Center	Reference/Website
Cognitive architecture and modeling framework	ACT-R Research Group at Carnegie Mellon University	http://act-r.psy.cmu.edu/
Model	SPAWARSYSCOM	http://www.movesinstitute.org/~shcalfee/index.html
Real-time game environment	U.S. Army	http://www.americasarmy.com/
Hybrid model incorporating social networks, disease models, dispersion models	Carnegie Mellon University, DARPA, CDC, NSF	http://www.casos.cs.cmu.edu/projects/biowar/index.html
Network modeling tool	U.S. Army, RDECOM	http://www.msrr.army.mil/index.cfm?RID=MNS_A_1001514
Modeling framework	Army Research Laboratory (ARL)	http://www.arl.army.mil/ARL-directorates/HRED/imb/imprint/References.pdf

continued

Appendix TABLE 2-A1 Continued

Acronym	Acronym Expansion	Description
C3TRACE	Command, Control, and Communications—Techniques for the Reliable Assessment of Concept Execution	Network modeling tool to support the evaluation of different organizational structures and communications network topologies to evaluate overall C3 system performance. Built on top of MicroSAINT.
CART	Combat Automation Requirements Testbed	Network modeling tool initially applied to single-pilot model operating JSF and subsequently applied to a nine-member time critical targeting (TCT) cell in an air operations center (AOC). Built on top of IMPRINT, it enabled one of the first instances of integrating MicroSAINT with an external world model simulation.
CLARION	Connectionist Learning with Adoptive Rule Induction ON-line	Cognitive architecture for connectionist/neural representation of implicit (subsymbolic or neural network) knowledge and semantic representation of explicit (symbolic chunks and rules) knowledge. Provides for explicit representation of static knowledge as well as acquisition of subsymbolic knowledge through learning over time.
COGNET, iGEN	Cognition as a Network of Tasks	COGNET is an executable cognitive architecture and iGEN is the associated development environment. Both have been applied in a number of DoD-sponsored modeling exercises, most notably in the AFRL/HE AMBR air traffic control human behavior modeling and simulation program and in the Navy TADMUS antiaircraft defense modeling effort. Little or no technical literature appears to be available describing the technical details and therefore used little outside CHI Systems, its commercial developer.
Construct		A multiagent model of group and organizational behavior in which the agents communicate, learn, and make decisions in a continuous cycle, dependent on the perceptions and goals of the individual and the goals and culture of the group. When agents interact they communicate and learn both task knowledge and cognitive knowledge. These dynamic relationships are grounded in structuration theory, which is the notion of construction and reconstruction of the social system through human interaction based on rules and resources.

Category	Sponsor/ Research Center	Reference/Website
Modeling framework	ARL	www.hfes.org/web/BulletinPdf/ bulletin0405.pdf
Modeling framework	USAF AFRL/HE	http://www.maad.com/MaadWeb/ ongoing_projects/onprojma.htm#Combat
Cognitive architecture for individual entity modeling	Dept. of Cognitive Science, Rensselaer Polytechnic Institute; Army Research Institute (ARI)	http://www.cogsci.rpi. edu/~rsun/clarion-ub.html
Cognitive architecture and model development environment	USAF AFRL/HE and Navy SPAWAR	http://www.chisystems.com/
Multiagent dynamic network model	Carnegie Mellon University, DARPA, ONR	http://www.casos.cs.cmu.edu/projects/ construct/index.html

continued

Appendix TABLE 2-A1 Continued

Acronym	Acronym Expansion	Description
CORES	Complex Organizational Reasoning System	A multiagent network simulation model that uses organizational, social, political, and economic dynamics to generate forecasts of the likely actions and responses of adversarial actors. Scenarios are represented in a framework consisting of actors, resources, goals, actions, effects, and relations. Based on these entities, the model generates forecasts of likely actions and responses of actors. Potential areas of application include military intelligence and learning, political and corporate negotiation, disaster relief and crisis management, and business intelligence.
DDD	Distributed Dynamic Decision-making	Focuses on team functions that drive performance, such as communications and coordination. Model users can specify allocations of people, equipment, and material, and specify performance objectives/ constraints, such as job/mission objectives, timing, and coordination requirements. DDD models the resultant team/environment interactions based on empirically observed team/organization interactions and provides a simulation environment for calculating team performance metrics, based on a team performance model embedded in the simulator.
DIAS	Dynamic Information Architecture System	Object-oriented framework for integrating disparate multidisciplinary simulation models, supporting legacy code reuse, and modeling of cooperative behaviors of agents.
EPIC	Executive-Process/ Interactive Control	Cognitive modeling architecture for human information processing that accurately accounts for the detailed timing of parallel human perceptual, cognitive, and motor activity, in multitasking situations. Primarily an academic tool for researchers interested in fine-level details of perception and cognition. Applied to operator-centered design of undersea ship systems and many other systems.
FLAMES	Flexible Analysis Modeling and Exercise System	A framework for developing constructive simulations and interfaces between constructive, virtual, and live simulations. It has applications that support scenario definition, scenario execution, scenario postprocessing and scenario visualization.

Category	Sponsor/ Research Center	Reference/Website
Multiagent network model, incorporating DIME/PMESII factors	Carnegie Mellon University, DARPA, NSF	http://www.casos.cs.cmu.edu/ (Kowalchuck, Singh, and Carley, 2004)
Team/organization performance modeling tool/environment	Aptima; AFOSR, AFRL, ARL, DOT, NASA, NavAir, Office of Naval Research (ONR)	http://www.aptima.com/a-sim.php
Generic simulation framework	Argonne National Lab, DIS Division	http://www.dis.anl.gov/DIAS/
Cognitive architecture for individual entity modeling	University of Michigan, ONR	http://www.eecs.umich.edu/~kieras/ epic.html
Generic framework	USAF AFRL/MN and NAIC	http://www.ternion.com

continued

Appendix TABLE 2-A1 Continued

Acronym	Acronym Expansion	Description
HOS	Human Operator Simulator	HOS V is a MicroSAINT-based task-level simulation language for the individual human operator. Invokes micro models for primitives like perception, decision, and action. The output of HOS V consists of task performance timelines, errors, user-defined system performance measures, and component, person, and other resource utilization. Has been incorporated into COGNET to support modeling of low-level operator activities.
IBC	Integrated Battle Command	The IBC framework provides a means of integrating disparate environmental and IOS behavior models to support the analysis of the potential effects that a given set of DIME actions will have across the full range of PMESII variables, at the nation-state level. Each model in IBC may represent its portion of the domain in a manner and level of fidelity quite different from other models. Modeling paradigms include such techniques as concept maps, social network models, influence diagrams, differential equations, causal models, Bayesian networks, Petri nets, event-based simulation, and agent-based simulation.
ICET	Integrated Concept Evaluation Tool	Addresses modeling, simulation, and analysis of advanced cross-weapons communications concepts. Built on top of FLAMES.
ICEWS	Integrated Crisis Early Warning System	Goal is "the development of state-of-the-art computational modeling capabilities that can monitor, assess, and forecast, in near-real time, a variety of phenomena associated with country instability." This is a relatively recent start with no publications as of this date.
IMPRINT	Improved Performance Research Integration Tool	A stochastic task network modeling tool for the individual soldier. Task analysis is used as a starting point to assess the interaction of soldier and system performance. A network is constructed representing the flow and performance time and accuracy for operational and maintenance missions. Workload profiles for crew members are generated so the workload distribution and peaks and valleys can be examined. The underlying engine is the MicroSAINT task network modeling environment.

Category	Sponsor/ Research Center	Reference/Website
Language for individual operator task modeling	ARL	http://www.dtic.mil/dticasd/ddsm/closed/ ddsm0023.html
Framework for integrating different DIME/PMESII models	DARPA	http://www.darpa.mil/sto/solicitations/IBC/ index.htm (Allen, 2004)
Generic framework	USAF AFRL/MN and NAIC	http://www.ternion.com
Decision aid with embedded models of nation-state behaviors	DARPA IXO	http://www.darpa.mil/ipto/solicitations/ open/07-10_PIP.pdf
Human task modeling environment	ARL	http://www.arl.army.mil/ARL-Directorates/ HRED/imb/imprint/imprint.htm

continued

Appendix TABLE 2-A1 Continued

Acronym	Acronym Expansion	Description
IUSS, IWARS	Integrated Unit Simulation System, Infantry Warrior Simulation	Constructive, force-on-force model for assessing the combat worth of systems and subsystems for both individuals and small-unit dismounted war-fighters in high-resolution combat operations. Early versions were labeled IUSS. Current version, IWARS, is being built on IUSS Version 4 and the AMSAA Infantry MOUT Simulation (AIMS) models.
JSAF	Joint Semi-Automated Force	With its roots in the DARPA Synthetic Theater of War (STOW) program and derived from ModSAF, the JSAF simulation provides entity-level simulation of air, ground, and maritime forces in support of command and staff training and mission rehearsal. The JSAF federation provides a distributed modeling and simulation (M&S) framework composed of multiple federates to represent a realistic synthetic environment; model C2, logistics, and weapon effects; provide automated reasoning of entities via simple task behaviors and more advanced pilot behavior modeling via TacAir Soar; and interface with simulation and real-world systems (e.g., DIS, HLA, C4I Gateways). Based on technology developed prior to OneSAF.
JSIMS	Joint Simulation Systems	A federation of service-unique models of service-specific entities, based on a high-level architecture, common standards, and common protocols. JSIMS was going to be the primary M&S tool to support future joint and service training, education, doctrine development, and mission rehearsal for the Army, Air Force, Navy, DIA, DISA, NASM, TRANSCOM, and SOCOM. JSIMS was going to be progressively developed into a robust, interactive joint synthetic battlespace (JSB) for training strategic national joint tasks and joint and service tactical tasks in all phases of operations (mobilization, deployment, employment, sustainment, and redeployment). After nearly 7 years and $2 billion of investment, it was cancelled in 2004.

Category	Sponsor/ Research Center	Reference/Website
Modeling framework and specific small unit models for infantry behaviors	Army Natick Soldier Center/AMSAA	http://nsc.natick.army.mil/media/fact/ss&t/ IWARS.PDF
Computer-generated force (CGF) application for simulating a wide range of cross-service military entities	Joint Forces Command (JFCOM) Training and Analysis Center	http://afmsrr.afams.af.mil/ index.cfm?RID=MDL_AF_1000066
M&S environment for all DoD needs	JFCOM	N/A

continued

Appendix TABLE 2-A1 Continued

Acronym	Acronym Expansion	Description
MAMID	Methodology for Analysis and Modeling of Individual Differences	An integrated symbolic architecture, models high-level decision making, with focus on the role of affective factors (emotions and traits). MAMID models the cognitive appraisal process to dynamically generate emotions in response to incoming stimuli and models the subsequent effects of these emotions on distinct stages of decision making. Its parametric methodology supports the modeling of multiple, interacting individual differences and facilitates the rapid creation of distinct agent profiles.
MATREX	Modeling Architecture for Technology Research & Experimentation	The RDECOM-supported MATREX STO (science and technology objective) enables the integration of interoperable component engineering-level simulations and models that conform to a common architecture specification. MATREX is a framework, not a model, designed to integrate existing models into a robust representation of the battlespace (terrain, dynamic environmental effects, and physics-based modeling). It will be used to support and augment testing and training in either human-in-the-loop or constructive simulations. It will also support the integration of human behavioral models, such as IWARS, but does not support the direct construction of such models.
MicroSAINT	Microprocessor-based Systems Analysis of Integrated Networks	A discrete-event network simulation language for developing task network models of humans performing well-defined sequential tasks. It combines the operator with the external world model entities (e.g., airplanes), making plug-in operator models difficult to implement. Many models have been developed for military simulations, and the basic language has been extended by development supported by ARL under the IMPRINT program and by subsequent extensions by AFRL under the CART program. The language is particularly popular with modelers having little background in human perceptual or cognitive processes because of its ease of use. More sophisticated researchers, particularly from the ACT-R community, have made efforts to integrate MicroSAINT models with more traditional cognitive architectures.

Category	Sponsor/ Research Center	Reference/Website
Cognitive architecture for individual entity modeling	Army Research Institute (ARI, NASA, AFOSR)	http://www.psychometrixassociates.com/ hudl_mamid.pdf
M&S environment for all Army needs	Army RDECOM	N/A
Simulation language and tools for developing task-network models of human behavior	Micro Analysis and Design, ARL	http://www.maad.com/index.pl/ micro_saint

continued

Appendix TABLE 2-A1 Continued

Acronym	Acronym Expansion	Description
MIDAS	Man-machine Integration Design and Analysis System	Developed to support helicopter cockpit design for the Army, with a primary focus on anthropometry, physical layout of instrumentation, and operator workload. Primitive sensory models drive a production rule system to drive activity selection effected by simple motor models of the operator's limbs. A Z-scheduler handles rule collisions, but its psychological basis is unclear. MIDAS was a standalone system with a single instantiation at NASA Ames, and little public documentation is available regarding the detailed cognitive structures employed.
MINDS	Modeling Individual Differences and Stressors	The MINDS Behavior Moderator Engine (BME) has been developed as a plug-in for other cognitive architectures (e.g., ACT-R, OMAR, SAMPLE, Soar), as a means for generating personality- or stress-based moderators that can moderate structures or parameters of the target cognitive architecture, to emulate, for example, the effect of fatigue level on perception or fear on cognitive task performance. MINDS has been integrated with the SAMPLE cognitive architecture and embedded in the IWARS simulation environment to model infantry squad leader decision making.
ModSAF	Modular Semi-Automated Forces	An outgrowth of the early semi-automated force (SAF) program to simulate red ground force entities (e.g., tanks) executing basic maneuvers and missions (attack, defend, etc.) while engaging blue forces commanding simulated ground force entities (e.g., tanks) in the simulation network (SIMNET) environment developed during the 1980s. The modular SAF (modSAF) was developed to support composable (red) SAF behaviors, to minimize recoding efforts needed for training under different battle conditions, tactics, etc. SAF behaviors can be operated by behind-the-scenes red entity operators.

Category	Sponsor/ Research Center	Reference/Website
End-to-end workstation for the design of multicrew helicopter cockpits, with embedded operator models	Army Aeroflight Dynamics Laboratory, NASA Ames Research Center	http://caffeine.arc.nasa.gov/midas/ New_MIDAS_Design.html
Generic behavior moderator engine for use in individual entity cognitive architectures	Army Natick Soldier Systems Center, ONR	(Neal Reilly, Bachman, Harper, Marotta, and Pfautz, 2007) (Neal Reilly, Harper, and Marotta, 2007)
Computer-generated force (CGF) application for simulating maneuvering ground entities	Army Simulation, Training, and Instrumentation Command (STRICOM)	N/A

continued

Appendix TABLE 2-A1 Continued

Acronym	Acronym Expansion	Description
OMAR, D-OMAR	Operator Model Architecture, Distributed OMAR	OMAR is a cognitive architecture and simulation environment to develop models of human operators interacting with a variety of other operators and nonhuman entities. The basic components are a production rule-based cognitive processor driven by inputs from production memory, long-term memory, and working memory, this last driven by auditory and visual inputs. The architecture relies heavily on a centralized, synchronous production rule framework. An initial version was programmed in LISP, limiting its usability; a more recent version is implemented in Java. OMAR has been used in a number of simulations, including military air traffic control.
OneSAF, OOS, OTB	One Semi-Automated Forces, OneSAF Objective System, OneSAF Testbed	OneSAF is a constructive modeling and simulation environment intended to replace entity-based simulations. OneSAF is designed for numerous M&S domain applications, including research, experimentation, training, COA analysis, and mission planning. OneSAF models automated and semi-automated behaviors for entities and units up to the brigade level and supports the full spectrum of military operations, including urban missions. Designed as an extensible architecture, the OneSAF distribution includes tools for creating new components and behaviors to meet future modeling and simulation requirements. OOS was the predecessor system for OneSAF. OTB was the predecessor program for developing new technologies for OOS, focusing on test, integration, and user feedback. ModSAF was an earlier predecessor of all the programs.
ORA	Organizational Risk Analyzer	A risk assessment tool for locating individuals or groups that are potential risks given social, knowledge, and task network information. After building the network by connecting the nodes (people) via links (relationships) to other nodes (people), ORA conducts a form of social network analysis (SNA) to assess risk of individuals in the network. ORA is essentially a network development and analysis tool.

Category	Sponsor/ Research Center	Reference/Website
Cognitive architecture for individual entity modeling	USAF AFRL/HE	http://omar.bbn.com/manual/index.html, http://omar.bbn.com/
Computer-generated force (CGF) application for simulating a wide range of military entities	Army's Program Executive Office for Simulation, Training, and Instrumentation (PEO STRI)	http://www.onesaf.org/, http://www.onesaf.net/
Social network model building and analysis tool	ONR, DARPA, ARL, NSF, AFOSR	http://www.casos.cs.cmu.edu/projects/ora/ software.html

continued

Appendix TABLE 2-A1 Continued

Acronym	Acronym Expansion	Description
PCAS	Pre-Conflict Anticipation and Shaping	A recently concluded DARPA program to investigate the effectiveness of different computational social science approaches to support forecasting the likelihood of a nation-state failure (e.g., Sudan). The PCAS architecture consists of four modules for data collection, modeling, gaming/shaping tools, and decision support tools. Computational modeling approaches include system dynamics, multiagent systems, Bayesian influence models, diffusion models, and regression modes.
PMFServ	Performance Moderator Function Server	An integrated framework that permits one to examine the impacts of stress, culture, and emotion on decision making. PMFServ has been used to create and simulate the people and objects of a number of scenarios, including crowd scenes (civil unrest in the United States, urban conflict in the Mideast), asymmetric threat leaders and followers, the Black Hawk Down recreation in the UnrealTournament™ game engine, and world leader modeling in a diplomacy and strategy game. Over the past 5 years the instructor has been sponsored by DMSO, ONR, IDA, GM, Army, DARPA, JFCOM, and others.
RAID	Real-time Adversarial Intelligence and Decision-making	Supports real-time forecast analysis of probable enemy actions in urban operations against irregular. RAID leverages novel approximate game-theoretic and deception-sensitive algorithms to continuously identify and update forecasts of likely enemy actions while continuously estimating likely deceptions in the available battlefield information. Significant effort in the program is being applied to evaluating the program's performance relative to that of human analysts unaided by RAID.
SAMPLE, GRADE	Situation Awareness Model for Person in the Loop Evaluation, Graphical Agent Development Environment	SAMPLE is a cognitive architecture comprised of modules for fuzzy rule-based perception, Bayesian belief network-based situation awareness, and production rule-based decision making. GRADE is an agent development environment for rapidly creating SAMPLE models for different domains/tasks. Both have been used with JSAF, IWARS, the EAAGLES air combat simulation, the FACET ATM simulation, and the UnrealTournament™ gaming engine.

Category	Sponsor/ Research Center	Reference/Website
Nation-state DIME/ PMESII modeling methodologies	DARPA/IXO	(Popp et al., 2006)
Cognitive architecture for individual entity modeling and associated agent development environment	University of Pennsylvania, DMSO, ONR, IDA, U.S. Army, DARPA, JFCOM	http://www.seas.upenn.edu/~barryg/ HBMR.html
Decision aiding tool with game-theoretic model for adversary behavior forecasting	DARPA IIXO	http://dtsn.darpa.mil/IXO/ programs.asp?id=43 (Kott and Ownby, 2005)
Cognitive architecture for individual entity modeling and associated agent development environment	AFOSR, AFRL, ARL, DARPA, NRC, NSSC, ONR	http://www.cra.com (Harper, Ton, Jacobs, Hess, and Zacharias, 2001)

continued

Appendix TABLE 2-A1 Continued

Acronym	Acronym Expansion	Description
SEAS	Synthetic Environments for Analysis and Simulation	An agent-based software development environment that incorporates seven behavioral primitives: initiate, search, decide, execute, communicate, update, terminate. No attempt is made to model fundamental cognitive or social behavioral models, but a capability is provided for representing entities at the individual, organizational, and institutional level. Developers claim that the SEAS environment integrates multiple theories from various disciplines to program behaviorally accurate agents, but little has been available in peer-reviewed journals to substantiate that claim. JFCOM has been a strong supporter, especially in the attempts to model large-scale, nation-state-level projections (DIME/PMESII input-output forecasts) and COA assessments.
SIAM	Situational Influence Assessment Model	A collaborative decision aiding tool to help multiple analysts and experts decompose and analyze complex problems. It consists of a user-friendly graphical interface that supports the development and exercising of influence networks, a utility function decision-theoretic approach that builds on belief networks. SIAM allows each factor or influencing relationship affecting a decision to be examined separately, yet it optimizes understanding of the overall impact of, and the interrelationships among, the contributing factors.
Soar, Soar-EPIC	Simulation of Adaptive Resource, Executive-Process/ Interactive Control	Soar is an operator modeling production rule system in which existing rules propose potential operators that might be used to solve the current goal or problem. It is focused on problem solving and has its roots in GOMS. Its lack of a perceptual front end and motor back end has motivated hybridization with EPIC to provide these services. Although its psychological basis is less well-developed than other research-oriented models, Soar has been applied to a number of military systems modeling efforts (notably TacAir Soar).

Category	Sponsor/ Research Center	Reference/Website
Software agent-based development environment	Simulex, JFCOM	http://www.simulexinc. com/products/case_studies/#seas-vis
Visualization and decision aiding tool	SAIC	http://www.saic.com/business/technologies/ license/it/siam.Pdf
Operator modeling production rule system	University of Michigan, SoarTech	http://sitemaker.umich.edu/soar/home, http://www.soartech.com

continued

Appendix TABLE 2-A1 Continued

Acronym	Acronym Expansion	Description
SPECTRUM		Provides an environment with multicolored, multisided icons in an effort to simulate realistic situations that is conducive to MOOTW (SASO). SPECTRUM portrays the graphics and terrain of this environment and adds the human dimension, to account for the impact of economics, politics, regional populations, nongovernmental agencies (NGOs), and humanitarian relief agencies.
SROM	Stabilization and Reconstruction Operations Model	Analyzes the organizational hierarchy, dependencies, interdependencies, exogenous drivers, strengths, and weaknesses of a country's PMESII systems using systems dynamics modeling techniques. SROM models a country in a holistic lumped parameter manner as a national submodel, which is then defined in terms of its n regions as a system of systems. Each regional submodel itself contains six functional submodels: demographics submodel, insurgent and coalition military submodel, critical infrastructure, law enforcement, indigenous security institutions, and public opinion.
STELLA		A simulation-based training environment to train soldiers in information operations. A cognitive model was constructed using Bayes inference nets and neural nets to guide combat models based on internal logic. At the time of this review, fuzzy set theory was being contemplated for modeling the propagation of rumors, and a mathematical submodel of IW was being developed using q-analysis and Boolean nets to study the structure and dynamics of IW.

Category	Sponsor/ Research Center	Reference/Website
Sociocultural training system	National Simulation Center (NSC)	http://www.msrr.army.mil/
DIME/PMESII regional or nation-state modeling environment	USAF AFRL/IF	(Robbins, Deckro, and Wiley, 2005)
Information warfare training system	DISA, AFAMS	http://www.disa.mil/

REFERENCES

Air Force Research Laboratory. (2004). *Urban sunrise. AFRL-IF-RS-TR-2004-22, Final technical report.* Rome, NY: Author.

Alberts, D.S., and Hayes, R.E. (2003). *Power to the edge: Command and control in the information age.* Washington, DC: Command and Control Research Press.

Allen, J.G. (2004). *Commander's automated decision support tools.* Briefing to proposers' symposium for DARPA's Integrated Battle Command program Defense Advanced Research Projects Agency (DARPA). Available: http://www.afcea.org/events/pastevents/documents/AFCEAIICPanel.ppt [accessed Feb. 2008].

Anderson, J.R. (1983). *The architecture of cognition.* Cambridge, MA: Harvard University Press.

Axelrod, R. (2004). *Modeling security issues of Central Asia.* Prepared for CMT International, U.S. Govt. Contract # 2003*H513400*000. Available: http://www.personal.umich.edu/~axe/research/Security_Central_Asia.pdf [accessed Feb. 2008].

Belanich, J., Sibley, D.E., and Orvis, K.L. (2004). *Instructional characteristics and motivational features of a PC-based game, research report 1822.* Alexandria, VA: U.S. Army Research Institute for the Behavioral and Social Sciences.

Brown, M.T. (1997). *Terrorist use of weapons of mass destruction within the United States: Asymmetric warfare paradigm in the 21st century.* Carlisle Barracks, PA: U.S. Army War College. Available: http://stinet.dtic.mil/cgi-bin/GetTRDoc?AD=ADA326609&Location=U2&doc=GetTRDoc.pdf [accessed Feb. 2008].

Defense and the National Interest. (2007). *Boyd and military strategy.* Available: http://www.d-n-i.net/second_level/boyd_military.htm#discourse [accessed July 2007].

Defense Modeling and Simulation Office. (2007). *Modeling and simulation resource repository (MSRR).* Available: http://www.msrr.dmso.mil/ [accessed April 2007].

The Drucker School, Claremont Graduate University. (2008). *Perspectives on Peter.* Available: http://www.cgu.edu/pages/4181.asp [accessed Feb. 2008].

Entin, E.E. (1999). Optimized command and control architectures for improved process and performance. In *Proceedings of the 1999 Command and Control Research and Technology Symposium,* Newport, RI.

Farrell, C.M., Klimack, W.K., and Jacquet, C.R. (2003). *Employing interactive multimedia instruction in military science education at the U.S. Military Academy.* Presented at the 2003 Interservice/Industry Training, Simulation, and Education (I/ITSEC 2003) Conference, Dec. 1–4, Orlando, FL.

Frost, R. (1998). *Simulation based acquisition.* Washington, DC: Office of the Secretary of Defense.

Glenn, R.W. (2000). *Heavy matter: Urban operations' density of challenges.* (RAND Monograph Report.) Santa Monica, CA: RAND.

Harper, K.A., Ton, N., Jacobs, K., Hess, J., and Zacharias, G.L. (2001). Graphical agent development environment for human behavior representation. In *Proceedings of the 10th Conference on Computer Generated Forces and Behavioral Representation,* Orlando, FL: Simulation Interoperability Standards Organization.

Johnson, D. (2004). Air Force modeling and simulation: "New Vector" updates. *Intercom,* 45(12).

Joint Chiefs of Staff. (2000). *Joint vision 2020: America's military: Preparing for tomorrow.* Washington, DC: U.S. Government Printing Office.

Joint Chiefs of Staff. (2002). *Joint publication 3-06: Doctrine for joint urban operations.* Washington, DC: U.S. Government Printing Office.

Kasales, M.C. (2002). The reconnaissance squadron and ISR operations. *Military Review,* May–June, 52–58 [English version].

Kipp, J., Grau, L., Prinslow, K., and Smith, D. (2006). The human terrain system: A CORDS for the 21st century. *Military Review*, *LXXXVI*(5), 8–15. Available: http://www.army. mil/professionalwriting/volumes/volume4/december_2006/12_06_2.html [accessed Feb. 2008].

Kott, A., and Ownby, M. (2005). *Tools for real-time anticipation of enemy actions in tactical ground operations*. Paper presented at Simulation Interoperability Standards Organization's Tenth Conference on Computer Generated Forces and Behavioral Representation, Orlando, FL.

Kowalchuck, M., Singh, S., and Carley, K. (2004). *CORES—Complex organizational reasoning system*. (Report No. CASOS Technical Report CMU-ISRI-04-131.) Pittsburgh, PA: Center for Computational Analysis of Social and Organizational Systems, Carnegie Mellon University.

Krulak, C.C. (1997). The three block war: Fighting in urban areas. *Vital Speeches of the Day*, *64*(5), 139–141.

Levchuk, G.M., Levchuk, Y.N., Luo, J., Pattipati, K.R., and Kleinman, D.L. (2002a). Normative design of organizations—Part I: Mission planning. *IEEE Transactions on Systems, Man, and Cybernetics. Part A: Systems and Humans*, *32*(3), 346–359.

Levchuk, G.M., Levchuk, Y.N., Luo, J., Pattipati, K.R., and Kleinman, D.L. (2002b). Normative design of organizations—Part II: Organizational structure. *IEEE Transactions on Systems, Man, and Cybernetics. Part A: Systems and Humans*, *32*(3), 360–375.

Levchuk, G.M., Levchuk, Y.N., Meirina, C., Pattipati, K.R., and Kleinman, D.L. (2004). Normative design of project-based organizations—Part III: Modeling congruent, robust, and adaptive organizations. *IEEE Transactions on Systems, Man, and Cybernetics. Part A: Systems and Humans*, *34*(3), 337–350.

Levchuk, G.M., Yu, F., Levchuk, Y., and Pattipati, K.R. (2004). Networks of decision-making and communicating agents: A new methodology for design and evaluation of organizational strategies and heterarchical structures. In *Proceedings of the 2004 International Command and Control Research and Technology Symposium*, June 15–17, San Diego, CA.

Levchuk, G.M., Yu, F., Levchuk, Y., and Pattipati, K. (2006). PERSUADE: Modeling framework for the design of modular Army organizations. In *Proceedings of the 2006 Command and Control Research and Technology Symposium*, June 22–22, San Diego, CA.

Lwin, M.R. (1997). *Great powers, weak states and asymmetric strategies*. Monterey, CA: Naval Postgraduate School. Available: http://stinet.dtic.mil/cgi-bin/GetTRDoc?AD=AD A340989&Location=U2&doc=GetTRDoc.pdf [accessed Feb. 2008].

MacMillan, J., Diedrich, F.J., Entin, E.E., and Serfaty, D. (2005). How well did it work? Measuring organizational performance in simulation environments. In W.B. Rouse and K.R. Boff (Eds.), *Organizational simulation* (pp. 253–272). Hoboken, NJ: John Wiley & Sons.

Mahoney, R. (2005). *Joint urban operations workshop on command and control and net centric environmental science and technology, executive summary*. (Report No. IDA Doc. D-3132.) Alexandria, VA: Institute for Defense Analyses.

McCrabb, M. (2001). *Explaining effects: A theory for an effects-based approach to planning, executing, and assessing operations, version 2.0.* Available: http://www.sci.fi/~fta/EBO. htm [accessed Feb. 2008].

National Research Council. (1998). *Modeling human and organizational behavior: Application to military simulations*. Richard W. Pew and Anne S. Mavor (Eds.). Panel on Modeling Human Behavior and Command Decision Making: Representations for Military Simulations. Commission on Behavioral and Social Sciences and Education. Washington, DC: National Academy Press.

Navy Modeling and Simulation Office. (2004). *Autonomous agent-based simulation of an AEGIS cruiser combat information center performing battle group air-defense commander operations.* S.H. Calfee, Master's thesis, Naval Postgraduate School, Monterey, CA. Available: http://stinet.dtic.mil/cgi-bin/GetTRDoc?AD=ADA414842&Location=U2 &doc=GetTRDoc.pdf [accessed Feb. 2008].

Neal Reilly, W.S. (2006). *Modeling what happens between emotional antecedents and emotional consequents.* Paper presented at the Agent Construction and Emotions (ACE 2006) Conference, April 18–19, Vienna, Austria.

Neal Reilly, W.S., Bachman, J., Harper, K.A, Marotta, S., and Pfautz, J. (2007). Modeling the effects of behavior moderators for simulation-based human-factors design. In *Proceedings of the 51st Annual Meeting of the Human Factors and Ergonomics Society,* Baltimore, MD.

Neal Reilly, W.S., Harper, K.A., and Marotta, S. (2007). *Modeling concurrent interacting behavior moderators for simulation-based acquisition tasks.* Paper presented at Spring Simulation Multiconference (SpringSim '07), Norfolk, VA.

Nieborg, D.B. (2004). *America's Army: More than a game.* Bridging the gap: Transforming Knowledge into Action through Gaming and Simulation. In *Proceedings of the 35th Conference of the International Simulation and Gaming Association* (ISAGA), Munich, Germany: SAGSAGA.

North, M.J., Howe, T.R. Collier, N.T., and Vos, R.J. (2005). *The repast simphony runtime system.* Paper presented at the Agent 2005 Conference on Generative Social Processes, Models, and Mechanisms, Argonne National Laboratory, October, Argonne, IL. Available: http://repast.sourceforge.net/papers/Agent_2005_Repast_Runtime.pdf [accessed Feb. 2008].

Office of the Secretary of Defense. (1996). *Study on the effectiveness of modeling and simulation in the weapon system acquisition process.* Washington, DC: Author.

Parsons, D., and Wittman, R. (2004). *OneSAF: Tools and processes supporting a distributed development environment for a multi-domain modeling and simulation community.* Presented at the European Simulation Interoperability Workshop (Euro SIW 04), June 28–July 1, Edinburgh, Scotland.

Pattipati, K.R., Meirina, C., Pete, A., Levchuk, G., and Kleinman, D.L. (2002). Decision networks and command organizations. In A.P. Sage (Ed.), *Systems engineering and management for sustainable development, Encyclopedia of life support systems.* Oxford, England: UNESCO–EOLSS.

Popp, R., Kaisler, S.H., Allen, D., Cioffi-Revilla, C., Carley, K.M., Azam, M., Russell, A., Choucri, N., and Kugler, J. (2006). *Assessing nation-state instability and failure.* Presented at the Aerospace Conference 2006 IEEE, Boston, MA.

Robbins, M., Deckro, R.F., and Wiley, V.D. (2005). *Stabilization and reconstruction operations model (SROM).* Paper presented at the Center for Multisource Information Fusion Fourth Workshop on Critical Issues in Information Fusion: The Role of Higher Level Information Fusion Systems Across the Services, University of Buffalo. Available: http://www.infofusion.buffalo.edu/ [accessed Feb. 2008].

Schaffer, P. (2005). *Mapping the human terrain: GIS in support of cultural intelligence, Track: Defense and intelligence.* Redlands, CA: ESRI GIS Mapping and Software.

Tatara, E., North, M.J., Howe, T.R., Collier, N.T., and Vos, J.R. (2006). An introduction to repast modeling by using a simple predator-prey example. In *Proceedings of the Agent 2006 Conference on Social Agents: Results and Prospects,* Argonne National Laboratory, Argonne, IL. Available: http://repast.sourceforge.net/papers/Agent_2006_Repast_Tutorial.pdf [accessed April 2008].

United Nations. (2002). *World urbanization prospects: The 2001 revision (ST/ESA/SER. A/216). Department of Economic and Social Affairs Population Division.* New York: Author.

U.S. Air Force Scientific Advisory Board. (2002a). *Predictive battlespace awareness to improve military effectiveness, volume 1: Summary.* (Report #SAB-TR-02-01—authors: G.B. Harrison, R.O. Johnson, M.R. O'Neill, M.W. Ganz, and T.F. Saunders.) Washington, DC: Author.

U.S. Air Force Scientific Advisory Board. (2002b). *Predictive battlespace awareness to improve military effectiveness, volume 2: Panel reports.* (Report #SAB-TR-02-01—authors: G.B. Harrison, R.O. Johnson, M.R. O'Neill, M.W. Ganz, and T.F. Saunders.) Washington, DC: Author.

U.S. Air Force Scientific Advisory Board. (2005). *Report on Air Force operations in urban environments, volume 1: Executive summary and annotated brief.* (Report #SAB-TR-05-01—Committee on Air Force Operations in Urban Environments.) Washington, DC: Author.

U.S. Army. (2002). *U.S. Army Training and Doctrine Command (TRADOC) pamphlet (Pam) #525-3-90, The United States Army objective force operational and organizational plan for maneuver unit of action.* Washington, DC: U.S. Army Training and Doctrine Command.

U.S. Department of Defense. (1995). *Modeling and simulation (M&S) master plan. DoD #5000.59-P.* Washington, DC: Office of the Undersecretary of Defense for Acquisition and Technology.

U.S. Department of Defense. (2006). *Quadrennial defense review report.* Washington, DC: Author.

Vick, A.J., Stillion, J., Frelinger, D.R., Kvitky, J., Lambeth, B.S., Marquis, J.P., and Waxman, M. (2002). *Aerospace operations in the urban environment.* (RAND monograph report #MR-1187.) Santa Monica, CA: RAND.

Wait, P. (2007). Defense domain, civilian awareness. Elder, Garcia walk two sides of the cyber-security beat. *Government Computer News,* January 22. Available: http://integrator. hanscom.af.mil/2007/January/01252007/01252007-25.htm [accessed March 2008].

Zyda, M., Mayberry, A., McCree, J., and Davis, M. (2005). From Viz-Sim to VR to games: How we built a hit game-based simulation. In W.B. Rouse and K.R. Boff (Eds.), *Organizational simulation* (pp. 553–590). Hoboken, NJ: John Wiley & Sons.

Part II

STATE OF THE ART IN
ORGANIZATIONAL MODELING

Part II

State of the Art in Organizational Modeling

Part II reviews the multitude of individual, organizational, and societal (IOS) modeling approaches, methods, and tools that are potentially useful for addressing the military modeling needs described in Chapter 2. Models take many forms, ranging from loose conceptual models to precise mathematical models (Lave and March, 1975). They include agent-based models, cognitive models, expert systems, dynamical systems, and input-output models. Here we survey and explore many different types of models relevant to our questions. We describe each, show their strengths and limitations, and discuss research and development efforts that could make the approaches more useful for addressing military modeling needs.

The diverse expertise of the committee members contributed greatly to the completeness of this review but also made it challenging to agree on an organizing framework for presenting the review results. Refined through multiple iterations, the organizing framework that we developed represents a significant product of the study.

CATEGORIES OF MODELS: INITIAL EMPIRICAL RESULTS

As a first step in organizing our review, we took an empirical approach to organizing the various terms and approaches used in IOS modeling. Using the methods of cultural domain analysis (see Chapter 3 for a description), we developed a perceptual map of the field of modeling based on committee members' perceptions.

Methodology

The first step in our investigation was to collect "free lists" of models from each member of the committee. Effectively, we asked, "What are all the kinds of models you can think of?" A large number of unique "kinds of models" were elicited with little overlap, implying that the domain itself lacks a high degree of cultural coherence. A total of 240 items were elicited, with approximately 35 items per member. Much of this lack of overlap was due to differences in the level of specificity for the kinds of models listed. For example, some of the items were specified at the level of named models, such as DyNet, EpiSims, NetWatch, etc., while others were at a very general level, such as conceptual models or verbal models. Aggregating across all lists, a master list of distinct terms was obtained after standardizing word forms. Also, an attempt was made to keep all items at the same level of specificity, in this case at a more general level.

The second step was to take the 38 most frequently mentioned items at a more general level of specificity and construct a pile-sorting task, in which each committee member was asked to sort the items into piles according to how similar the kinds of models are. They could use as many or as few piles as they wished. The task was conducted online using interview software that simulated cards and allowed the virtual cards to be placed into piles. When they were done, the program recorded the membership of each pile. Then, an aggregate proximity matrix X, whose rows and columns corresponded to "kinds of models," was constructed such that each cell X_{ij} of the matrix recorded the number of respondents that placed the ith kind of model in the same pile as the jth kind of model.

The final step was to visualize this proximity matrix using a standard network visualization package called Netdraw (Borgatti, 2002). In this approach to visualization, a line is drawn to connect two items if the similarity of the two items exceeds a certain user-defined threshold (DeJordy, Borgatti, Roussin, and Halgin, 2007; Johnson and Griffith, 1998).

Results

The resulting map is shown in Figure II-1. In the map, a line is drawn between two modeling techniques if at least 28 percent of the respondents placed the items together in the same pile. (A cutoff of 28 percent was chosen because above that level the main section of the network becomes disconnected.)

The results show three basic clusters of modeling techniques. The first cluster, at the top left of the map, consists of multiagent models in which the agents are connected to each other by social ties or interactions. In these models, the combination of agents and ties forms a single interconnected

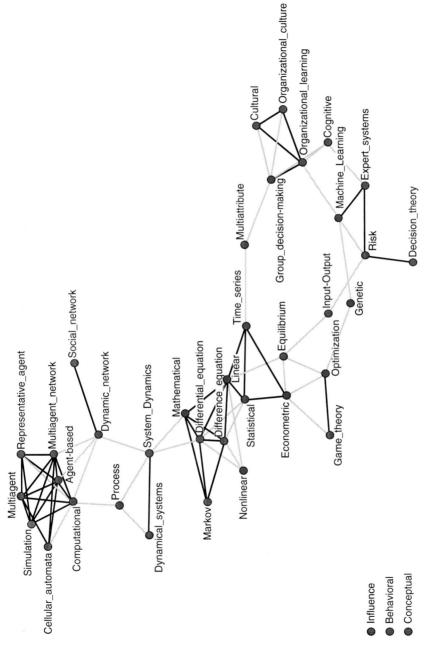

FIGURE II-1 Perceived similarities among types of models.

- Influence
- Behavioral
- Conceptual

and interdependent system. For convenience, we refer to the models in this cluster as the *computational network models* cluster (although it contains some models for which this would not be the ideal name).

The computational network models cluster is connected to the next cluster via the system dynamics node. This new cluster consists of low-level statistical and mathematical techniques that have broad application across many different settings. Although these techniques are often thought of as tools rather than models, statisticians would recognize that they do indeed constitute models. For convenience, we refer to this cluster as the *mathematical systems models* cluster.

At the bottom right of Figure II-1, there is another cluster of models focused on the cognition or culture of the agents. We call these the *cognitive models*. The difference between these and the computational network models at the top of the map is one of emphasis rather than substance. The cognitive models are defined by their focus on the details of cognition. The objective of the cognitive models is to understand the patterns of who believes or chooses what. In contrast, the computational network models are defined by the processes that the modeler builds into the system and may not represent cognition at a detailed level. The outcomes of the computational network models may well be the same as those of the cognitive models, and the processes of the cognitive models often involve the same multiagent interactions of the computational network models: it is only the focus of the investigation that is different. Finally, as noted at the bottom left, three model types—influence, behavioral, and conceptual—did not cluster with other types.

FOUR-PART ORGANIZING FRAMEWORK FOR MODELS

On the basis of the empirical clustering results described above and further discussion, the committee developed a four-part categorization for reviewing modeling approaches: (1) macro models, (2) micro models, (3) meso models, and (4) integrated, linked micro-meso-macro models. No single one of these approaches is the correct one, and the best modeling approach depends on the nature of the problem to be solved. It is a common theme throughout this book that models constitute "use-driven research" (Stokes, 1997) and cannot be developed or evaluated without an in-depth understanding of the uses to which they are to be put.

A macro model considers interactions between macro-level variables, such as unemployment, crime, education, poverty, and resources. Macro modeling approaches like system dynamics enable one to identify feedbacks and to see system-level effects without getting bogged down in details. At the other extreme, one can model the cognitive or affective processes of individual actors or at least their outcomes—individual decisions and actions. These more micro modeling approaches include cognitive models from psychology, expert systems models, and rational choice models, which include game theory and decision theory.

Fifty years ago, this distinction between micro and macro would have been thought sufficient. One can look at the trees, or one can look at the forest. Over time, social scientists have come to appreciate the importance of the level in between—the meso level (Miller and Page, 2007). To complete the metaphor, one can think of stands of aspen trees with a shared root system. The stand is a part of the forest, yet it does not function merely as the sum of its individual trees, given the sharing of resources. The social analog of a shared root system is social capital. People join movements, participate in riots, and support government in part based on the actions of their friends and peers. Predictions based on individual attributes can almost always be improved by adding in social factors.

We highlight two types of meso models: network models and agent-based models. Both modeling approaches have produced flurries of attention over the past decade. Network models allow one to formalize, measure, and test loose conceptions of social capital, centrality, and connectedness. Agent-based models allow one to include diverse, purposive agents who interact in space and time. As the name suggests, agent-based models originate with the agents, but these agents can self-organize, creating emergent meso-level structures that take on meaning and have predictive value.

The fourth category links micro, meso, and macro models. Agent-based models, and to some extent game theory and network models, achieve this double linkage. Yet only recently have researchers begun to create hybrid models that include agents who employ sophisticated psychological models and whose macro effects link to a system dynamics model. These hybrid models have great potential for addressing the needs identified in Chapter 2. Thus, we make this linkage between levels explicit. Although we do not have a separate chapter on integrated models, agent-based models and network models are discussed in Chapter 6, and the challenges of achieving such multilevel model integration are discussed in Chapter 8.

PART II GUIDE

Chapter 3 discusses conceptual models and cultural models. Adequate conceptual models provide the foundation for development of computational and mathematical models. Cultural models occupy a special position in our review because our interest is in understanding people at multiple levels of aggregation. The questions that concern us require the ability to model individuals, teams, communities, and entire societies. At each of these levels, cultural factors are at work, so we first explain what we mean by culture and cultural effects.

The discussion next turns to a review of formal modeling approaches (Chapters 4, 5, and 6) organized using the four-part framework described above. For each modeling approach we describe the current state of the art, the most common applications of the approach, and its strengths and limitations for the problems described in Chapter 2, and we provide suggestions for further research and development.

In Chapter 7, we turn to online games as a methodology. Gaming, the creation of an environment in which real people can play against one another or against artificial players, can be thought of as a methodology, but as these gaming models apply many of the other types of models, and as they involve people interacting with the games, we set them apart. Online gaming environments are both consumers of models—to create artificial players and the social effects of player actions—and potential testbeds for generating data to develop and test models of the communications and actions of large numbers of individuals interacting in a simulated world.

Chapter 8 discusses important methodological issues that are common across many modeling approaches, including modeling frameworks, tools, and data, and it includes a discussion of model verification and validation. Model validation is a key issue for complex social models, and we argue for a "validation for action" approach that considers how the model is to be used rather than attempting to evaluate model accuracy or model fidelity without considering context of use. Chapter 9 summarizes the state of the art in IOS modeling and its applicability to the requirements and uses discussed in other chapters.

REFERENCES

Borgatti, S.P. (2002). A statistical method for comparing aggregate data across a priori groups. *Field Methods, 14*(1), 88–107.

DeJordy, R., Borgatti, S.P., Roussin, C., and Halgin, D.S. (2007). Visualizing proximity data. *Field Methods, 19*(3), 239–263.

Johnson, J.C., and Griffith, D.C. (1998). Visual data: Collection, analysis, and representation. In V. de Mjmck and E. Sabo (Eds.), *Using methods in the field*. Walnut Creek, CA: Altamira Press.

Lave, C.A., and March, J.G. (1975). *An introduction to models in the social sciences*. New York: Harper and Row.

Miller, J.H., and Page, S.E. (2007). *Complex adaptive social systems: An introduction to computational models of social life*. Princeton, NJ: Princeton University Press.

Stokes, D.E. (1997). *Pasteur's quadrant: Basic science and technological innovation*. Washington, DC: Brookings Institution Press.

3

Verbal Conceptual and Cultural Models

In this chapter we discuss models that are not instantiated in formal algorithms or software, but in words. Verbal conceptual models are presented first, followed by verbal cultural models.[1] These models are important for their attempts to apply theoretical constructs to the behavior of individuals and groups. The models and the terms and constructs they encompass may provide a foundation for some of the more applied and formal models discussed later. These models have been developed in the disciplines of social psychology, sociology, anthropology, and organizational behavior studies.

VERBAL CONCEPTUAL MODELS

What Are Verbal Conceptual Models?

Verbal conceptual models characterize entities, variables, or events/ processes/mechanisms and the relations among them in words, not in equations or other mathematical or operational formulations. Although they may use mathematical terms—for example, Kurt Lewin's statement that all behavior is a "function" of the person and the situation (1951)—the nature

---

[1]Note that, in general, both conceptual models and cultural models can be articulated in formal logical, mathematical, algorithmic, or computational forms. Our focus in this chapter is on verbal representations of conceptual and cultural models, as an initial stepping stone toward computational implementations in individual, organizational, and societal (IOS) models and simulations.

and form of relations described in a verbal model are commonly under-specified compared with formal models. Verbal conceptual models include very general classifications or broad characterizations that provide the foundation for a new discipline, such as the brain-as-computer metaphor on which modern cognitive science was founded, and mid-level frameworks, such as "images" of organizations (Morgan, 1997) as machines, or organisms, or brains. Typologies or taxonomies, such as a taxonomy of emotional states (Borgatti, 1994) or a typology of small groups (Arrow, McGrath, and Berdahl, 2000), are another form of verbal conceptual model. Most numerous of all are the small-scale models that characterize relations among variables or processes relevant for understanding a specific phenomenon. The "progression of withdrawal" and "compensatory behaviors withdrawal" models, for example, are alternate models of job withdrawal (quitting and absenteeism) (Hanisch, 2000); another example is a two-variable model of how social norms emerge in a newly formed group, based on whether or not new members' characterizations of the situation and the "scripts" they retrieve to guide behavior match (Bettenhausen and Murnighan, 1985).

The use of such terms as "theory," "framework," "model," and "paradigm" in psychology and the social sciences is as informal as the models themselves.[2] One person's conceptual model is another person's theory or framework. In this chapter, we use the term "conceptual model" (and, for brevity, sometimes just "model") as a way to group theories, frameworks, and paradigms into rough classification systems based on common features in structure (for example, dual process models, dynamic models, threshold models) or relevant domain (group development models, organizational withdrawal models, visual attention models). What verbal conceptual models have in common is that they tend to be "highly informal constructions, use the natural language system, are rich in metaphor, and use lavishly nuanced statements" (Davis, 2000, p. 218). If rendered as diagrams instead of in straight prose, they tend to be represented via two-by-two tables, labeled boxes with arrows drawn between them, or perhaps a flowchart for a process model.

In psychology and the social sciences, theorizing about a problem typically begins with verbal conceptual models, which then may be elaborated and adjusted over time as relevant empirical data accumulate. Formal mathematical models, computational models, statistical models, etc. rely on verbal conceptual models to specify variables and relations among them, although a host of extra assumptions and plausible estimates are typically needed to translate a verbal theory into a workable implementation. Hence

---

[2]This is also true in the behavior modeling and simulation community, which is why we attempted in Chapter 1 to identify and differentiate four levels of representation: theory, architecture (here, framework), model, and simulation (here, paradigm).

a computational model of emotional response relies on a conceptual taxonomy of emotional states (Ekman and Davidson, 1995), and a process simulation model of jury decision making, such as DISCUSS (Stasser, 1988), relies on the guiding metaphor of "interacting minds" engaged in "collective information processing" (which represents the mind-as-computer metaphor generalized to groups-as-networked-computers).

State of the Art for Verbal Conceptual Models

Sophisticated verbal conceptual models (whose authors often call them theories) are typically more specific about the nature of relations among variables or about the nature of processes described than are ad hoc models or global metaphor models. They may also be more sophisticated in incorporating contingencies, dynamics, and multiple levels of analysis. For example, in the study of leadership effectiveness, a very simple model, the leadership grid (Blake and Mouton, 1982), proposes that leadership effectiveness is explained by two dimensions—concern for people and concern for production—and the more a leader has of both, the better. This model focuses entirely on the leader (single level), entertains no contingencies, assumes linear additive components, and has no dynamic elements. A more sophisticated model, situational leadership theory (Hersey and Blanchard, 1988), proposes that the optimal mix of task-oriented and relation-oriented behavior by leaders depends on the level of maturity and corresponding skill level of the subordinate, which is expected to change over time. In a heterogeneous group of members at different levels, effective leadership will require that the leader tailor her style to individual members and adjust that style as each member progresses through four successive levels of maturity and autonomy. This model incorporates three levels (individual, dyad, and group), contingencies (different levels of member development), and change over time.

Computational models are often used to model complex processes that unfold over time, so verbal conceptual models that include attention to dynamics are particularly useful as a resource for the implementation of more formal models. Verbal conceptual models of groups and organizations can be arrayed along a continuum of increasing complexity using the four different levels of complexity in time research of Ofori-Dankwa and Julian (2001).[3] First-level models focus on mean differences in, for example, how much time a process takes and assume stationarity (sometimes implicitly rather than explicitly). Second-level models add change as a possibility, so that the rate of a process may speed up or slow down across time. Third-

---

[3]The following summary is adapted from Arrow, Henry, Poole, Wheelan, and Moreland (2005, pp. 313–368).

level models incorporate more than one hierarchical level of a system. For example, the rate of change at the member, small group, and organizational levels may be expected to differ systematically. Fourth-level models allow for multiple simultaneous and potentially nonstationary processes at different levels.

Zaheer, Albert, and Zaheer (1999) introduced the concept of "time-scale completeness" for a process model. In essence, they define what a process model needs to specify to provide sufficient guidance in designing a research program. Although their focus was on empirical data collection, the same desiderata apply for implementing a verbal model in a computational form. A theory is time-scale complete if it specifies time scale for all of its variables, relationships, and boundary conditions. For example, it needs to specify the time needed for a complete instance of the phenomenon to occur, the nature and rate of change in variables, and the duration and sequence of any subphases in the process. Otherwise researchers cannot make theory-driven choices of observation, recording, and aggregation intervals, and the criteria for evidence either in support of or contrary to model predictions remain unclear.

Finally, state-of-the-art conceptual models allow for conceptual "docking" with other models by clarifying how the terms used relate to other, closely related (or synonymous) terms in the literature, and note where other models might "plug in" (for example, a structural model might refer to possible plug-in models that address processes or mechanisms not included in but relevant to the structural model) and clearly specifying boundary conditions.

Relevance to Modeling Requirements

One way to demonstrate the relevance of verbal models is to give an example of how a well-developed verbal conceptual model could be used for rapid cultural awareness training. The conceptual model is the cross-cultural framework of Fiske (1991, 2000), which proposes that human beings in all cultures coordinate their social interactions using a mix of fundamental relational models: communal sharing, authority ranking, equality matching, and market pricing. The four models are organized sets of associated concepts and rules that serve as a generative grammar for thinking about and coordinating relationships.[4] When following the communal sharing model, people emphasize the common identity of group members and focus on what is good for the group as a whole. The preferred model of decision making is consensus and people pool resources and draw on the pool without keeping track of individual contributions and withdrawals.

---

[4]The following summary is adapted from Arrow and Burns (2004, pp. 176–178).

Prototypical contexts and domains in which this model is used are family and food. Families commonly share food resources freely, and people who are defined as the "in group" in a particular context (such as invited guests at a party) are expected to help themselves to whatever food and drink they want. Violations of the rules occur when out-group members attempt to access in-group resources (for example, someone crashes a party).

In relationships organized by the authority ranking model, people structure their interactions according to status, position, and dominance hierarchy. Military organizations commonly use this model, and personnel wear insignias of rank to signal status. How people behave is strongly governed by whether they have the higher or lower rank of the two people in a given interaction. In distributing resources, high-status members get more, and low-status members get less. Rank also comes with obligations: superiors are expected to provide for or take care of inferiors. Violations occur when lower status members are insubordinate, treating a higher ranked person as an equal, for example, or when higher status members abuse their rank and power and betray their obligations to lower status followers or dependents.

When a relationship is governed by the equality matching model, people reciprocate favors after some delay and maintain a balance between giving and receiving. This model is commonly applied among people who consider themselves to be of equal status, such as friends, classmates, or colleagues. People in equality matching relationships often respond to favors by saying "I owe you one" or "I'll pay next time." Note the difference from authority ranking, in which a lower status person responds to favors with gratitude and loyalty, rather than reciprocating in kind. If the relationship is using the equality matching model, however, the failure to reciprocate a favor (or to express one's understanding of this obligation) would be a violation.

In market pricing relationships, people seek the best deal for themselves and expect that others will do the same. This model commonly governs trade and other social exchanges among strangers or acquaintances and is guided by the equity principle of proportionality—so that price, for example, should be proportional to value. Self-interested or selfish behavior is not a violation (it is expected), but cheating or stealing (which violates the equity rule) is.

Particular cultural implementations of these models organize social exchange, distribution, contribution, decision making, social influence, moral judgment, aggression, and conflict (Fiske, 1991). There is no practical limit, for example, to the types of objects or services that might be deemed appropriate or inappropriate to reciprocate the gift of a chicken or a radio. This sort of idiosyncratic and culturally specific content can be provided only by an informant who is very familiar with a culture. However, more important to practical application in a field situation is simply detecting whether the context in which a chicken is received is one that calls for gratitude and acknowledgment of an in-group bond (communal sharing), expression of deference and humility (authority ranking), reciprocation with a favor of roughly equal value (equality matching), or direct payment (market pricing). Mistakes involving specific cultural content (reciprocating with an odd sort of gift) may invoke humor or surprise; violation of relational models (paying someone for a gift, failing to reciprocate) are more likely to give offense and damage the relationship.

Major Limitations

The strengths of verbal models are also their weakness. Natural language is a flexible and nuanced instrument in which one can express highly sophisticated ideas, including multiple overlapping metaphors and embedded narratives. However, because natural language is an encompassing sea in which we all swim, shadings of meaning and idiosyncratic clouds of associations allow four people to encounter the "same" verbal model and understand it in four different ways. Some of the ambiguities and gaps that are common in verbal models may become evident when designing an experiment, and they are highlighted most sharply when one attempts to extract a set of formal relations from a natural language model.

In psychology and the social sciences, the grand metaphors of conceptual models often govern the whole direction of a field, but metaphors always direct attention to some features and lead to the neglect of others. Once a broad conceptual framework such as this becomes pervasive, scholars tend to forget that a metaphor is involved. For example, the information-processing metaphor for the brain, and the researchers who focused on it, probably contributed to a pervasive neglect of research on emotional and social processes for the first several decades of cognitive science.[5] Computers don't have emotions and are not social beings. So if the mind is not simply *like* a computer in some ways (simile with boundary conditions), but *is* a biological computer (unreflective metaphor), the ways in which minds are decidedly *not* like computers get overlooked, even by researchers engaged in intensively social activities about which they have strong feelings. The same curious social blindness is evident in the early era of organizational research dominated by the organization-as-machine metaphor. The related notion that workers are cogs in the machine led researchers to study the impact of physical conditions, such as lighting, on worker productivity while completely ignoring the possible impact of one human being on another (Mayo, 1960).

---

[5] A small pocket of researchers clearly forged ahead in this important area; see, for example, Ortony et al. (1988).

Verification and Validation Issues

Verbal conceptual models are sometimes specific enough that they can be tested and plausibly falsified, using empirical field studies or controlled experiments. For example, in studies of subjects from Bengali, Chinese, Korean, Vai (Liberia and Sierra Leone), and U.S. cultures (Fiske, 1992), Fiske and colleagues have used social cognition experiments to demonstrate that people organize acquaintances in memory according to the dominant model that organizes the relationship and that for many subjects this classification accounts for more variance in recall and substitution errors than such personal attributes as gender, race, and age.

In contrast to such well-developed conceptual frameworks, broad metaphors (brains as information-processing devices, organizations as cultures) are not really subject to verification or falsification. Whether or not they are used in a particular domain is likely to depend largely on face validity and established precedent. In evaluating the usefulness of a verbal model of this nature, the yardstick is often not how well supported the model is, but how much interesting research it inspires. Even when a verbal model seems, in principle, to be subject to falsification, the underspecification of relations and processes often means that a rather broad array of different outcomes can be presented as "consistent with" the theory. As Harris (1976) noted in "The Uncertain Connection Between Verbal Theories and Research Hypotheses in Social Psychology," theoretical terms often are not defined, boundary conditions are unspecified, and, under various plausible interpretations of assumptions or conditions, several well-known theories include internal contradictions and inconsistencies (as cited in Davis, 2000).

Future Research and Development Requirements

Verbal conceptual models can be highly influential and generative and do not require intensive funding or technology to develop, yet the development of such models is often overlooked as a funding priority. The scarce resource in improving verbal theory is intellectual time and energy. Motivation may also be an issue when grant funding is available primarily for doing (conducting experiments, writing code, designing games, collecting reams of data) and not for thinking. This can encourage the proliferation of low-level, poorly specified, ad hoc conceptual models that get spawned in discussion sections of journal articles to explain the results of a single set of studies and, if they survive, are later herded together in introduction sections of subsequent articles without actually being systematically integrated into more comprehensive integrated models. That work is generally left for the writers of literature reviews who are trying to make sense of a mountain of facts and ideas and find a deeper order.

Stronger theory is needed for domains that social scientists still don't know how to think about and those in which numerous weak conceptions have not been integrated. Verbal conceptual models are essential building blocks for theory building. Bringing people together for conferences and funding edited books and special issues that explore themes and issues in depth are useful. Measurable advances in theory should also be specified as a valuable deliverable for grants. Think tanks could be funded for scholars to come together and work intensively for an extended period (three to six months) on theory development and integration for issues and areas in which it is increasingly clear not only that there are not enough data, but also that it is difficult to know how to conceptualize the problem. Of course this sort of conversation *is* going on in labs and institutes around the country, but the focus on generating data (at least in psychology) seems to eclipse or marginalize the systematic development and integration of theory that goes beyond the highly specific area in which people tend to do research.

CULTURAL MODELING

What Is Cultural Modeling?

The term "cultural modeling" encompasses two broadly different areas of research. One area is concerned with modeling growth and distribution of cultural phenomena, such as the evolution of norms or the diffusion of beliefs. Research in this tradition typically treats culture (or, more accurately, some characteristic of culture) as an outcome and concentrates on the factors shaping those outcomes. This kind of cultural modeling is distinguished from other kinds of modeling surveyed in this volume only by the domain of study—namely, an element of culture. It does not imply a particular modeling technique. For example, the evolution of norms may be studied using a variety of methods, including multivariate statistics, agent-based models, system dynamics models, event history models, and so on. This kind of cultural modeling is discussed in several chapters in this volume and is not discussed further here.

The other kind of cultural modeling, which is discussed here, is concerned with describing (and often formally representing) a group's culture. Work in this tradition typically does not concern itself with how the culture came to be but rather with how it is distributed in the population and, in the best cases, what the consequences of having that culture might be.

Finally, it is appropriate to note that perhaps the most fundamental verbal cultural models are those that are implicit in a region's or society's language and history. It is abundantly clear—from Laurence of Arabia's exploits to today's attempts to "democratize" Iraq—that deep and broad knowledge of the local history and language are still fundamental for the kind of high-level understanding of societal dynamics that is the main focus of this report and of today's military. This committee acknowledges the importance of both language and history as the foundational knowledge base for any cultural model development, and as perhaps the starting point for identifying "implicit" models embedded in the language and history—models that can be built on in successive formalization efforts.

What Is Culture?

Culture can be defined in a number of different ways. Indeed, over 200 scholarly definitions have been documented (Kroeber and Kluckhohn, 1952). Researchers have defined culture in normative, historical, biological, cognitive, functional, structural, categorical, and symbolic terms. Definitions typically make use of some combination of the following elements: beliefs, behaviors, values, customs, artifacts, organizational orientations, preferences, experiences, attitudes, meanings, hierarchies, religions, perceptions, conceptions, material objects, possessions, symbols, motives, traditions, strategies, ideals, rules, habits, reasoning, identities, conventions, customs, and institutions, among others. While definitions of cultures differ on which of these elements constitute culture, most view a group's culture as an essential factor in problem solving, coping, and adapting to environmental changes. In addition, they generally agree that culture is something possessed by groups (such as societies, organizations, occupations, teams) and that it is learned, transmitted, and shared (albeit imperfectly and unevenly). At the same time, scholars regard culture as being held in individual minds and do not consider it an oxymoron to talk about an individual's culture.

State of the Art of Culture Models

There are four basic types of descriptive culture models popular today: cultural inventory models, dominant trait models, semantic models, and cultural domain models.

Cultural Inventory Models

Cultural inventory models are a way of describing cultures by listing which of a list of traits they do or do not possess. Thus cultures are conceived of as distinctive bundles of features that can be represented as a string of 1s and 0s indicating the presence or absence of a trait. A number of anthropologists have undertaken a compilation of cultural traits across human societies. The best and most relevant example of such a compilation across cultures is the Standard Cross-Cultural Survey developed by George

Murdock and others. The database consists of 186 societies and 22 cultural categories involving almost 1,000 standard coded variables derived from ethnographic sources (Murdock and Morrow, 1970). Essentially, a team of researchers has combed through ethnographies written by anthropologists and coded the cultures described using a universal codebook. The ability to compare features across societies is critical for both developing models and testing theories concerning patterns of and associations among cultural traits, categories, and features. Table 3-1 provides examples of some of the 22 cultural categories and associated variables and their codes.

For military purposes (McFate, 2005), many of these traits may be irrelevant, while others would need to be gathered, such as information on cultural gestures (e.g., meaning of certain hand gestures), cultural greeting etiquettes (e.g., rules for properly entering a village), cultural norms surrounding conflict (e.g., cultural notions of courage, honor, and revenge), etc. Such a database would considerably improve the ability to interact in a satisfactory manner with natives and to accurately predict their reactions to stimuli.

The key difficulty with cultural inventories is obtaining the necessary data. Data need to be collected on an ongoing basis to ensure the quality and timeliness of information. Also, it is important to recognize cultural boundaries and subcultures. For example, a nation like China may form a single political unit but may contain many different cultures. Furthermore, collecting new cultural information can be particularly difficult during periods of conflict, which means that the data need to be collected on an ongoing basis regardless of whether it has immediate utility.

Another approach is the cultural classification system developed by Karabaich, which is intended to cover the possible group types that might be encountered in a military, business, or political context (Karabaich, 2004). These group types are summarized in Table 3-2.

TABLE 3-1 Examples of Cultural Categories and Coded Variables from the Standard Cross-Cultural Survey

Examples of Cultural Categories	Examples of Labels for Variables Within Categories

Subsistence economy and supportive practices	Marital residence
	  Matrilocal or uxorilocal—with wife's kin
	  Avunculocal—with husband's mother's brother's kin
	  Patrilocal or virilocal—with husband's kin
	  Ambilocal—with either wife's or husband's kin
	  Neolocal—separate from kin

TABLE 3-1 Continued

Examples of Cultural Categories	Examples of Labels for Variables Within Categories
Political organization	Political power—most important source Direct subsistence production Warfare wealth Tribute or taxes Slaves Contributions of free citizens Large landholdings Political office Foreign commerce Capitalistic enterprises Priestly services
Cultural complexity	Fixity of residence Nomadic Seminomadic Semisedentary Sedentary, impermanent Sedentary
Sexual attitude and practice	Frequency of premarital sex—male Universal Moderate Occasional Uncommon
Relative status of women	Mythical founders of the culture All male Both sexes, but the role of men more important Both sexes, and the role of both sexes fairly equal Both sexes, but female role more important, or solely female
Cultural theories of illness	Theories of soul loss Absence of such a cause Minor or relatively unimportant cause An important auxiliary cause Predominant cause recognized by the society
Female power and male dominance	Female economic control of products of own labor Absent Present
Political decision making and conflict	Conflict between communities of the same society Endemic: high physical violence, feuding, and/or raiding occur regularly Moderately high, often involving physical violence Moderate: disputes may occur regularly but tendency to manage them in a more or less peaceful manner Mild or rare
Nature of warfare	Value of war: violence/war against nonmembers of the group Enjoyed and considered to have high value Considered to be a necessary evil Consistently avoided, denounced, not engaged in

SOURCE: Adapted from Murdock and Morrow (1970).

TABLE 3-2 Summary of Karabaich Group Stereotype Taxonomy

Group Stereotype	Description
Social	Shared interest, but no clear political agenda
Religious	Shared beliefs and goals based on shared faith in particular dogma
Economic	Seek advances in their economic objectives
Professional	Shared interests, problems, and objectives concerning their livelihood and profession
Political	Shared goals in addressing particular grievance or advance specific set of rights or benefits. Success requires interaction with existing societal/governmental power structure and group works "within the system"
Militant	Shared sense of threat to fundamental values, rights, or benefits and a desire to fight against the existing power structure that they blame. Group is willing to use violence and therefore generally operates in opposition to the power structure or without its overt support
Military	Goal is to defend existing system by threat of or actual physical force. Individual members may have joined voluntarily, due to social pressure, or been forced

SOURCE: Hudlicka (2004, Table 5.1-1 from Psychometrix Technical Report 0412, p. 48).

Each group and its culture are characterized by a series of attributes, which were derived in part from Karabaich's extensive experience in Army psychological operations and in part from the work of several political psychologists, most notably the work of Alexander George on operational codes (op-codes) (1979, 1998), who in turn built on the work of Leites. Op-codes capture the role of internal, subjective schemas (the operational codes) that guide individual (and group) behavior. They include values, beliefs, perceptions, and goals and jointly define what the group considers important, its view of the world, what motivates its behavior, and how it goes about accomplishing its goals. The assumption then is that these attributes will influence the individual members of the group in a manner similar to, but more powerful than, the influence of the national and ethnic groups to which the individual belongs.

Table 3-3 shows a subset of the key attributes used to characterize groups, listing examples of the specific values for these attributes for three of the group categories: a political group, a religious group, and a militant group.

TABLE 3-3 Examples of Group Stereotypes

Attribute	Group Categories		
	Political	Religious	Militant
Goals	Influence	Acceptance, validation, advance dogma	Protect livelihood and values, influence, defy authority
Goal scripts	Propaganda, street demonstrations	Good works, proselytizing, humanitarian/ education	Propaganda, attack symbols of power, attack infrastructure
Acceptable means	Work within existing power structure	Work within existing power structure	Work against existing power structure, violence
Historical data (past behavior)			
Values/beliefs			
World view	Neutral	Neutral/friendly	Hostile
Demographics	Heterogeneous	Homogeneous in belief	Homogeneous in religion, ethnicity, or socioeconomic status
Motivation for joining			

SOURCE: Hudlicka (2004, Table 5.1-2 from Psychometrix Technical Report 0412, pp. 48–49).

Dominant Trait Models

Dominant trait models are similar to cultural inventories but differ in the fundamental unit of analysis. Whereas cultural inventory models are based on ethnographic assessments of the culture as a whole, dominant trait models are based on individuals' responses to survey questions about themselves. This approach is based on the concept of modal personality developed by the cultural and personality school of psychological anthropology (Benedict, 1934; LeVine, 1982; Hsu, 1972). In this approach, culture is seen as "personality writ large." Cultures are described by the dominant psychological traits of the members of the culture. If certain traits are more prevalent in one society than another, the cultures are said to be different in this respect.

Perhaps the most famous advocate of this approach in modern times is Hofstede, who has identified five dimensions of culture that he regards as fundamental and that are thought to vary widely across cultures. The five dimensions are power distance (degree of tolerance for uneven distribution of power), individualism-collectivism, femininity-masculinity (task versus process/people orientation), uncertainty avoidance, and short- versus long-term orientation.

This trait set was recently augmented by Klein and colleagues and termed the "cultural lens" model (Klein, Pongonis, and Klein, 2000; Klein and McHugh, 2005). The cultural lens model adds several cognitively oriented factors to the Hofstede dimensions, including counterfactual thinking versus hypothetical reasoning and dialectical reasoning. Other cultural trait sets have also been identified, including that of Schwartz, which consists of conservatism (degree of preference for status quo and established order); intellectual autonomy (independence of intellectual pursuits); affective autonomy (desirability for individual's positive affective experience); hierarchy (same as power distance); egalitarianism (similar to Hofstede's collectivism); mastery (getting ahead through active self-assertion); and harmony (fitting into the environment) (Schwartz, 1999).

Another trait-based approach focuses on the characteristic cognitive styles of a group's members (Hudlicka, 2004; Hudlicka et al., 2004). For example, one of the more striking findings in cross-cultural cognition research is the recognition that the "fundamental attribution error" (Ross, 1977) is in fact dependent on culture, and more common in Western, individualistic cultures, than Eastern, more group-oriented cultures. Fundamental attribution error refers to the individual's tendency to attribute the behavior of others to individual dispositions rather than to environmental influences. Similarly, Western subjects exhibit a greater focus on isolated objects than Asian subjects, who attend more to the gestalt of the situation and the interrelationships among the objects.

Examples of findings from cross-cultural cognition research are listed in Tables 3-4 and 3-5. The tables follow the categorization of inference types of Peng, Ames, and Knowles (2001).

Finally, a large number of cross-cultural studies focus on emotion: its expression, recognition, and elicitation across cultures. The specific data of interest for a particular modeling effort depend on the objective of the model (a training system designed to teach cultural awareness should provide information about acceptable expressions of particular emotions; a decision aid designed to improve behavior prediction needs to represent emotion elicitors, etc.). As might be expected, there are significant commonalities across cultures in both emotion recognition and expression, particularly in the case of the more fundamental (basic) emotions, such as fear, anger, sadness, and happiness (Ekman and Davidson, 1995). For

TABLE 3-4 Findings Regarding Cultural Differences in Human Inference: Inductive Reasoning (ability to generalize from limited data) (Hudlicka, 2004)

Category of Inference	Findings
Covariation judgment (identifying correlations between cues)	Ji, Peng, and Nisbett (2000) Chinese versus Americans Simple stimuli presented on computer screen Chinese more confident about judgments Chinese more correct in judgments Chinese showed no primacy effect Americans showing strong primacy effect "East Asian cognition has been held to be relatively holistic; that is, attention is paid to the field as a whole. Western cognition, in contrast, has been held to be object focused and control oriented. In this study East Asians (mostly Chinese) and Americans were compared on detection of covariation and field dependence. The results showed the following: (a) Chinese participants reported stronger association between events, were more responsive to differences in covariation, and were more confident about their covariation judgments; (b) these cultural differences disappeared when participants believed they had some control over the covariation judgment task; (c) American participants made fewer mistakes on the Rod-and-Frame Test, indicating that they were less field dependent; (d) American performance and confidence, but not that of Asians, increased when participants were given manual control of the test"
Causal attribution (identifying causal relations between cues) Social	Miller (1984) Americans versus Hindu Indians Fundamental attribution error evident in Americans Hindu Indians attribute behavior to social roles, obligations, physical environment Attributed to different beliefs regarding causality (content difference) Morris, Nisbett, and Peng (1995) Americans versus Chinese Fundamental attribution (mass murderers, computer animations of fish) Americans attributed behavior to individual dispositions Chinese attributed behavior to environment Lee, Hallahan, and Herzog (1996) Americans and Hong Kong Chinese Sportswriters' descriptions of events American writers focus on individuals Hong Kong writers focus on situational factors Nisbett (2003); Jones and Harris (1967) Americans and Koreans Judgment of another person's attitude Americans assume due to disposition Koreans assume due to contextual influences

continued

TABLE 3-4 Continued

Category of Inference	Findings
Causal attribution Physical	Asian folk physics is relational, emphasizing fields and force over distance Western folk physics focuses on nature of object itself, rather than its relation to the environment Peng et al. (2001, p. 252) Peng and Knowles (2003) Chinese versus Americans Force-over-distance explanations (aerodynamic, hydrodynamic, magnetic) Americans referred more to nature of object Chinese referred more to the field
Person perception	Chiu, Hong, and Dweck (1997) Hong Kong Chinese versus Americans Judgment of self as fixed versus changing Americans assume fixed, enduring traits Chinese assume changing self Peng et al. (2001) Chinese versus Americans Type of information used in person perception judgments Americans focused on evidence provided by target Chinese focused on evidence provided about the target by others
Inference of mental states	Americans prefer "what you see is what you get" norm of authenticity Asians would consider this impolite Knowles, Morris, Chiu, and Hong (2001) Chinese versus Americans Judgment of mental states (thoughts, feelings, desires) Americans: focus on what "they say" Chinese: focus on what "they don't say"
Categorization	General findings: • Some categories are more stable across cultures than others. Examples of stable categories are: basic emotions, colors, basic shapes • Westerners tend to categorize objects by color at an early age and by function later. Africans tend to use color throughout their life. (This finding may be related to formal education more than culture.) • More cultural influence for goal-based categories than for environment-based categories • More salient categories for a given culture are more highly differentiated (culture directs attention) • Culture determines types of features used in defining categories • Asians may be less attuned to categories in their inferences and category learning • Some evidence that Asians tend to use relational features as basis for categorization • Differences in "chronic accessibility" (less for Koreans than for Americans) • Possible differences in category acquisition (exemplar based versus rule based) • Self-descriptions (Americans in terms of fixed traits; Asians in terms of roles; more "socially diffused")

SOURCE: Hudlicka (2004, Table 3.2.2-1 from Psychometrix Technical Report 0412, pp. 28–30).

TABLE 3-5 Findings Regarding Cultural Differences in Human Inference: Deductive Reasoning

Category of Inference	Findings
Syllogisms	Luria (1931, Russia); Cole (1996, Africa)
	Subjects did not engage syllogistic problems at the theoretical level (i.e., if asked to deduce something based on a presented syllogism, they would frequently think out of the box and suggest that the experimenter go find out for himself; why would x be true, etc.)
	Real-world (culturally relevant) grounding of topic makes a large difference in success on task
Dialectical reasoning	Asians: changing nature of reality and enduring presence of contradictions versus Western: linear epistemology built on notions of truth, identity, and noncontradiction
	Resolving contradiction: Chinese seek compromise; Americans seek exclusionary (either-or) truth and resolution (Peng and Nisbett, 1999)
	Assumption in Eastern dialectical epistemology:
	• Principle of change—everything is always in flux (thus x may not be identical with itself because it may change over time)
	• Principle of contradiction—opposing qualities coexist
	• Principle of holism—everything is linked to everything else and isolating phenomena may lead to misleading conclusions
	• Folk wisdom: greater frequency and preference for dialectical (apparently contradictory proverbs) among Chinese than Americans
Social contradictions/ conflicts	Americans tended to blame one side versus Chinese tended to see fault in both

SOURCE: Hudlicka (2004, Table 3.2.2-2 from Psychometrix Technical Report 0412, p. 31).

behavior prediction, the most significant differences are those in emotion elicitation; that is, in the specific situations and stimuli triggering particular emotions. Variations were found both in the nature of the emotions elicited and the intensity of those emotions. Some of these findings are summarized in Table 3-6.

Semantic Models

Semantic models are not researcher-based models but rather the models that ordinary people use to understand their worlds. The models are often tacit, in the sense that individuals are not aware they have them. Anthropologists discover the models by interviewing people and listening to their accounts of daily life. They typically consist of chains of prototypical events

TABLE 3-6 Differences in Emotion Elicitors Across Cultures: Summary
of Findings

Situation	Emotion Elicited
Birth of new family member	More intense joy for Europeans/Americans than Japanese
Body-centered basic pleasures	More intense joy for Europeans/Americans than Japanese
Achievement	More intense joy for Europeans/Americans than Japanese; more fear for Americans
Death of loved one	More frequent triggers of sadness for Europeans/Americans than Japanese
Physical separation from a loved one	More frequent triggers of sadness for Europeans/Americans than Japanese
World news	More frequent triggers of sadness for Europeans/Americans than Japanese
Strangers	More frequent trigger of anger for Japanese than for Europeans/Americans; more fear for Americans
Novel situations	More fear for Japanese
Negative developments in relationships	More sadness for Japanese than Europeans/Americans

SOURCE: Hudlicka (2004, Table 3.2.2-3 from Psychometrix Technical Report 0412, p. 33).

that constitute plans of action. D'Andrade defined these sorts of models as "a cognitive schema that is intersubjectively shared by a social group" (D'Andrade, 1989, p. 809). Semantic models are qualitative or conceptual rather than computational models.

As an example of a semantic model, Naomi Quinn (1987) has analyzed hundreds of hours of interviews to discover concepts underlying American marriage and to show how these concepts are tied together. She began by looking at patterns of speech and at the repetition of key words and phrases, paying particular attention to informants' use of metaphors and the commonalities in their reasoning about marriage. For example, one of her informants said that "marriage is a manufactured product." This metaphor paints marriage as something that has properties like strength and staying power and as something that requires work to produce. Some marriages are "put together well," while others "fall apart" like so many cars or toys or washing machines (Quinn, 1987, p. 174).

The objective is to look for metaphors in rhetoric and deduce the schemas, or underlying principles, that might produce patterns in those metaphors. Quinn found that people talk about their surprise at the breakup of a marriage by saying that they thought the couple's marriage was "like the Rock of Gibraltar" or that they thought the marriage had been "nailed in cement." People use these metaphors because they assume that their listeners know that cement and the Rock of Gibraltar are things that last forever (i.e., they are intersubjectively shared).

Quinn reasons that if schemas or scripts are what make it possible for people to fill in around the bare bones of a metaphor, then the metaphors must be surface phenomena and cannot themselves be the basis for shared understanding. Quinn found that the hundreds of metaphors in her corpus of texts fit into just eight linked classes that she calls lastingness, sharedness, compatibility, mutual benefit, difficulty, effort, success (or failure), and risk of failure. For example, Quinn's informants often compared marriages (their own and those of others) to manufactured and durable products ("it was put together pretty good") and to journeys ("we made it up as we went along; it was a sort of do-it-yourself project"). Quinn sees these metaphors, as well as references to marriage as "a lifetime proposition," as exemplars of the overall expectation of lastingness in marriage.

Other examples of the search for cultural schemas in texts include a study of the reasoning that Americans apply to interpersonal problems (Holland, 1985), a study of ordinary Americans' theories of home heat control (Kempton, 1987), and a study of what chemical plant workers and their neighbors think about the free enterprise system (Strauss, 1997).

Cultural Domain Analysis

Cultural domain analysis refers to perspectives on and methods for analyzing culture drawn from cognitive anthropology (Borgatti and Everett, 1992). A cultural domain is a collection of items that in some sense go together or are all examples of a kind of x (e.g., animals, plants). Such domains are often linguistic categories (e.g., semantic domains or concepts) in that there is a simple name for the set of items, like fruit or vegetables. What makes these domains cultural is that they are consensual. There is general agreement on the part of cultural actors regarding membership of most items in the domain. However, like all human things, the boundaries of a domain can be porous or fuzzy. There are items that are clearly in the domain, and items that are clearly outside, and many items that are in-between. The general objective of this type of analysis and modeling is to understand the cultural domain, which means to know what items belong in it and how these items are perceived to relate to one another (i.e., the extent to which they are similar or different). The data are collected and analyzed in a systematic manner using data collection techniques, such as pile sorts, sentence completion tasks, and triads tests (similar methods are referred to as repertory grid analysis in psychology; see Johnson and Weller, 2002), and analytical methods, such as hierarchical clustering and multidimensional scaling, to identify the conceptual organization and shared dimensions among concepts. Analysis of domain items can also include their attributes (e.g., diseases and their symptoms). A good example of a general principle stemming from this form of analysis comes from Stefflre (1972) in his proposition that people will behave similarly toward things they perceive as being similar.

The importance of this approach lies in the ability to quickly assess the nature of cultural beliefs and conceptions, albeit for a rather narrowly delineated set of cultural items. However, such an understanding can facilitate the ability to alter or change cultural beliefs and ultimately human behavior. These types of methods have been used in consumer research for both product development and marketing (Stefflre, 1972). Johnson, Griffith, and Murray (1987) and Murray, Griffith, and Johnson (1987), for example, have used this approach in changing people's beliefs about underutilized fish species, leading to increased consumption of fish that were traditionally considered "trash" fish.

Another branch of cultural domain analysis is the cultural consensus model (CCM) of Romney, Weller, and Batchelder (1986). The model originated as a theoretical exploration of the formal conditions under which similarity of beliefs would imply cultural knowledge. It was shown that, in the context of a true/false questionnaire asking respondents to react to propositions of fact, the degree of knowledge of each respondent could be inferred when three conditions held. First, that a single culturally correct right answer exists that is valid for all respondents in the sample. Second, that conditional on the underlying cultural answer key, the responses of subjects are independent (i.e., when they did not know the answer to a question, their responses were uncorrelated). Third, that the questionnaire contained questions about only one domain of knowledge (that is, a single competence level for each person sufficed to characterize their probability of answering any question correctly). When these three conditions held, the model was capable of deriving both the culturally correct answer key and the cultural competence of each respondent. The model allows for a test of the degree to which cultural knowledge is shared, who has more or less of this cultural knowledge, and how it varies among a group of people in terms of, for example, gender, levels of human capital, and social class. It also allows for the construction of the culturally correct answers by working backward via Bayesian statistical techniques from the patterns of agreement concerning a series of related cultural propositions or statements.

This approach has a number of advantages in terms of understanding and modeling culture, particularly with respect to modeling aspects of intra-cultural variation.[6] CCM has been used in a variety of contexts, but it has been applied practically to solving policy and management issues, modeling indigenous ecological knowledge, and understanding people's cultural beliefs concerning various aspects of health and illness. It has recently been used to measure cultural consonance (i.e., the correspondence between cultural beliefs and actual behavior) that has been shown to correlate with health outcomes (e.g., low consonance is related to high blood pressure; see Dressler and Bindon [2000]).

The CCM approach can be used to empirically determine shared beliefs and knowledge that can be used in models incorporating cultural variables. In addition, the approach can also be used to more finely tune an understanding of cultural beliefs and their variation that may be patterned in terms of different social attributes (e.g., gender, age). Thus, cultural knowledge (the correct cultural response) or individual cultural competency can be treated as either a dependent or an independent variable in a model at various levels of analysis.

Relevance to Modeling Requirements and Major Limitations

For the purposes of this study, a key limitation of all the models reviewed in this section is that they were not built for military purposes. The variables and dimensions they have focused on (such as power distance) have not been shown to be relevant for any given military situation. More generally, different aspects of culture are relevant for different situations, and as a result a new model must be built for each substantially different military purpose and for each group of people (who have distinct cultures).

Another limitation of these models is that they do not explicitly link culture and behavior and therefore do not provide direct guidance on how to intervene in a group in order to change the culture. A partial exception is cultural domain analysis, which posits that people behave similarly to

---

[6]Understanding *intercultural* variations has benefited significantly from the approach taken by Heinrich et al. (2004), in which an economic "game" (such as the Ultimatum Game, Güth, Schmittberger, and Schwarze, 1982) is introduced across a number of different societies, and the resulting behaviors correlated (or "normalized") with respect to the interaction pattern norms found in each society. As the authors note: "We draw two lessons from the experimental results: first, there is no society in which experimental behavior is even roughly consistent with the canonical model of purely self-interested actors; second, there is much more variation between groups than has been previously reported, and this variation correlates with differences in patterns of interaction found in everyday life" (p. 5). Clearly there are implications for war games, understanding cultural biases with respect to aggression across cultures, and anticipating adversary tactics for a range of Department of Defense IOS modelers.

similar stimuli. As a result, it is possible to predict that people's behavior toward a new course of action will be similar to their reactions toward other courses of action that are similar.

Another difficulty with predicting behavior is that the behavior of interest to predict may often be that of individuals. However, some models, such as the semantic models, unless they were based on a single individual, are not intended to apply to any single individual. Other models, such as the trait-based models of Hofstede, are based on individuals but then aggregated to the group level. Cultures are then described by the traits of the majority.

Data, Verification, and Validation Issues

Cultural inventory models rely on ethnographic observation and are therefore both time-consuming to develop and highly subjective. Having multiple independent observers helps ameliorate the subjectivity problem but is expensive.

Dominant trait models, such as the Hofstede dimensional models, can involve two sets of data. The first set of data is used to derive the dimensions. These can be validated by a number of different statistical methods, such as factor analysis. Once these are fixed, another set of data is obtained to score each new culture on the dimensions. These data have to be obtained from willing natives of the culture, and the data have to be updated over time because cultures change.

Future Research and Development Needs

In a certain sense, cultural models are critical for all the computational models discussed in this volume, because the cultural models provide the principles to be embedded in those models. For example, an agent-based model of crowd behavior needs to know the cultural rules for behavior that will govern the agents' interactions.

The biggest limitation of cultural models at present is that existing models were not designed with military purposes in mind. As a result, a key research need is to develop models applicable to military needs. This would include semantic models of how natives think about land, nation, war, foreigners, and so on, as well as cultural inventory models that include relevant variables. Note that different models are needed for different cultures.

The semantic models are particularly powerful for military applications. However, they are currently not formal models, meaning that they are expressed verbally and not in ways that are immediately amenable to computational analysis. As a result, another key research direction is to develop formal ways of expressing semantic models that are simple enough to be used by field researchers and subject matter experts.

REFERENCES

Arrow, H., and Burns, K.L. (2004). Self-organizing culture: How norms emerge in small groups. In M. Schaller and C. Crandall (Eds.), *The psychological foundations of culture* (pp. 171–199). Mahwah, NJ: Lawrence Erlbaum Associates.

Arrow, H., Henry, K.B., Poole, M.S., Wheelan, S.A., and Moreland, R.L. (2005). Traces, trajectories, and timing: The temporal perspective on groups. In M.S. Poole and A.B. Hollingshead (Eds.), *Theories of small groups: Interdisciplinary perspectives* (pp. 313–367). Thousand Oaks, CA: Sage.

Arrow, H., McGrath, J.E., and Berdahl, J.L. (2000). *Small groups as complex systems: Formation, coordination, development, and adaptation.* Thousand Oaks, CA: Sage.

Benedict, R. (1934). *Patterns of culture.* New York: Houghton Mifflin.

Bettenhausen, K., and Murnighan, J.K. (1985). The emergence of norms in competitive decision-making groups. *Administrative Science Quarterly, 30*, 350–375.

Blake, R.R., and Mouton, J.S. (1982). How to choose a leadership style. *Training and Development Journal, 36*(2), 38–48.

Borgatti, S.P. (1994). Cultural domain analysis. *Journal of Quantitative Anthropology, 4*, 261–278.

Borgatti, S.P., and Everett, M.G. (1992). Notions of position in social network analysis. *Sociological Methodology, 22*, 1–35.

Chiu, C.Y., Hong, Y.Y., and Dweck, C.S. (1997). Lay dispositionism and implicit theories of personality. *Journal of Personality and Social Psychology, 73*, 19–30.

Cole, M. (1996). *Cultural psychology: A once and future discipline.* Cambridge, MA: Harvard University Press.

D'Andrade, R.G. (1989). Cultural cognition. In M.I. Posner (Ed.), *Foundations of cognitive science* (pp. 795–830). Cambridge, MA: MIT Press.

Davis, J.H. (2000). Simulations on the cheap: The Latané approach. In D.R. Ilgen and C.L. Hulin (Eds.), *Computational modeling of behavior in organizations: The third scientific discipline* (pp. 217–220). Washington, DC: American Psychological Association.

Dressler, W.W., and Bindon, J.R. (2000). The health consequences of cultural consonance: Cultural dimensions of lifestyle, social support, and arterial blood pressure in an African American community. *American Anthropologist, 102*(2), 244–260.

Ekman, P., and Davidson, R.J. (1995). *The nature of emotion: Fundamental questions.* Oxford: Oxford University Press.

Fiske, A.P. (1991). *Structures of social life: The four elementary forms of human relations.* New York: Free Press.

Fiske, A.P. (1992). The four elementary forms of sociality: Framework for a unified theory of social relations. *Psychological Review, 99*(4), 689–723.

Fiske, A.P. (2000). Complementarity theory: Why human social capacities evolved to require cultural complements. *Personality and Social Psychology Review, 4*(1), 76–94.

George, A. (1979). Cognitive beliefs about decision-making behavior. In L. Falkowski (Ed.), *Psychological models in international politics.* Boulder, CO: Westview Press (Praeger).

George, A., and George, J. (1998). *Presidential personality and performance.* Boulder, CO: Westview Press.

Güth, W., Schmittberger, R., and Schwarze, B. (1982). An experimental analysis of ultimatum bargaining. *Journal of Economic Behavior and Organization, 3*(4), 367–388.

Hanisch, K.A. (2000). The Impact of organizational interventions on behaviors: An examination of different models of withdrawal. In D.R. Ilgen and C. L. Hulin (Eds.), *Computational modeling of behavior in organizations: The third scientific discipline* (pp. 33–60). Washington, DC: American Psychological Association.

Harris, R.J. (1976). The uncertain connection between verbal theories and research hypotheses in social psychology. *Journal of Experimental Social Psychology, 12*(2), 210–219.

Heinrich, J., Boyd, R., Bowles, S., Camerer, C., Fehr, E., and Gintis, H. (Eds.). (2004). *Foundations of human sociality: Economic experiments and ethnographic evidence from fifteen small-scale societies.* New York: Oxford University Press.

Hersey, P., and Blanchard, K.H. (1988). *Management of organizational behavior: Utilizing human resources.* Englewood Cliffs, NJ: Prentice-Hall.

Holland, D. (1985). From situation to impression: How Americans get to know themselves and one another. In J.W.D. Dougherty (Ed.), *Directions in cognitive anthropology* (pp. 389–412). Urbana: University of Illinois Press.

Hsu, F.L.K. (1972). *Psychological anthropology.* Cambridge, MA: Schenkman.

Hudlicka, E. (2004). *Technical Report 0412. Using cultural and group affiliations to infer individual characteristics and predict behavior.* Blacksburg, VA: Psychometrix Associates.

Hudlicka, E., Karabaich, B., Pfautz, J., Jones, K., and Zacharias, G. (2004). Predicting group behavior from profiles and stereotypes. In *Proceedings of Behavior Representation in Modeling and Simulation (BRIMS) '04 Conference,* May 17–24, Arlington, VA.

Ji, L.-J., Peng, K., and Nisbett, R.E. (2000). Culture, control, and perception of relationships in the environment *Journal of Personality and Social Psychology, 78*(5), 943–955.

Johnson, J.C., and Weller, S. (2002). Elicitation techniques in interviewing. In J. Gubrium and J. Holstein (Eds.), *Handbook of interview research* (pp. 491–514). Thousand Oaks: Sage.

Johnson, J.C., Griffith, D.C., and Murray, J.D. (1987). Encouraging the use of underutilized marine fishes by southeastern U.S. anglers. Part I: The research. *Marine Fisheries Review, 49*(2), 122–137.

Jones, E.E., and Harris, V.A. (1967). The attribution of attitudes. *Journal of Experimental Social Psychology, 3*, 1–24.

Karabaich, B. (2004). Towards a working taxonomy of groups. In *Proceedings of Behavior Representation in Modeling and Simulation (BRIMS) '04 Conference,* May 17–24, Arlington, VA.

Kempton, W. (1987). Two theories of home heat control. In D. Holland and N. Quinn (Eds.), *Cultural models in language and thought* (pp. 222–243). New York: Cambridge University Press.

Klein, H.A., and McHugh, A.P. (2005). National differences in teamwork. In W.B. Rouse and K.R. Boff (Eds.), *Organizational simulation* (pp. 229–252). Hoboken, NJ: Wiley.

Klein, H.A., Pongonis, A., and Klein, G. (2000). Cultural barrier to multinational C2 decision making. In *Proceedings of 2000 Command and Control Research and Technology Symposium,* Monterey, CA. Available: http://stinet.dtic.mil/cgi-bin/GetTRDoc?AD=ADA 461631&Location=U2&doc=GetTRDoc.pdf [accessed Feb. 2008].

Knowles, E., Morris, M.W., Chiu, C., and Hong, Y. (2001). Culture and the process of person perception: Evidence for automaticity among East Asians in correcting for situational influences on behavior. *Personality and Social Psychology Bulletin, 27*, 1344–1356.

Kroeber, A.L., and Kluckhohn, C. (1952). *Culture: A critical review of concepts and definitions.* Cambridge, MA: Peabody Museum.

Lee, F., Hallahan, M., and Herzog, T. (1996). Explaining real life events: How culture and domain shape attributions. *Personality and Social Psychology Bulletin, 22*, 732–741.

LeVine, R.A. (1982). *Culture, behavior, and personality: An introduction to the comparative study of psychosocial adaptation.* New York: Aldine.

Lewin, K. (1951). *Field theory in social science: Selected theoretical papers.* New York: Harper.

Luria, A.R. (1931). Psychological expedition to central Asia. *Science, 74*, 383–384.

Mayo, E. (1960). *The human problems of an industrial civilization*. New York: Viking Press.

McFate, M. (2005). The military utility of understanding adversary culture. *Joint Force Quarterly, 38,* 42–48.

Miller, J.G. (1984). Culture and the development of everyday social explanation. *Journal of Personality and Social Psychology, 46,* 961–978.

Morgan, G. (1997). *Images of organization, second edition*. Thousand Oaks, CA: Sage.

Morris, M.W., Nisbett, R.E., and Peng, K. (1995). Causality across domains and cultures. In D. Sperber and A.J. Pemack (Eds.), *Causal cognition*. New York: Oxford University Press.

Murdock, G.P., and Morrow, D.O. (1970). Subsistence economy and supportive practices: Cross-cultural codes I. *Ethnology, 9*(3), 302–330.

Murray, J.D., Griffith, D.C., and Johnson, J.C. (1987). Encouraging the use of underutilized marine fishes by southeastern U.S. anglers. Part II: Educational objectives and strategy. *Marine Fisheries Review, 49*(2), 138–142.

Nisbett, R.E. (2003). *The geography of thought*. New York: Free Press.

Ofori-Dankwa, J., and Julian, S.D. (2001). Complexifying organizational theory: Illustrations using time research. *The Academy of Management Review, 26*(3), 415–430.

Ortony, A., Clore, G., and Collins, A. (1988). *The cognitive structure of emotions*. New York: Cambridge University Press.

Peng, K., and Knowles, E. (2003). Culture, ethnicity and the attribution of physical causality. *Personality and Social Psychology Bulletin, 29*(10), 1272–1284.

Peng, K., and Nisbett, R.E. (1999). Culture, dialecticism, and reasoning about contradiction. *American Psychologist, 54,* 741–754.

Peng, K., Ames, D.R., and Knowles, E.D. (2001). Culture and human inference: Perspectives from three traditions. In D. Matsumoto (Ed.), *Handbook of culture and psychology* (pp. 243–263). New York: Oxford University Press.

Quinn, N. (1987). Convergent evidence for a cultural model of marriage. In D. Holland and N. Quinn (Eds.), *Cultural models in language and thought* (pp. 173–195). New York: Cambridge University Press.

Romney, A.K., Weller, S.C., and Batchelder, W.H. (1986). Culture as consensus: A theory of culture and informant accuracy. *American Anthropologist, 88*(2), 313–338.

Ross, L. (1977). The intuitive psychologist and his shortcomings: Distortions in the attribution process. In L. Berkowitz (Ed.), *Advances in experimental social psychology* (vol. 10, pp. 173–220). New York: Academic Press.

Schwartz, S.H. (1999). A theory of cultural values and some implications for work. *Applied Psychology: An International Review, 48*(1), 23–47.

Stasser, G. (1988). Computer simulation as a research tool: The DISCUSS model of group decision making. *Journal of Experimental Social Psychology, 24*(5), 393–422.

Stefflre, V.J. (1972). Some applications of multidimensional scaling to social science problems. In R.N. Shepard, A.K. Romney, and S.B. Nerlove (Eds.), *Multidimensional scaling: Theory and applications in the behavioral sciences*. New York: Seminar Press.

Strauss, C. (1997). Research on cultural discontinuities. In C. Strauss and N. Quinn (Eds.), *A cognitive theory of cultural meaning* (pp. 210–251). New York: Cambridge University Press.

Zaheer, S., Albert, S., and Zaheer, A. (1999). Time scales and organizational theory. *The Academy of Management Review, 24*(4), 725–741.

4

Macro-Level Formal Models

This chapter presents modeling approaches for representing the behavior of humans in groups and organizations. It discusses system dynamics models first, followed by a discussion of several approaches to organizational modeling.

SYSTEM DYNAMICS MODELS

What Is System Dynamics Modeling?

System dynamics modeling is a method of modeling the dynamic behavior of complex systems by breaking down these systems into simpler interconnected components ("blocks") connected together via links or "wires" that connect one block's outputs to another block's inputs. This breaking down or recursive modeling continues until simple blocks can be defined in terms of well-understood interactions between the block's inputs, outputs, and its "internal state." Within any given block, this state is defined by the associated state variables, which are usually related by a set of differential equations that underlie the dynamics of that block.[1]

To provide a quick illustration of the basic concepts involved, if one were to model the dynamics of two cars traveling down a straight road, one behind the other, one might specify four blocks: one for each car and one for each driver. Each car would have (a) two states: a speed and a

---

[1]The use of differential equations reflects the history of system dynamics modeling and its roots in electrical and mechanical engineering and control systems theory.

position/location down the road; (b) a single input (or control) of acceleration, determined by the driver's application of the gas or brake pedal; and (c) a single output, the position/location down the road.[2] Simple differential equations, based on the laws of physics (and the vehicle acceleration/braking dynamics) would then be used to define the relation of the input (control) of the driver's use of the gas pedal or brake to the car's output, the position down the road. The second car would be modeled similarly. The trailing car driver would be likewise modeled as a block, with perhaps two inputs, distance and closing speed to the front car, and a single output, gas/brake pedal usage. The differential equations or "control law" relating driver inputs to driver outputs would be specified by well-understood manual control dynamics (see, for example, McRuer and Krendel, 1974). The lead driver could be modeled in "open-loop" fashion, as a block with no input but with a randomly varying output of gas pedal pressure, leading to random speed behavior. By specifying each individual block's behavior (via the inputs, the outputs, and the differential equations underlying the internal dynamics) and by linking up the appropriate inputs to the appropriate outputs of the four-block system, one then has a general system dynamics representation of the dynamics of the two-car, two-driver "system."

The fundamental power of this approach lies in four areas:

1. System dynamics concepts are tightly bound to the twin notions of (1) the dynamic behavior of systems over time and (2) feedback and cross-connectivity between different elements of the system. Dynamic behavior can evolve simply because of a system's internal dynamics and its initial conditions (e.g., a frictionless swing set to infinite harmonic oscillation by an initial offset from the vertical). But the dynamic behavior is considerably more interesting when it is driven by the dynamics of yet some other system (e.g., someone pumping the swing ever higher and eliciting nonlinear swing behaviors), through a cross-coupling or feedback loop involving real physics or abstract information. And when these loops are contaminated by noise (an erratic "pumper"), time delays (a slow-to-respond pumper), and/or distortion in the form of frequency- or amplitude-selective feedback channels, then the opportunity exists for often unanticipated and sometime surprising behaviors across the system as a whole. These are often the characteristics of com-

---

[2]Two states suffice for a simple kinematic representation of the longitudinal (fore-aft) control of vehicle location; additional states would be added for finer grained representation of the situation if one were interested in modeling the effect of the detailed dynamics of the brake calipers, for example. The approach would be the same, however, via the introduction of yet another block placed between the driver's brake pedal and the block representing the vehicle kinematics.

plex human-machine and human-human systems that modelers are
dealing with.

2. The use of blocks, which can be made up of subblocks ad infinitum, so that any level of detail can be examined in a given model,
within practical computational limits. Literally millions of state
variables can be introduced—in a structured manner—to allow
the finest grained examination of the impact of very small components (e.g., O-ring brake failure) on overall system behavior
(e.g., a 20-car pileup on the Los Angeles freeway). In essence, this
approach provides one means of modeling the "butterfly effect,"
as an alternative to chaos theory, which models how small changes
in the initial state (or initial conditions) of a nonlinear system can
lead to large changes of the system state (or system trajectory) at
some later point in time.[3] The systems dynamics approach takes
a bottom-up building block approach, which is appealing in its
dependence on well-understood domain-specific theory and laws,[4]
whereas chaos theory takes a broader systems level view that, if
more abstract, is well grounded mathematically.

3. The use of interconnected blocks ensures that the fundamentals of
feedback are (nearly) always present. In the example above, the
driving behavior of the lead driver clearly will affect the behavior
of the trailing driver.[5] Thus, subtle interactions can be accounted
for, as one element of the system accounts for and accommodates
to others. It is often these feedback loops that give rise to unanticipated "emergent" behaviors (pilot-induced oscillations in aircraft
handling, stock market crashes, etc.).

4. The use of blocks with "internals" that can be elaborated as the
need arises. Generally, differential equations serve as the basis for
a block's dynamics, but it is straightforward to elaborate, via either
the addition of subordinate blocks as just described or the addition of, for example, nonlinear characteristics (e.g., a limit on the
acceleration obtainable via a fully pressed-down gas pedal in the
above example). However, any such nonlinear additions often tend

---

[3]The term "butterfly effect" was introduced by one of the pioneers of chaos theory, Edward
Lorenz, in a paper given by him in 1972 to the American Association for the Advancement of
Science in Washington, D.C., entitled *Predictability: Does the Flap of a Butterfly's Wings in
Brazil Set Off a Tornado in Texas?*
[4]See later comments on the limits to the system dynamics approach of building, from the
ground up, models that seem plausible at each level, until they are actually run and compared
with dramatically different real-world results.
[5]And to explore the impact of the trailing driver's behavior on the lead driver's, one would
merely need to add in a rear-view mirror into the model of the lead driver, and postulate the
dynamics of lead driver behavior as a function of, say, trailing driver tailgating activity, thus
fully "closing the loop" between the two drivers.

to make the theoretical analysis of such systems intractable, so that system dynamics analysts must then rely on simulation execution and analysis in order to understand or predict system behavior.

A specialized version of system dynamics modeling, and the main focus of this section, focuses on a fairly explicit representation of the system states, called "stocks" (entities that accumulate or deplete over time) and their associated "flows" (the rates of change of stocks) (Forrester, 1968). In essence, Forrester[6] transformed the generic nth order differential equations characterizing general system dynamics theory into n first-order differential equations that are intuitively simple to understand and, via the associated programming language Dynamo, into a transparent graphic representation of the key interrelationships among variables (Richardson and Pugh, 1981). Using Dynamo to implement these first-order relations, it becomes a relatively simple exercise in computational model development by the nonspecialist who may not have been schooled in differential equations and their specification or solution. Feedback and interconnections are introduced by defining how the level of one stock controls the flow of another. Nonlinearity is introduced via simple limits on stock levels and flow rates.

A simple example is given in Box 4-1, which illustrates how two states (birth rate and death rate) define the flow of a third state (net growth rate). This is a simple open-loop example with no feedback, but it is not a difficult exercise to close the loop, for example, by postulating how population growth rate might influence economic growth rate, which could induce consumer confidence and, through that, cause birth rates to increase.

An example showing this level of loop closure is given in Figure 4-1, which illustrates one component of a larger system dynamics model of the spread of an epidemic (Sage and Armstrong, 2000). The three state variables (stocks) are X_1, the population susceptible to infection (susceptible population), X_2, the population that is actually infected (infected population), and X_3, the population that has developed an immunity to the infection (immune population). Note that boxes are used to represent these states graphically. The associated flows are LR (loss of immunity rate), IR (infection rate), and RR (recovery rate). Note that the valve symbols are used to indicate how the flows control the stock levels, via the following intuitive graphic analogy: flow into a block increases the stock level, while

---

[6]Although Jay Forrester's name is the one most closely associated with the system dynamics concept, his work owes much to the electrical engineering pioneers at Bell Laboratories working with feedback circuits and notions of system stability in the 1920s and 1930s (see, e.g., Black, 1977); the discipline of *cybernetics* developed at the Massachusetts Institute of Technology by Norbert Weiner and colleagues during the 1940s and 1950s (Weiner, 1948); and, more recently, practitioners who have done much to popularize its application to important problems in the social sciences, most notably Richardson and colleagues (see, e.g., Richardson and Pugh, 1981; Richardson, 1991).

BOX 4-1
The Equation, Variables, and Mathematical Representations for Birth and Death Used in Population Modeling

Description of variables:
$\beta(t)$: Average birth rate per unit person in the population at time t
$\Delta(t)$: Average death rate per unit person in the population at time t
$\mu_n(t)$: Expected value

Mathematical representation of birth rate, death rate, and average rate of population growth:
$\beta(t)\mu_n(t)$: Total average birth rate
$\Delta(t)\mu_n(t)$: Total average death rate
$\dfrac{d\mu_n(t)}{dt} = [\beta(t) - \Delta(t)]\mu_n(t)$: Average rate of population growth (the difference between the total average birth rate and death rate)

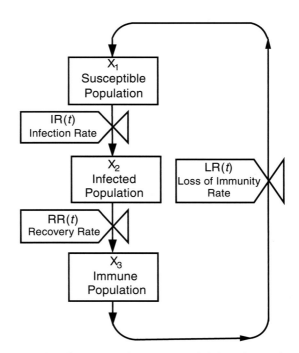

FIGURE 4-1 Example of a system dynamics model that shows the partial system dynamics description for propagation of a potential epidemic.
SOURCE: Adapted from Sage and Armstrong (2000, p. 235).

flow out decreases it.[7] The diagram captures the following qualitative and, for the mathematically inclined, quantitative notions:[8]

- For the states:
  - The susceptible population X_1 will *increase* as the recovered lose immunity (LR) and *decrease* as the susceptibles become infected (IR). Or[9]
    - $d(X_1)/dt = LR - IR$
  - The infected population X_2 will *increase* as the susceptibles become infected (IR) and *decrease* as the infected recover (RR). Or
    - $d(X_2)/dt = IR - RR$
  - The immune population X_3 will *increase* as the infected recover (RR) and *decrease* as the immune lose immunity (LR). Or
    - $d(X_3)/dt = RR - LR$
- For the flows (not illustrated for simplicity):
  - The infection rate (IR) increases both as the susceptibles (X_1) increase and as the infected (X_2) increases, due to the networked nature of spreading infections. Or[10]
    - $IR = a*X_1*X_2$
  - The recovery rate (RR) is directly proportional to the infected (X_2). Or
    - $RR = b*X_2$
  - Likewise, the loss of immunity rate (LR) is directly proportional to the infected (X_3). Or
    - $LR = b*X_3$

Note the complete loop closure relating the three states, and the potential for continuing growth and decay of an infected population over time. Note also the potential for nonlinear behavior over time, because of the fundamental nonlinearity introduced via the infection rate equation ($IR = a*X_1*X_2$).

The structure of system dynamics models can be characterized by four hierarchical levels, as shown in Figure 4-2.[11] All interactions and impacts

---

[7]Not explicitly shown is how the flows are influenced by the stock levels.

[8]Note that in this set of equations and in subsequent sets, the asterisk (*) is not meant to represent a convolution operation or function composition, but rather a simple multiplication, in line with DYNAMO code conventions, as well as FORTRAN syntax, which was a popular computational language at the time of DYNAMO's introduction.

[9]d( )/dt is used to denote the first-order derivative of the associated variable.

[10]The constants (a,b,c) are chosen on the basis of underlying knowledge of dynamics of infection, recovery, etc.

[11]This description borrows heavily from Sage and Armstrong (2000, p. 237).

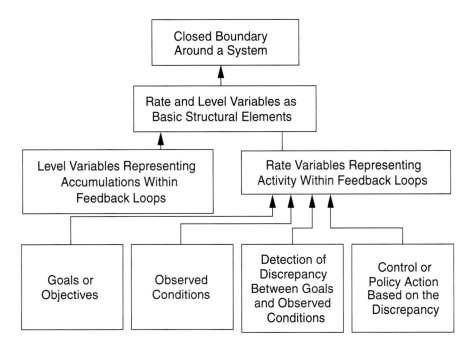

FIGURE 4-2 The four hierarchical levels of system dynamics modeling.
SOURCE: Sage and Armstrong (2000, p. 237).

in the system dynamics model take place inside a boundary. Within this boundary, variables are chosen to represent the key states that define overall system behavior. A derivative variable is chosen to control a flow into the state or level variable, which integrates or accumulates this level. Information concerning the level is used to control the rate variable (state feedback to the same associated state). In other words, we define a rate variable as the time derivative of a level or state variable and determine rate variables as functions of level variables.

Some useful readings on system dynamics modeling methodology are Roberts, Anderson, Deal, Garet, and Shaffer (1983); Sterman (2000); Ogata (2003); and Karnopp, Margolis, and Rosenberg (2006). A more detailed description of system dynamics modeling and the equations it uses is available in Sage (1977) and Sage and Armstrong (2000). Comprehensive approaches to modeling complex projects—including industrial and military—are described by Williams (2002).

State of the Art in System Dynamics Modeling

Early History of System Dynamics

Jay W. Forrester created this focused version of system dynamics in the mid- to late 1950s at the Massachusetts Institute of Technology's Sloan School of Management, basing it on the more traditional modeling used at the time, implementing differential equation models on analog computers. Forrester brought these concepts to the digital domain, codified them in the stocks and flows paradigm described above, and used this approach to model highly complex systems such as organizations and the urban environment (Forrester, 1961; see also Forrester, 1969). This novel approach of developing computational dynamic models of hitherto unmodeled phenomena led to the founding of the System Dynamics Group at the Massachusetts Institute of Technology in the early 1960s (see http://web.mit.edu/sdg/www/what_is_sd.html).

Forrester wrote several books on system dynamics methodology that provide the foundations of the field. The first was *Industrial Dynamics* (Forrester, 1961), providing a computational foundation for understanding the dynamics of organizations and processes in industry. Forrester then published *Urban Dynamics* (1969), which was the first noncorporate application of system dynamics (Radzicki, 1997). Shortly thereafter Forrester published *World Dynamics* (1971) in which he applied system dynamics methodology to the behavior of the highly interrelated forces of global dynamics (Sage and Armstrong, 2000). Forrester's student, Dennis Meadows, and colleagues expanded on *World Dynamics* in *The Limits to Growth* (Meadows, Meadows, Randers, and Behrens, 1972) and a follow-up, *Beyond the Limits* (1992) (Radzicki, 1997). The Malthusian projections that came from these early models not only alienated the growth-oriented policy makers of the West, but also brought severe criticism from many of the academics in the field (e.g., economists), because of the glaring mismatch between model "predictions" and what was actually occurring on the world stage. This became more apparent as time went on, and it is fair to say that this failure to meet empirical validation standards considerably dampened the initial enthusiasm that met the system dynamics viewpoint toward understanding the complex interrelations of complex systems.[12]

---

[12]However, system dynamics modeling has been applied to several other areas, including software project dynamics (Abdel-Hemid and Madnick, 1991), organizational learning (Senge, Kleiner, Roberts, Ross, and Smith, 1994; Morecroft and Sterman, 1994), agriculture (Elmahdi, Malano, and Khan, 2006), health care management (Rohleder, Bischak, and Baskin, 2007), and transportation (Springael, Kunsch, and Brans, 2002).

More Recent Applications of System Dynamics Modeling

More recently, there has been a resurgence of interest in system dynamics modeling, most particularly in public policy and business areas. Sterman's text on Business Dynamics (2000) presents a number of case studies that demonstrate successful applications across a number of areas, including global warming, the war on drugs, reengineering the supply chain of a major computer firm, developing a marketing strategy in the automobile industry, and planning process improvements in the petrochemicals industry. The Department of Defense (DoD) has also taken a keen interest in this approach, particularly for modeling diplomatic, information, military, and economic (DIME) actions, and political, military, economic, social, information, and infrastructure (PMESII) interactions. It is not our intent here to survey all of these efforts, but merely to provide a few illustrative examples to indicate the potential of system dynamics modeling in this area.

For example, Robbins' Stabilization and Reconstruction Operations Model (SROM) (Robbins, Deckro, and Wiley, 2005) analyzes the organizational hierarchy, dependencies, interdependencies, exogenous drivers, strengths, and weaknesses of a country's PMESII systems using a complex set of interdependent system dynamics representations. SROM models a country system in a holistic manner as a national model, which, as shown in Figure 4-3, is then defined in terms of its *n* regional submodels that interact with each other and the national model. Each regional submodule contains six functional submodels: the demographics submodel, the insurgent and coalition military submodel, critical infrastructure, law enforcement, indigenous security institutions, and public opinion. Each submodel is comprised of approximately 600 model parameters, 90 random variables, 80 states (stocks), and 190 rates of change (flows).

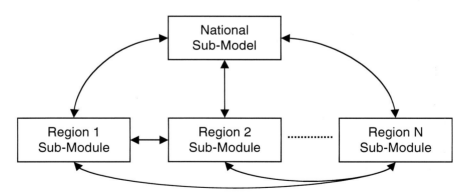

FIGURE 4-3 Top-level nation SROM.
SOURCE: Robbins et al. (2005, p. 19).

Figure 4-4 shows a portion of the critical infrastructure model of SROM. The model captures a sequence of influences among variables, starting from the power supply at an electrical substation. The generated power is fed into an industrial water plant, which produces water consumed by oil field work. An oil field produces crude oil to be refined by a refinery. Finally, refined fuel is used to generate power, which in turn is supplied to various power substations, thus forming a closed loop.

SROM has been demonstrated in modeling and analysis of Iraqi reconstruction and recruiting efforts (Robbins et al., 2005). Parameters were set to reflect prevailing conditions in Iraq on May 1, 2003, including

- Regional makeup (governorates)
- Regional population
- Population subgroup distribution
- Population support for coalition
- Oil and gas infrastructure
- Power infrastructure
- Transportation infrastructure
- Economic—regional gross domestic product

Robbins (2005) claims that the SROM allows analysts to more precisely investigate the multifaceted process that is nation building: "[Because] the complexities of nation-building involve many different but interrelated systems and institutions, understanding the significance of the dynamic relationships between these systems and institutions is paramount to operational success. The system dynamics model proposed in this study allows decision-makers and analysts to investigate different sets of decision approaches at a sub-national, regional level" (p. 135).

The Pre-Conflict Anticipation and Shaping (PCAS) program (Popp et al., 2006) was an attempt to evaluate alternative DIME/PMESII modeling efforts to predict nation-state collapse and to anticipate instabilities that might lead to conditions necessitating military intervention. One of the approaches, led by Nazli Choucri, developed a "state stability model" using a system dynamics approach; a high-level view of the model is given in Figure 4-5.

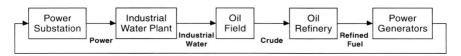

FIGURE 4-4 SROM infrastructure model.
SOURCE: Robbins, Deckro, and Wiley (2005).

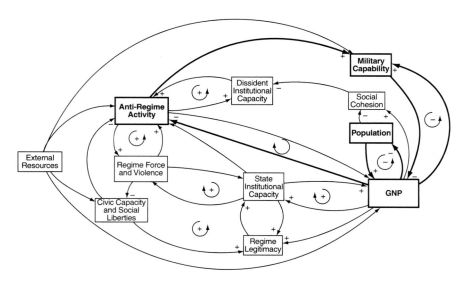

FIGURE 4-5 High-level view of system dynamics implementation of state stability model.
SOURCE: Popp (2005).

According to Popp (2005, p. 18), it "shows loads, demands and stresses on state and the causal dependencies; shows feedback loops, tipping points and unintended consequences; [and] shows the internal and lateral pressures that can lead to conflict." By looking at the loads (demands) placed on the system (nation-state) and evaluating those demands in terms of the system's capabilities, an assessment of stability can be made based on how much demands exceed capacity.

Finally, O'Brien's Integrated Crisis Early Warning System (ICEWS) is a new program at DARPA/IPTO aimed at following on from the PCAS exploration just described. According to the announcement of the research program, its goal "is to develop a comprehensive, integrated, automated, generalizable, and validated system to monitor, assess, and forecast national, sub-national, and international crises in a way that supports decisions on how to allocate resources to mitigate them. ICEWS will provide Combatant Commanders (COCOMs) with a powerful, systematic capability to anticipate and respond to stability challenges in the Area of Responsibility (AOR); allocate resources efficiently in accordance to the risks they are designed to mitigate; and track and measure the effectiveness of resource allocations toward end-state stability objectives, in near-real time" (see http://www.arpa.mil/ipto/solicitations/open/07-10_PIP.pdf [accessed July 2007]).

Environments for System Dynamics Modeling

The earliest computer-based system dynamics simulations were created by Richard Bennett, who developed the SIMPLE (Simulation of Industrial Management Problems with Lots of Equations) compiler in 1958 (Forrester, 1989). In 1959, Phyllis Fox and Alexander Pugh used the SIMPLE compiler to form the DYNAMO simulation package, which was used as the standard system dynamics language for over 30 years (Radzicki, 1997).

There are several computer-based simulators that are used to model system dynamics problems. The DYNAMO system dynamics simulation language is described in Richardson and Pugh (1981) and a personal computer–based language, STELLA, is discussed in Richmond and Peterson (1992). Other software packages that are used for system dynamics modeling include Powersim, Vensim, MapSys, Simile, and Evolución. The tradition of easy model development is also carried on via the Ptolemy systems modeling language developed by Buck, Ha, Lee, and Messerschmitt (1994).

Relevance, Limitations, and Future Directions

The relevance of the system dynamics approach to the problems addressed by this panel is manifest both by the early work by Forrester and colleagues in attempting to model organizations, cities, nations, and overall world dynamics and by the current resurgence in interest by DoD in retackling these very hard problems, in recognition of the "soft" nature of warfare now dominating current conflicts. The fundamental appeal of this methodology is due to the strengths noted earlier:

- System dynamics concepts provide a means of representing critical dynamic behavior of systems over time, as well as feedback, and cross-connectivity between different elements of the system.
- The use of blocks that can be made up of subblocks ad infinitum, so that any level of detail can be examined.
- The use of interconnected blocks that ensures that the fundamentals of feedback are (usually) always present, enabling emergent behavior.
- The use of blocks with internals that can be elaborated as the analysis need arises, in terms of both resolution and modeling fundamentals.

One of the major limitations of the system dynamics approach is that its strong grounding in a mathematical description of the organizational dynamics (namely, first-order differential equations) tends to preclude participation by researchers and modelers who are more linguistically and semantically oriented, for example, those working in causal networks, expert systems, or the like. Attempting to bring these different communities together is not a trivial task, as evidenced by the experience of the PCAS program, and attempting to integrate across these different methodologies is likewise problematic, as described in Chapter 8.

Another limitation is verification and validation, since these models are particularly easy to build by making simple assumptions about structures, feedback path, and parameter values, without ever relying on "real" data.[13] As noted by Sage and Armstrong with respect to the urban modeling effort of Forrester: "Forrester's interest in modeling the city is a somewhat abstract one in that he does not fit the data and parameters for his city to any particular city. Effort is primarily directed at discovering the essential features of the city and expressing relationships between these features in mathematical terms as difference equations" (Sage and Armstrong, 2000, p. 253).

A system dynamics model must have both behavioral and structural validity (Quadrat-Ullah, 2005). Forrester and Senge (1980) presented some tests for determining if a model has structural validity:

- Boundary adequacy: whether the important concepts and structures for addressing the policy issue are endogenous to the model.
- Structure verification: whether the model structure is consistent with relevant descriptive knowledge of the system being modeled.
- Parameter verification: whether the parameters in the model are consistent with relevant descriptive and numerical knowledge of the system.
- Dimensional consistency: whether each equation in the model dimensionally corresponds to the real system.
- Extreme conditions: whether the model exhibits a logical behavior when selected parameters are assigned extreme values.

---

[13]Or, as one of the report reviewers so aptly noted, "Expressing a social relationship as a differential equation (or any other kind of equation for that matter) does not make it so. There are such things as accounting identities, but mathematically exact descriptions of social or human processes generally do not exist." We agree, but, of course, this is not a shortcoming that is specific to computational models built on system dynamics concepts; it is a general issue with any "model" that can be reduced to executable software. Adequate verification and validation is always a critical issue for any modeling paradigm, as we discuss at length in Chapter 8.

Finally, there are a number of potential directions for future research and development (R&D) efforts, including bridging the gap to models and simulations that are not so formally mathematically defined, improving model composability from smaller component libraries (Davis and Anderson, 2004), and ensuring that the difficult problem of verification and validation does not outstep the progress being made in developing development environments that are easy to use.

ORGANIZATIONAL MODELING

An organization consists of a number of individuals who must work together to achieve a goal. Corporations, governmental bureaus, religious organizations, the Armed Forces, divisions, squads, and teams are all organizations; they are everywhere and each of us can be members of many organizations. A fundamental aspect of an organization is that the organizational task requires the efforts of many individuals who must work together to accomplish this overall task. The big task is broken down into smaller subtasks or jobs which must then be coordinated in order to achieve the organizational goal. Each organization is the context or structure for the individuals to achieve their own smaller tasks. Individuals are linked through the organizational structure.

What Is Organizational Modeling?

Organization theory is a study of the structure, behavior, and performance of the organization (Scott, 1998) in order to describe, explain, and predict. Basic questions include: Under what circumstances is decentralization better than centralization? When should an organization be highly formalized with many rules and when should it be more informal? When should the information and communications follow the hierarchy, and when should there be many cross-hierarchy exchanges? When should tasks be grouped together by task specialization and when by purpose—all for better performance?

There are many ways to describe an organization. One generally thinks of an organization in terms of its task assignment and its hierarchy of command and control. Another more basic description is: Who does what, when, and where?—the four Ws of an organization. The "who" is the individual—the task assignment is the "what"—the "when" introduces the time of action—and the "where" is the location. This is a rather complete description of an organization, leaving out only the how (with what resources and knowledge) and the why (with what goals). Beyond a few individuals, it is difficult to think about an organization at this level, without the use of dynamic network analysis models, so we introduce organizational properties, such as decentralization and formalization, or rules for how decisions are made and the information communicated, and the behavioral patterns or routines that are repeated. In information-processing terms, the Ws take on a slightly different aspect: who talks to whom about what (communication networks, which may be hierarchical) and who decides what to do (decision making or command structure). In modern information-intensive organizations, the information-processing view of an organization is a frame for both describing an organization and designing it (Burton and Obel, 2004).

Organizational design is the complement of organization theory in which one specifies what the organization should be, beginning with the purpose of the organization and then specifying the tasks and coordination mechanisms or communications and decision-making structures. The focus goes beyond what is or has been to what might be and then what should be (Burton, 2003). A good design requires a good understanding of organization theory, as the theory indicates what is feasible and not feasible in the circumstances. Organizational design can be the specification of what the organization should be: its hierarchy, its formalization, its decentralization, its communications, and its coordination and control mechanisms—an information-processing view. Or the term "organizational design" may describe the process of finding a good design and the issues of change or redesign. Organizational design is used both as a noun for what the design should be and as a verb for the process of determining the design. We discuss both below.

In the social and managerial sciences, the research on organizations is deeper in organization theory than organizational design, although the opposite tends to be true in engineering. In general, however, the focus has been on the "what is" description of organizations and explaining what has happened—to enhance understanding of how organizations behave. There are theories of bureaucracy, routines as rules and patterns, decentralization, coordination, and control; all yield hypotheses for confirmation or rejection. For the most part, these hypotheses have been examined empirically. Researchers have used field data and, to a much lesser extent, lab experimental data to test the hypotheses and add to the understanding of organizations. In addition, some researchers take an ethnographic approach in which they gather more detailed data about an organization, describe it in great depth, and in doing so generate emergent hypotheses and theories. For both approaches, there is an emphasis on the data gathered from organizations using an inductive approach for understanding.

There has been a smaller effort on formal mathematical modeling of organizations using a deductive approach, in which the analysis of the models yields insights and hypotheses that can be tested using field or lab data. This includes optimization approaches, in which the structure of the organization is optimized to meet the demands of the mission and tasks to be performed (Pattipati, Meirina, Pete, Levchuk, and Kleinman, 2002).

Simulation or computational models offer a third approach (Axelrod, 1997). Simulation modeling is distinctively different from both approaches described above yet also has characteristics of both. First, a simulation model of an organization, which includes its structure and agents, generates behavioral and performance data on the organization, which can be analyzed *as if* they were field data.[14] These are frequently called virtual experiments. Agent-based models explicitly model both the agents or individuals and the decision-making structure of the organization, which includes the communication and authority links among the agents. In these experiments, the simulation model parameters can be varied beyond what can be observed from field or lab data to explore what might be; for example, new structures and decision procedures can be created, and even the information-processing characteristics of the agents (human or machine) can be varied. In both situations, the simulation models generate a larger set of possibilities from which to gain insight and understanding. Second, simulation models can be similar to mathematical models, but they are more complex and not amenable to closed-form solutions. Here, the simulation model can be used to explore and generate hypotheses for further investigation in the field or lab.

Simulation models free one from the constraints of mathematical models in which closed-form equilibrium solutions are required. Simulation models free one from the size and scale restrictions of lab experiments and from the limitations of field data, which necessarily are historical and limited to what did happen—not what could have happened. For action, we are interested in the alternatives available and what might happen—which is broader than what has happened in the past.

---

[14]We emphasize "as if" for two reasons: First, the simulation-generated data can be processed and analyzed via the same methods and toolsets used for real-world data. This is clearly an advantage both in terms of economical reuse of methods and software already developed for real-world data, and in terms of ease of comparing *processed and analyzed* data collected from the two domains (simulated and real-world). This latter case is particularly important, since it is often impossible to compare, for example, single-time histories of organizational "state" recorded from real and virtual experiments (because of "noise," for example), whereas it is possible to compare processed data, obtained from time- or ensemble-averaged statistics calculated over many "runs," both virtual and real.

The second reason for the emphasis on "as if" is not so positive. Because simulation-generated data can be made to look so much like real data, they are often confounded, and researchers can be led to overinterpret the results of a simulation, coming to conclusions as if they had been looking at real data generated by real-world experiments (or at real-world data generated by an experiment which grew out of a hypothesis created by a simulation-based virtual experiment), rather than at data generated by a simulation that has not been adequately validated.

State of the Art in Organizational Modeling

Here, we focus on simulation or computational organizational models. A number of books contain overviews and examples of many models in this area (Carley and Prietula, 1994; Carley and Gasser, 1999; Lomi and Larsen, 2001). Some models consider organization theory questions; others are more oriented to organizational design questions; and some can be used for both purposes. We begin with the theory models and then consider the design models, with comments when the models can be used both ways.

Organization Theory Models

There are numerous organization simulations or computational organizational models; here we review a few of them. Most, but not all, are agent-based models in which the organization is represented as agents that are linked together by communication or authority structures or both.

The earliest computational organizational model was a behavioral theory of the firm in which the organization was modeled in terms of goals, expectations, and choice (Cyert and March, 1963). Simple systems were used to demonstrate how nonrational behavior could generate behavior similar to that observed in real organizations. This was then extended in the now canonical model, the garbage can model of organizational choice (Cohen, March, and Olsen, 1972). This was a simple Fortran program in which basic matching and accumulation functions were combined to show how variations in the problem access, salience of problems, and energy of the participants altered the level of work and the quality of outcomes.

The Lin and Carley models look at organizations as networks of communication linkages among agents, such that agents learn only from the information that they get from the outside world or that is provided to them by another agent in the organization (Lin and Carley, 2003; Lin, Zhao, Ismail, and Carley, 2006). Using these models, they investigated questions of crisis response. They conducted a "matched-set" validation experiment, in which they compared the behavior of 69 real-world organizations faced with industrial crises with the behavior of the simulated versions of those same 69 companies. Using what-if analysis, they were then able to show that the type of decision making employed by the organization—for example, following standard operating procedures or following the dictates of historically based experience—often led organizations to false conclusions about their performance.

This work was generalized and extended to produce the OrgAhead model. OrgAhead is a multiagent model of organizational design and the examination of the impact of learning and strategic adaptation on that design (Carley and Svoboda, 1996). In this model, learning occurs at the operational and structural levels, using experiential and expectation-based learning models. From a technical standpoint, the model uses simulated annealing[15] to alter the communication and authority lines and number of agents. The agents are information-processing units with a simple learning component. OrgAhead can be thought of as an operationalized grounded theory. The basis for OrgAhead is the body of research, both empirical and theoretical, on organizational learning and organizational design. The model has built into it several theories of different aspects of organizational behavior. From the information-processing tradition comes a view of organizations as information processors composed of collections of intelligent individuals, each of whom is boundedly rational and constrained in actions, access to information by the current organizational design (rules, procedures, authority structure, communication infrastructure, etc.), and his or her own cognitive capabilities. Organizations are seen as capable of changing their design (DiMaggio and Powell, 1983; Romanelli, 1991; Stinchcombe, 1965) and as needing to change if they are to adapt to changes in the environment or the available technology (Finne, 1991). Different organizational designs are seen as better suited to some environments or tasks than others (Hannan and Freeman, 1977; Lawrence and Lorsch, 1967). Aspects of the model have been tuned to reflect the findings of various empirical studies related to these theories. The set of theories that are unified into a single computational theory of organizational behavior interact in complex fashions to determine the overall level of organizational performance.

Harrison and Carroll (1991) investigated the effect of turnover on organizational culture for different prototypical organizations and policies. Their model is stated as a set of mathematical functions, which are then simulated and yield data that are analyzed as if they were field data. The model is essentially a cultural diffusion model operating at the group level. On the basis of "virtual experiments" conducted with the model and a follow-on analysis of the resulting simulation-based data, they found that some employee turnover can help stabilize the culture of the organization, suggesting that some previously held truths about turnover are not general.

An alternative information diffusion model is Construct, developed by Carley to examine the coevolution of structure and culture that results from individual information exchange and the formation and dissolution

---

[15]Simulated annealing is a technique to find a good solution to an optimization problem by trying random variations of the current solution. A worse variation is accepted as the new solution with a probability that decreases as the computation proceeds. The slower the cooling schedule, or rate of decrease, the more likely the algorithm is to find an optimal or near-optimal solution (see http://www.nist.gov/dads/HTML/simulatedAnnealing.html [accessed August 2007]).

of social networks (Carley, 1991). Construct has been used to examine the impact of new technologies on the workplace (Carley and Schreiber, 2002), performance under diverse leadership styles (Schreiber and Carley, 2004), and the emergence of organizational vulnerabilities (Carley, 2004).

NK models, originally suggested by Kauffman, are simple optimization models, often operationalized using genetic algorithms, in which N is the number of actors and K is degree of connectedness among the actors (Kauffman and Weinberger, 1989). NK models have been applied to organization theory questions of adaptation (Levinthal, 1997), search and stability (Rivkin and Siggelkow, 2003), modularity and innovation (Ethiraj and Levinthal, 2004), imitation and benchmarking (Rivkin, 2000), and other basic questions about organizations. The explicit modeling of rugged landscapes permits one to understand the limitations of organization explanations that implicitly assume smooth performance surfaces. It also yields greater insights into the persistence of variety among organizations.

The SimVision model (earlier called VDT) is a project organization model (Levitt, Thomsen, Kunz, Jin, and Nass, 1999) which explicitly models the project tasks (similar to a critical path method network) and the hierarchical organization structure. In essence, this model is the merger of Gantt chart technology with a limited information-processing model for the agents. The project tasks are linked by the project network, and each task is assigned directly to an agent in the hierarchy. SimVision has been used as a laboratory for organization experiments.[16] For example, Carroll, Burton, Levitt, and Kiviniemi (2006) found that "fast tracking" or concurrent engineering of projects quickly leads to increased coordination demands that do not reduce total project time; additional personnel can also increase project time as they require time to manage; and decentralization increases coordination demands. Earlier, Kim and Burton (2002) found that decentralization reduces project time but may also decrease quality. Long, Burton, and Cardinal (2002) demonstrated that three simultaneous control approaches are better than any single control approach. These studies began with organizational questions and observations of real organizations as base models. The simulation experimental manipulations ("virtual experiments") went beyond real-world observations to investigate plausible conditions of what could happen for a better understanding of potential outcomes. Field obser-

---

[16]In the studies cited here it must be remembered that the conclusions drawn from analysis of the simulation-based data (in turn generated by virtual experiments in the simulation domain) are not to be confounded with conclusions drawn from an analysis of homologous real-world data. This is in keeping with our earlier footnote regarding how simulation-based data can be analyzed as if it were real-world data. It often can, but the fundamental issue still remains regarding the validity of applying the simulation-based conclusions to real-world organizational behavior. Naturally, the more validated the model, the more likely one is to be correct in cross-applying one's conclusions.

vations and generalizations are limited in their applicability and should be used with caution in the design of future organizations. Simulation studies provide deeper insight into what is possible and what is desirable for organizational redesign and change. SimVision can also be applied as an organizational design model.

Organizational Design Models

The term "organizational design" is used both to mean the design of the organization and the process of design. The two meanings are different but closely related. In a special issue of *Organization Science*, Dunbar and Starbuck (2006) focus on the process of organizational design in its many facets. The articles give insight into how design can be accomplished and the challenges encountered.

SimVision was applied to investigate organization theory questions. But it was originally created as an organizational design tool to help project managers optimize projects and project management implementation (Levitt, 2004) This included avoiding unforeseen bottlenecks and finding options to compress project time. One of the insights is that project managers adapted quite well to minor variations from the normal base case but less well when there were large changes in requirements. The simulations were extremely useful in helping project managers reframe the project and redesign the project.

Pattipati and colleagues (Pattipati et al., 2002; Levchuk, Levchuk, Luo, Pattipati, and Kleinman, 2002a, 2002b; Levchuk, Levchuk, Meirina, Pattipati, and Kleinman, 2004) have used multiobjective optimization algorithms to develop organizational designs optimized to meet mission requirements for military command and control organizations, focusing specifically on Joint Task Force command teams. These designs specify both structure and process by specifying roles in the organization defined in terms of control of resources, responsibility for tasks, and requirements for coordination. Designs are then tested in simulations of organizational performance and finally tested in field experiments in which military officers play the roles that were designed using the model. Studies have shown that optimized organizational designs based on the model result in performance that exceeds that observed under more traditional designs suggested by military subject matter experts (Entin, 1999). A key finding of this work is that sufficient training is essential for the officers to function effectively in the innovative organizational structures developed using the model.

Carroll, Gormley, Bilardo, Burton, and Woodman (2006) describe an organizational design process at the National Aeronautics and Space Administration (NASA), where SimVision and other organizational design tools were used as decision aids in creating a new organization. The challenge was to create an organization that had multiple functional experts, was geographically disperse, and had severe resource constraints in which project time and quality were paramount. The design team began with the construction of the design structure matrix; it gave a good beginning but generated questions as well as answers. Next, they used OrgCon—an expert system organizational diagnosis and design tool—to model the proposed organization at a high level in terms of structural properties, such as formalization and decentralization. One purpose of this modeling was to identify "misfits" (Burton and Obel, 2004) that suggested a need for change; they found few of them. But many questions remained. Then they created a SimVision of the proposed design to obtain greater detail and better understanding of how the organization would actually work. Using variations in the design, they confirmed that the design developed with the aid of the tools was reasonable. Perhaps most importantly, the usual organizational design approach would have resulted in an organization that would have failed to meet the goals and would have incurred delays and unanticipated costs. The results indicate that the tools can make a difference and lead to better designs; furthermore, the theory-based notion of organizational misfits aids in the process. It can be a bridge between theory and design and theory and practice, as managers find the identification of misfits and their correction both intuitive and practical. NASA had been accustomed to using simulations in engineering design but not in organizational design. Nonetheless, the culture was amenable to the application of such tools for organizational design.

Similarly, OrgAhead was built to explore the relative effectiveness of different organizational designs. For example, it was used to determine the adaptability and performance characteristics of different designs under consideration by the Naval Strategic Studies Group. Construct, referred to earlier, has also been used to evaluate various organizational designs under different turnover regimes. Moreover, when data are collected on the who, what, where, and how of organizations, such data can first be assessed for points of vulnerability in ORA and then Construct can be applied to the same empirical description of the real organization to forecast its behavior in terms of information diffusion and performance with or without turnover (Carley, Diesner, Reminga, and Tsvetovat, 2005).

Levis and Wagenhals (2000) and the subsequent work with Shin, Kim, Bienvenu, and Shin led to the development of a Petri net model for designing and assessing organizational architectures (Bienvenu, Shin, and Levis, 2000; Wagenhals, Shin, Kim, and Levis, 2000). Modeling agents, their resources, and the decision process, this overall approach makes possible the fine tuning of detailed designs of core groups in organizations. This approach has been used consistently to evaluate command and control structures. The key advantage of this approach is that designs can be optimized to the specific communication and timing requirements.

Relevance, Limitations, and Future Directions

The relevance of organizational models to the requirements outlined in Chapter 2 is obvious. Representative tasks, such as designing effective organizations and disrupting adversary organizations, are clear candidates for the use of such models. If it were possible to accurately assess the probable effectiveness of various organizational options before implementing them, much effort could be saved and many potentially catastrophic mistakes avoided.

Limitations of such models as they now exist include requirements for data that may be totally unavailable or unavailable in appropriate formats and structures, the need for culturally appropriate information on which to base assumptions and algorithms, especially for non-Western organizations, and technical issues requiring further development and refinement of the models themselves.

R&D requirements include better methods for obtaining and using organizational performance data to provide leaders and managers with better tools for restructuring their organizations as necessary. The vast majority of current model-based organizational design methods are static. That is, they use prior performance data about the organization to develop future designs, but they do not use "streaming" performance data as it comes in to understand or modify the organization's structure and processes in real time. Organizational models that could accept and use real-time data could provide a tool for making organizations more flexible and able to adapt to changing conditions and missions more quickly.

An additional area in need of research is the ability to combine models at different levels of granularity and detail to represent large organizations, as well as the advantages and drawbacks of including more or less detail. Including detail for all of the individuals in a large organization can quickly lead to intractable size and computational infeasibility, but system-level models may not be able to represent the detail that leads to emergent behavior. For example, system dynamics models could be developed at the level of the entire organization, with individual agents developed to represent key individuals or groups in the organization. Data could flow in both directions between the detailed agent-based models and the organization-level system model. Challenges and existing approaches for developing such integrated multilevel models are discussed in Chapter 8.

Finally, innovative experimentation approaches are needed to advance the state of the art in organizational modeling. Systematic controlled experiments are not feasible for organizations of any size—team experiments rarely include more than six to eight team members. However, the development of agents that can represent the behaviors of members of the organization in a realistic way opens the door for "hybrid" experiments in which most roles in the organizations are played by agents, with only a few played by live subjects. Research is needed on the best ways to use this hybrid experimentation capability to advance organizational science: the types of questions that can best be addressed in such experiments, the best ways to "control" such experiments in the classical sense of experimental control, the level of fidelity needed in the agents, and the statistical techniques needed for analysis of the results.

REFERENCES

Abdel-Hemid, T., and Madnick, S.E. (1991). *Software project dynamics: An integrated approach*. Englewood Cliffs, NJ: Prentice-Hall.

Axelrod, R. (1997). Advancing the art of simulation in the social sciences. In R. Conte, R. Hegselmann, and P. Terna (Eds.), *Simulating social phenomena* (pp. 21–40). Berlin, Germany: Springer.

Bienvenu, M.P., Shin, I., and Levis, A.H. (2000). C4ISR architectures III: An object-oriented approach for architecture design. *Systems Engineering, 3*(4), 288–312.

Black, H.S. (1977). Inventing the negative feedback amplifier. *IEEE Spectrum, 14*, 54–60.

Buck, J.T., Ha, S., Lee, E.A., and Messerschmitt, D.G. (1994). Ptolemy: A framework for simulating and prototyping heterogeneous systems. *International Journal of Computer Simulation, Special Issue on Simulation Software Development Component Development Strategies, 4*.

Burton, R.M. (2003). Computational laboratories for organization science: Questions, validity and docking. *Computational and Mathematical Organization Theory, 9*(2), 91–108.

Burton, R.M., and Obel, B. (2004). *Strategic organizational diagnosis and design: The dynamics of fit, third edition*. Boston: Kluwer Academic.

Carley, K.M. (1991). Designing organizational structures to cope with communication breakdowns: A simulation model. *Industrial Crisis Quarterly, 5*, 19–57.

Carley, K.M. (2004). Estimating vulnerabilities in large covert networks using multi-level data. In *Proceedings of the North American Association for Computational Social and Organizational Science (NAACSOS) 2004 Conference*, June 27–29, 2004, Pittsburgh, PA.

Carley, K.M., and Gasser, L. (1999). Computational organization theory. In G. Weiss (Ed.), *Multiagent systems: A modern approach to distributed artificial intelligence* (pp. 299–330). Cambridge, MA: MIT Press.

Carley, K.M., and Prietula, M.J. (1994). *Computational organization theory*. Hillsdale, NJ: Lawrence Erlbaum Associates.

Carley, K.M., and Schreiber, C. (2002). Information technology and knowledge distribution in C³I teams. In *Proceedings of the 2002 Command and Control Research and Technology Symposium Conference*, Naval Postgraduate School, Monterey, CA.

Carley, K.M., and Svoboda, D. (1996). Modeling organizational adaptation as a simulated annealing process. *Sociological Methods and Research, 25*(1), 138–168.

Carley, K.M., Diesner, J., Reminga, J., and Tsvetovat, M. (2005). Toward an interoperable dynamic network analysis toolkit. *Decision Support Systems, Special Issue on Cyberinfrastructure for Homeland Security: Advances in Information Sharing, Data Mining, and Collaboration Systems, 43*(4), 1321–1323.

Carroll, T.N., Burton, R.M., Levitt, R.E., and Kiviniemi, A. (2006). *Fallacies of fast track heuristics: Implications for organization theory and project management.* Working paper, Duke University Fuqua School of Business.

Carroll, T.N., Gormley, T.J., Bilardo, V.J., Burton, R.M., and Woodman, K.L. (2006). Designing a new organization at NASA: An organization design process using simulation. *Organization Science, 17,* 202–214.

Cohen, M.D., March, J.G., and Olsen, J.P. (1972). A garbage can model of organizational choice. *Administrative Sciences Quarterly, 17*(1), 1–25.

Cyert, R.M., and March, J.G. (1963). *A behavioral theory of the firm.* Englewood Cliffs, NJ: Prentice-Hall.

Davis, P.K., and Anderson, R.H. (2004). *Improving the composability of Department of Defense models and simulations.* Santa Monica, CA: RAND.

DiMaggio, P.J., and Powell, W.W. (1983). The iron cage revisited: Institutional isomorphism and collective rationality in organizational fields. *American Sociological Review, 48*(2), 147–160.

Dunbar, R.L., and Starbuck, W.H. (Eds.). (2006). Learning to design organizations and learning from designing them. *Organization Science, 17*(2), 171–178.

Elmahdi, A., Malano, H., and Khan, S. (2006). Using a system dynamics approach to model sustainability indicators for irrigation systems in Australia. *Natural Resource Modeling, 19*(4), 465–481.

Entin, E.E. (1999). Optimized command and control architectures for improved process and performance. In *Proceedings of the 1999 Command and Control Research and Technology Symposium,* Newport, RI.

Finne, H. (1991). Organizational adaptation to changing contingencies. *Futures, 23*(10), 1061–1074.

Forrester, J.W. (1961). *Industrial dynamics.* Cambridge, MA: MIT Press.

Forrester, J.W. (1968). *Principles of systems, second edition.* Waltham, MA: Pegasus Communications.

Forrester, J.W. (1969). *Urban dynamics.* Cambridge, MA: MIT Press.

Forrester, J.W. (1971). *World dynamics.* Cambridge, MA: Wright-Allen Press.

Forrester, J.W. (1989). *The beginning of system dynamics.* Banquet speech at the International Meeting of the System Dynamics Society, July 13, Stuttgart, Germany. Available: http://sysdyn.clexchange.org/sdep/papers/D-4165-1.pdf [accessed May 2007].

Forrester, J.W., and Senge, P. (1980). Tests for building confidence in system dynamics models. In A.A. Legasto, Jr., J.W. Forrester, and J.M. Lyneis (Eds.), *System dynamics* (pp. 209–228, Studies in the Management Sciences, vol. 14). Amsterdam: North-Holland.

Hannan, M.T., and Freeman, J. (1977). The population ecology of organizations. *American Journal of Sociology, 82*(5), 929–964.

Harrison, J.R., and Carroll, G.R. (1991). Keeping the faith: A model of cultural transmission in formal organizations. *Admininistrative Science Quarterly, 36,* 552–582.

Karnopp, D.C., Margolis, D.L., and Rosenberg, R.C. (2006). *System dynamics: Modeling and simulation of mechatronic systems, 4th edition.* Hoboken, NJ: John Wiley & Sons.

Kauffman, S.A., and Weinberger, E.D. (1989). The N-K model of rugged fitness landscapes and its application to maturation of the immune response. *Journal of Theoretical Biology, 141,* 211–245.

Kim, J., and Burton, R.M. (2002). The effect of task uncertainty and decentralization on project team performance. *Computational and Mathematical Organization Theory, 8*(4), 365–384.

Lawrence, P.R., and Lorsch, J.W. (1967). *Organization and environment: Managing differentiation and integration.* Boston: Graduate School of Business Administration, Harvard University.

Levchuk, G.M., Levchuk, Y.N., Luo, J., Pattipati, K.R., and Kleinman, D.L. (2002a). Normative design of organizations—Part I: Mission planning. *IEEE Transactions on Systems, Man, and Cybernetics. Part A: Systems and Humans, 32*(3), 346–359.

Levchuk, G.M., Levchuk, Y.N., Luo, J., Pattipati, K.R., and Kleinman, D.L. (2002b). Normative design of organizations—Part II: Organizational structure. *IEEE Transactions on Systems, Man, and Cybernetics. Part A: Systems and Humans, 32*(3), 360–375.

Levchuk, G.M., Levchuk, Y.N., Meirina, C., Pattipati, K.R., and Kleinman, D.L. (2004). Normative design of project-based organizations—Part III: Modeling congruent, robust, and adaptive organizations. *IEEE Transactions on Systems, Man, and Cybernetics. Part A: Systems and Humans, 34*(3), 337–350.

Levinthal, D.A. (1997). Adaptation on rugged landscapes. *Management Science, 43*, 934–950.

Levis, A.H., and Wagenhals, L.W. (2000). C4ISR Architectures I: Developing a process for C4ISR architecture design. *Systems Engineering, 3*(4), 225–247.

Levitt, R.E. (2004). Computational modeling of organizations comes of age. *Computational and Mathematical Organization Theory, 10*(2), 127–145.

Levitt, R.E., Thomsen, J.C., Kunz, T.R., Jin, Y., and Nass, C. (1999). Simulating project work processes and organizations: Toward a micro-contingency theory of organizational design. *Management Science, 45*(11), 1479–1495.

Lin, Z., and Carley, K.M. (2003). *Designing stress resistant organizations: Computational theorizing and crisis applications.* Boston: Kluwer.

Lin, Z., Zhao, X., Ismail, K., and Carley, K.M. (2006). Organizational design and restructuring in response to crises: Lessons from computational modeling and real world cases. *Organizational Science, 17*(5), 598–618.

Lomi, A., and Larsen, E.R. (2001). *Dynamics of organizations: Computational modeling and organization theories.* Menlo Park, CA: AAAI Press/MIT Press.

Long, C.P., Burton, R.M., and Cardinal, L.B. (2002). Three controls are better than one: A computational model of complex control systems. *Computational and Mathematical Organization Theory, 8*(3), 197–220.

McRuer, D.T., and Krendel, E.S. (1974). *Mathematical models of human pilot behavior.* Hawthorne, CA: Systems Technology.

Meadows, D.H., Meadows, D.I., Randers, J., and Behrens, W.W.I. (1972). *The limits to growth.* New York: Universe Books.

Morecroft, J.D.W., and Sterman, J. (1994). *Modeling for learning organizations.* Portland, OR: Productivity Press.

Ogata, K. (2003). *System dynamics, fourth edition.* Englewood Cliffs, NJ: Prentice-Hall.

Pattipati, K.R., Meirina, C., Pete, A., Levchuk, G., and Kleinman, D.L. (2002). Decision networks and command organizations. In A.P. Sage (Ed.), *Systems engineering and management for sustainable development, Encyclopedia of life support systems.* Oxford, England: UNESCO–EOLSS.

Popp, R. (2005). Briefing to Committee on Organizational Modeling from Individuals to Societies, National Research Council, Keck Center, Washington, DC.

Popp, R., Kaisler, S.H., Allen, D., Cioffi-Revilla, C., Carley, K.M., Azam, M., Russell, A., Choucri, N., and Kugler, J. (2006). *Assessing nation-state instability and failure. Aerospace Conference, 2006 IEEE, 18,* 4–11. Available: http://ieeexplore.ieee.org/iel5/11012/34697/01656054.pdf?isNumber= [accessed Feb. 2008].

Quadrat-Ullah, H. (2005). Structural validation of system dynamics and agent-based simulation models. In *Proceedings of the 19th European Conference on Modeling and Simulation.* Available: http://www.econ.iastate.edu/tesfatsi/EmpValidSDACE.Hassan.pdf [accessed Feb. 2008].

Radzicki, M.J. (1997) *Introduction to system dynamics: Origin of system dynamics.* Available: http://www.systemdynamics.org/DL-IntroSysDyn/start.htm [accessed Feb. 2008].

Richardson, G.P. (1991). *Feedback thought in social science and systems theory.* Philadelphia: University of Pennsylvania Press.

Richardson, G.P., and Pugh, A.L. III. (1981). *Introduction to system dynamics modeling with DYNAMO.* Cambridge, MA: MIT Press.

Richmond, B., and Peterson, S. (1992). *STELLA II. An introduction to systems thinking.* Hanover, NH: High Performance Systems Inc.

Rivkin, J.W. (2000). Imitation of complex strategies. *Management Science, 46*(6), 824–844.

Rivkin, J.W., and Siggelkow, N. (2003). Balancing search and stability: Interdependencies among elements of organizational design. *Management Science, 49*(3), 290–311.

Robbins, M. (2005). *Investigating the complexities of nationbuilding: A sub-national regional perspective.* Master's thesis, Department of the Air Force Air University. Available: http://stinet.dtic.mil/cgi-bin/GetTRDoc?AD=ADA435214&Location=U2&doc=GetTRDoc.pdf [accessed Feb. 2008].

Robbins, M., Deckro, R.F., and Wiley, V.D. (2005). *Stabilization and reconstruction operations model (SROM).* Presented at the Center for Multisource Information Fusion Fourth Workshop on Critical Issues in Information Fusion: The Role of Higher Level Information Fusion Systems Across the Services, University of Buffalo. Available: http://www.infofusion.buffalo.edu/ [accessed Feb. 2008].

Roberts, N.D.F., Anderson, R.M., Deal, R.M., Garet, M.S., and Shaffer, W.A. (1983). *Introduction to computer simulation: A systems dynamics modeling approach.* Reading, MA: Addison-Wesley.

Rohleder, T.R., Bischak, D.P., and Baskin, L.B. (2007). Modeling patient service centers with simulation and system dynamics. *Health Care Management Science, 10*(1), 1–12.

Romanelli, E. (1991). The evolution of new organizational forms. *Annual Review of Sociology, 17,* 79–103.

Sage, A.P. (1977). *Methodology for large-scale systems.* New York: McGraw-Hill.

Sage, A.P., and Armstrong, J.E. (2000). *Introduction to systems engineering.* (Wiley Series in Systems Engineering and Management). Hoboken, NJ: John Wiley & Sons.

Schreiber, C., and Carley, K.M. (2004). Going beyond the data: Empirical validation leading to grounded theory. *Computational and Mathematical Organization Theory, 10*(2), 155–164.

Scott, W.R. (1998). *Organizations: Rational, natural, and open systems, fourth edition.* Upper Saddle River, NJ: Prentice-Hall.

Senge, P.M., Kleiner, A., Roberts, C., Ross, R., and Smith, B. (1994). *The fifth disipline fieldbook.* New York: Doubleday.

Springael, J., Kunsch, P.L., and Brans, J.-P. (2002). A multicriteria-based system dynamics modeling of traffic congestion caused by urban commuters. *Central European Journal of Operations Research, 10*(1), 81–97.

Sterman, J.D. (2000). *Business dynamics: Systems thinking and modeling for a complex world.* New York: McGraw-Hill.

Stinchcombe, A. (1965). Organization-creating organizations. *Trans-Actions, 2*, 34–35.

Wagenhals, L.W., Shin, I., Kim, D., and Levis, A.H. (2000). C4ISR Architectures II: Structured analysis approach for architecture design. *Systems Engineering, 3*(4), 248–287.

Weiner, N. (1948). *Cybernetics: Or the control and communication in the animal and the machine.* Cambridge, MA: MIT Press.

Williams, T. (2002). *Modeling complex projects.* Hoboken, NJ: John Wiley & Sons.

5

Micro-Level Formal Models

In this chapter we discuss several micro-level formal models of human behavior, models that most often are concerned with the behavior of individuals. We begin with cognitive architectures, followed by cognitive-affective models that consider the effect of human emotions on cognition and behavior, as well as of behavior on emotions. We then discuss expert systems, a legacy modeling approach that provides a framework for representing human expertise, and that now is often used as a programming paradigm in decision aiding systems. Finally we discuss decision theory and game theory and their limited applicability to individual, organizational, and societal modeling in general.

For each model or approach, we follow the same discussion framework as in Chapters 3 and 4: we present the current state of the art, the most common applications of the approach, its strengths and limitations for the problems described in Chapter 2, and suggestions for further research and development.

COGNITIVE ARCHITECTURES

Cognitive architectures are simulation-based models of human cognition. Their distinguishing feature is the broad focus on modeling the full sequence of information processing (stimulus-to-behavior) mediating adaptive, intelligent behavior. Cognitive architectures are built both for basic research and for applied purposes. Different architectures typically emphasize distinct aspects of human cognition (e.g., memory, multitasking, attention, learning, etc.), depending on their research objectives or application goals.[1]

Typically, cognitive architectures are used to model *individual* cognition. Less often, the applicability of this approach for modeling collective behavior has also been explored, that is, using a cognitive architecture to model the behavior of a group, team, or organization. The utility and appropriateness of this approach to modeling group cognition has yet to be demonstrated, however,[2] and so we have restricted our discussion here to covering the use of individual cognitive architectures to the modeling of individual behavior.

Cognitive architectures have their roots in the early artificial intelligence (AI) models of human problem solving developed in the 1950s. These models combined a number of key ideas emerging from observations of human problem solving and behavior, including symbolic processing, hierarchical organization of goals, problem spaces, rule- and heuristic-based behavior, and parallel and distributed representation and computation.

A number of cognitive models were developed in the 1970s and 1980s, such as the Model Human Processor (MHP) and Goals, Operators, Methods, and Selection rules (GOMS) (Card, Moran, and Newell, 1986), focusing on modeling a single function in the context of a single task and most often applied to models of human-computer interaction and, in particular, to the design and evaluation of user interfaces. Although limited in scope, these models provided the necessary methodological foundations for the more broadly scoped cognitive architectures of today, by demonstrating the feasibility and benefits of computational cognitive models, primarily in the context of human-computer interface design.

What Are Cognitive Architectures?

Cognitive architectures are computational, simulation models of human information processing and behavior. Cognitive architectures are also referred to as agent architectures, computational cognitive models, and human behavior models.[3] These simulation-based models aim to implement

---

[1]Indeed, this report's focus on models and simulations that can contribute to some element of improving forecasting or explanation in a Department of Defense context may limit the ultimate utility of applying some of the models described herein (and elsewhere in the report) in a broader nonmilitary context. Some researchers may argue that this is not the case because of inherent model generality, but this general issue goes beyond the original scope of the study and clearly deserves further study.

[2]Researchers are beginning to suggest future work in this area; see, for example, MacMillan (2007).

[3]Specific connotations may exist with each of these terms regarding the motivation and use of the cognitive architecture.

some version of a unified theory of cognition (Newell, 1990) by modeling the entire end-to-end human information-processing sequence, beginning with the current set of stimuli and ending with a specific behavior.

Cognitive architectures are typically classified into three broad categories, depending on their approach to knowledge representation and inferencing: symbolic, subsymbolic (also referred to as parallel-distributed), or hybrid (combining elements of the former two). Symbolic architectures use one or more propositional knowledge representation formalisms, such as rules, belief nets, or semantic nets. Subsymbolic, parallel-distributed architectures typically use some type of a connectionist representation and inferencing (e.g., recurrent neural networks), in which the mapping between conceptual entities and the representation is not one-to-one, because the knowledge is distributed over multiple representational elements (e.g., nodes within the network). Hybrid architectures use elements of both representational formalisms and are becoming increasingly common, as the benefits of the combined symbolic-subsymbolic knowledge representation and inferencing are recognized.

The specific functions represented in a particular architecture depend on its objective, level of resolution, and theoretical underpinnings. These also determine the specific modules that make up a given architecture. In most symbolic architectures, the modules and process structure correspond to (a subset of) the functions comprising human information processing. Most architectures thus contain some subset of the following broad cognitive and perceptual processes: attention, situation assessment, goal management, planning, metacognition, learning, action selection, and necessarily some form of memory (or memories), such as sensory, working, and long-term.

Thus, for example, an architecture attempting to model recognition-primed decision making (RPD) would have a module dedicated to situation assessment, since that is a core component of the RPD theory (Klein, 1997); an architecture focusing on models of learning would have corresponding modules responsible for such functions as credit assignment and creation of new schemas in memory. It should be noted here that most existing cognitive architectures are not capable of learning (Morrison, 2003). While some architectures, such as Soar, do contain elements of learning (e.g., creation of new operators by combining existing operators), typically, there is no *direct* learning resulting from the agent's interactions with the environment. However, the cognitive modeling community is beginning to recognize the limitations of human-constructed long-term memories in these models, and researchers are beginning to address the problem of automatic knowledge acquisition and learning in cognitive architectures (e.g., Anderson et al., 2003; Langley and Choi, 2006).

Depending on the architecture's control structure, the modules may execute in a fixed sequence, or in parallel, or anywhere between these two extremes. Figure 5-1 illustrates the module structure of a notional sequential cognitive architecture, frequently referred to as a "see-think-do" control structure. An alternative to this sequential approach is a parallel-distributed control structure, in which a number of parallel processes access a common memory structure (frequently referred to as a blackboard and hence the term "blackboard architectures," Corkill, 1991). As with the sequential architectures, the specific processes represented, as well as the structure of the memory blackboard, depend on the architecture objectives, the level of resolution, and theoretical foundations. Figure 5-2 shows an example of a blackboard architecture, illustrating examples of possible associated processes. Historically, cognitive architectures have focused on the middle stage of the see-think-do metaphor, frequently simplifying the perceptual input and motor output components. However, as cognitive architectures expand in model complexity and desired functionality (e.g., operating in a real-world environment), they increasingly incorporate sensory and motor models to become full-fledged agent architectures, capable of autonomous, intelligent, and adaptive behavior in a real or a simulated world.

Cognitive architectures thus contrast with the more narrowly scoped cognitive models (also referred to as micro models of cognition), which

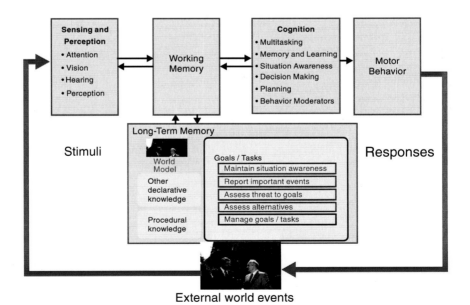

FIGURE 5-1 Example of a notional sequential cognitive architecture.

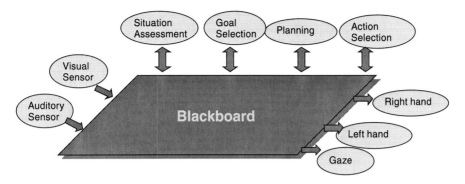

FIGURE 5-2 A blackboard architecture.

focus on a single function, such as attention, visual search, visual perception, language acquisition, or memory recall and retrieval, and implement micro theories of cognition, rather than unified theories of cognition.

This figure shows a high-level view of a parallel-distributed cognitive architecture, which represents an alternative to the sequential see-think-do model. In parallel-distributed models, processing occurs in multiple, concurrent processes, and coordination among these processes is achieved through the intermediate results posted on the blackboard, which represents the architecture memory. The structure of the blackboard varies, depending on a particular architecture, to represent the desired types of distinct memories.

State of the Art

A large number of cognitive architectures have been developed in both academic and industrial settings, and new architectures are rapidly emerging due to increasing demand, particularly in human-computer interaction (HCI) and decision support contexts, with emphasis on training, decision aiding, interactive gaming, and virtual environments. Three recent reviews provide a comprehensive catalogue of a number of established or commercially available cognitive architectures: a report focusing on U.S.-developed systems (Andre, Klesen, Gebhard, Allen, and Rist, 2000, pp. 51–111), a supplementary report focusing on systems developed in Europe, primarily in the United Kingdom (Ritter et al., 2003), and a review by Morrison that covers architectures in both the United States and Europe and includes some of the lesser known systems (Morrison, 2003). All three reviews provide detailed descriptions of the architectures in terms of the cognitive processes modeled, their historical context, applications, and implementation languages and any validation studies. A large number of research-oriented architectures also exist in laboratories around the world. The best sources for information regarding these architectures are conferences and workshops, such as the International Conference on Cognitive Modeling, the annual meeting of the Cognitive Science Society, symposia and conferences of the American Association for Artificial Intelligence, Autonomous Agents and Multi-Agent Systems, Human Factors, and BRIMS. See Table 2-1 for an overview of cognitive architectures used in military contexts.

Existing cognitive architectures are being used to support research on both human cognition and, more recently, emotion (see the next section on cognitive-affective models). They are also used in applied settings to control the behavior of synthetic agents and robots in a variety of contexts, including gaming and virtual reality environments, to enable user modeling in adaptive systems, and as replacements for human users and subjects for training, assessment, and system design purposes.

It is beyond the scope of this chapter to describe in detail the large number of architectures that have been developed over the past 25 years. The three reviews mentioned above are excellent sources of in-depth information regarding a number of architectures that are sufficiently established to be included in comprehensive reviews. Below we briefly discuss a subset of these, to provide a sense of the breadth of theoretical orientations, representational formalisms and modeling methodologies, and applications.

It should be noted that each architecture elaborates a particular subset of cognitive processing and that the architectures vary in their ease of transition to other domains and ease of use. These factors must be taken into consideration when a particular architecture is being considered as a modeling tool for a specific problem in a particular domain. For example, ACT-R's focus is on relatively low-level processing, and is particularly concerned with memory modeling. EPIC emphasizes models of multitasking. Soar emphasizes a particular model of learning, cast in relatively high-level symbolic terms. Thus, before a particular architecture is adopted for a specific modeling effort, it is necessary to carefully assess its ability to model the processes of interest at the desired level of resolution.

The most established architectures in the United States are ACT-R and Soar, each having a large and active academic research community, with annual workshops and tutorials, and each having an increasing presence in industry, primarily the defense industry. These are described below, followed by several other prominent architectures.

ACT-R

The historical focus of ACT-R (Atomic Components of Thought or Adaptive Character of Thought) has been on basic research in cognition and modeling of a variety of fundamental psychological processes, such as learning and memory (e.g., priming) (Anderson, 1983, 1990, 1993). ACT-R combines a semantic net representation with rule-based representation to support declarative and procedural memory representation and associated inferencing. ACT-R is probably the cognitive architecture that is "best grounded in the experimental research literature" (Morrison, 2003, p. 24). Primary early applications were tutoring in mathematics and computer programming (see www.carnegielearning.com). Gradually, ACT-R evolved into a full-fledged cognitive architecture, with increasing emphasis on sensory and motor components and applications in military settings (e.g., modeling adversary behavior in military operations on urban terrain, MOUT, tactical action officers in submarines, radar operators on ships; Andre et al., 2000; Anderson et al., 2004).

Soar

Soar (State, Operator, and Results) development was initially motivated by the desire to demonstrate the ability of generalized problem spaces, rules, and heuristic search capabilities to solve a wide range of problems and by the desire to develop an implementation of the unified theory of cognition of Newell (1990). Soar uses production rules to implement this problem-solving paradigm, via application of "operators" to states within a problem space. Soar represents all three types of long-term memory (declarative, procedural, and episodic) in terms of rules. A distinguishing feature of Soar is its ability to form new operators (rules) from existing operators (rules), when it reaches an impasse in its problem solving (impasse being defined as either no applicable operators selected or conflict among operators). It is thus one of the few architectures that explicitly addresses learning, albeit in the limited context of combining existing elements within its own knowledge base, rather than the bona fide acquisition of new knowledge from its interaction with the environment. Soar models both reactive and deliberative reasoning and is capable of planning (Hill, Chen, Gratch, Rosenbloom, and Tambe, 1998).

While Soar was in part motivated by theoretical considerations, particularly Newell's unified theory of cognition, the architecture has become a more traditional AI system, in its increasing emphasis on performance, rather than accurate emulation of human information processing. A frequent criticism of Soar is its large number of free variables, which enables a large number of specific models to match empirical data, thereby making it difficult to unequivocally establish the validity of a given model. This is the case with most computational cognitive architectures.

Soar's capabilities progressed from simple toy tasks (puzzles), through expert systems applications (medical diagnosis, software design), to architectures capable of controlling autonomous agents. Soar represents the more extensively applied cognitive architecture and includes a number of training installations or exercises in which it has replaced human role players or autonomous air entities: TacAir-Soar at the Air Force Research Laboratory (AFRL) training laboratory and at Williams Air Force Base (fixed-wing missions), Joint Forces Command (JFCOM) J9 exercises, MOUTBot (soldier models) VIRTE MOUT at the Office for Naval Research, JCATS at the Defense Modeling and Simulation Office; SOFSoar at JFCOM, RWA-Soar (rotary wing missions), STEVE for training simulations, and Quakebot for interactive computer games (Jones et al., 1999; Laird, 2000). The applications in the military are being developed by Soar Technology, Inc. (http://www.soartech.com). Soar also serves as the core technology at the Institute for Creative Technologies at the University of Southern California, where it acts as an agent architecture, controlling synthetic characters in virtual environments, primarily applied to training and game-based training environments. Soar has also been applied in a nondefense context, to develop a decision support system for businesses (KB Agent, developed by ExpLore Reasoning Systems, Inc.).

While the emphasis in Soar applications has been on individual models, it has also been applied in modeling multiagent environments, in which explicit representations exist of shared structures among team members (e.g., goals, plans). The STEAM model (Shell for TEAMwork) (1996) implements these enhancements and has been applied to military simulations (models of helicopter pilots) and to modeling soccer players in the RoboCup competition (Tambe et al., 1999).

EPIC

EPIC (Executive-Process/Interactive Control), developed from the MHP (Card et al., 1986), focuses on models of human behavior in multitasking contexts, in human-computer interaction. A distinguishing feature is its emphasis on integrating cognition with perceptual and motor processes. EPIC's sensorimotor capabilities have motivated its inclusion in some Soar models, to provide an interface with the real world. EPIC uses production rules to represent both its long-term memory and the control of processing within the architecture. It is primarily focused on research and is a good example of a more constrained architecture with a strong focus on validation against human performance data. Recently EPIC has also been used in more applied settings, for the design of undersea ship systems.

COGNET

COGNET (COGnition as a Network of Tasks) architecture was developed by CHI Systems and combines several knowledge representation formalisms in a blackboard-oriented framework. It was initially applied in user interface design (Zachary, Jones, and Taylor, 2002) but has been expanded to include models of multitasking in the context of air traffic control (Zachary, Santarelli, Ryder, Stokes, and Scolaro, 2001) and intelligent tutoring (Zachary et al., 1999). COGNET has an associated development environment iGEN, which is commercially available from CHI Systems.

OMAR

OMAR (Operator Model Architecture) is a task-goal network model with a focus on multitasking developed by BBN, Inc. (Deutsch, Cramer, Keith, and Freeman, 1999), from an earlier conceptual prototype, the CHAOS model (Hudlicka, Adams, and Feehrer, 1992). OMAR and its later distributed version, D-OMAR, have been used to model air traffic control and pilot error (Deutsch et al., 1999; Deutsch and Pew, 2001). It was one of the systems participating in the AMBR (Agent-based Modeling and Behavior Representation) validation project, in which its performance was compared with other cognitive architectures and with human subjects in the context of air traffic control (Gluck and Pew, 2005). Recent versions of OMAR were expanded with models of auditory and visual inputs, and the system was reimplemented in Java (from the original LISP version), to improve performance.

MIDAS

MIDAS (Man-machine Integrated Design and Analysis System) uses a goal-task network model to model simple, reactive decision making. It includes sensory inputs (visual and auditory) and simple motor outputs and has been applied in human-computer interaction to model pilot behavior in support of cockpit design (Corker and Smith, 1992; Corker, Gore, Fleming, and Lane, 2000; Laughery and Corker, 1997), air traffic control, the design of emergency communication systems, and the design of automation systems for nuclear power plants. MIDAS is also capable of modeling multiple, interacting agents.

SAMPLE

SAMPLE (Situation Awareness Model for Pilot-in-the-Loop Evaluation) is a sequential hybrid model developed by Charles River Analytics, using several knowledge representational mechanisms, including fuzzy logic and belief nets and rules. It has been applied to model air traffic control, pilot behavior, unmanned aerial vehicles, and soldier behavior in MOUT operations (Zacharias, Miao, Illgen, and Yara, 1995; Harper, Ton, Jacobs, Hess, and Zacharias, 2001). SAMPLE implements the recognition-primed decision-making model (Klein, 1997) and does not include complex planning. Sensorimotor components are represented at highly abstracted levels. SAMPLE has a drag-and-drop development environment GRADE, for rapid application prototyping, and is available commercially.

APEX

APEX is an architecture supporting the creation of intelligent, autonomous systems and serves also as a development environment. One of its goals is to reduce the effort required to develop agent architectures. Its primary applications are in human-computer interaction, to help design user interfaces and human-machine systems (Freed, Dahlman, Dalal, and Harris, 2002), and it has been applied in air traffic control.

Other Architectures

Several other architectures should be mentioned briefly. D-COG (Distributed Cognition) was developed at AFRL (Eggleston, Young, and McCreight, 2000) to model complex adaptive behavior. It was one of the architectures evaluated in the AMBR experiment (see Validation below). BRAHMS (Business Redesign Agent-Based Holistic Modeling System) is an environment developed by the National Aeronautics and Space Administration (NASA) for modeling multiple, interacting entities (Sierhuis and Clancey, 1997; Sierhuis, 2001) and emphasizes the interaction among entities rather than individual cognition.

Several well-established cognitive architectures have been developed in Europe. COGENT (Cognitive Objects within a Graphical EnviroNmentT) is a development environment for construction cognitive models developed by Cooper and colleagues (Cooper, Yule, and Sutton, 1998; Cooper, 2002). It supports the construction of cognitive architecture from individual, independent "modules," each responsible for a particular cognitive (or perceptual) function, and includes explicit support for systematic evaluation of the resulting models. COGENT offers a number of representational formalisms, including connectionist formalisms supporting the representation of distributed, subsymbolic knowledge. It has been applied to model medical diagnosis, models of memory, and models of concept learning.

The architectures outlined above are primarily symbolic and represent the most common approach to the development of integrated cognitive architectures. There are also examples of architectures that use connectionist formalisms, either exclusively or in combination with symbolic representations. We briefly mention two of these below. An example of the former is the ART (Adaptive Resonance Theory) architecture, developed by Grossberg (1999, 2000). ART emphasizes learning and parallel processing, both being key benefits of connectionist formalisms. An example of a hybrid connectionist-symbolic architecture is CLARION (Connectionist Learning with Adaptive Rule Indication On-Line), developed to support research in combined representations of symbolic knowledge (via rules) and subsymbolic knowledge (via connectionist networks) and inductive learning (Sun, 2003, 2005).

Current Trends

Several current trends in cognitive architecture development promise to contribute to more efficient development of these complex simulation systems, as well as more effective applications:

- Efforts to incorporate individual differences and behavior moderators, such as personalities and emotions, both to support basic research and to produce more realistic and robust agents (see next section).
- Efforts to provide broadly scoped end-to-end architectures, with increasing emphasis on sensory and motor processes, to enable the associated synthetic agent or robot to function in a virtual or actual environment (e.g., variety of Soar-based agents being developed at the Institute for Creative Technologies).
- Use of shared ontologies to facilitate the labor-intensive effort of cognitive task analysis and domain-specific model construction.
- Use of development environments to facilitate cognitive architecture construction, which may include automatic KA/KE facilities, visualizations, and model performance assessment and analysis tools.
- Increasing emphasis on empirical validation, frequently with respect to human performance data, and the development of validation methodologies and metrics (e.g., Gluck and Pew, 2005).

Verification and Validation Issues

As stated above, verification refers to ensuring that the architecture functions as intended, that is, that the model has been implemented according to the specifications. Validation refers to the degree to which the model specifications reflect the reality, at the desired level of resolution. We focus here on model validation and, more broadly, on model evaluation. While there is increasing emphasis on validation of cognitive architectures, validation remains one of the most challenging aspects of cognitive architecture research and development. "HBR [human behavior representation] validation is a difficult and costly process [and] most in the community would probably agree that validation is rarely, if ever done" (Campbell and Bolton, 2005, p. 365). Campbell goes on to point out that there is not a general agreement on exactly what constitutes an appropriate validation of a cognitive architecture. Since cognitive architectures are developed for a wide variety of reasons, there is a correspondingly wide set of validation (and evaluation) objectives and metrics and associated methods. Lack of established benchmark problems and criteria exacerbates this problem. It is interesting to note that a set of recommendations for model accreditation and validation was made in the 1998 National Research Council report on modeling human and organizational behavior, but these have yet to be implemented. The same report also emphasizes that a general validation of these complex models is not possible, and the models must be evaluated in the specific context for which they were developed.

Within these constraints, several approaches exist for cognitive architecture validation, varying in the data requirements, time, and effort required and the quality of the validation results. We list these below, in order of decreasing overall quality.

- *Comparative empirical studies:* the architecture's performance is compared with human performance on the same task and in the same context. Both outcome and process measures can be used: the former include time, mean time between errors, accuracy and error types, and behavioral choices. The latter include assessments of internal and intermediate states, such as emotions, workload, situation assessments, etc. The empirical data used can be obtained from a variety of sources. The ideal sources are parallel empirical studies, conducted in the same task context as the model development. As these types of studies become more common, guidelines are emerging regarding the methods (and criteria) for establishing the goodness of fit between the human and the model performance.
- *Performance-based evaluation:* the architecture's effectiveness is assessed with respect to selected performance criteria, which are defined on the basis of the architecture's role and objectives (e.g., improved training, degree of agent realism, improved prediction of the modeled decision maker's behavior, more robust and effective behavior).
- *Heuristic evaluation:* the architecture performance is evaluated by a panel of experts (or users). This is the weakest form of validation

but is frequently used because of resource limitations. Even with this weak method of validation, certain principles must be followed (e.g., judgments must be collected from individuals who were not involved in system development, data should be collected independently). When these guidelines are not followed, this approach is sometimes referred to as BOGSAT: "bunch of guys sitting around a table"—clearly to be avoided (National Research Council, 2003; Campbell and Bolton, 2005).

Validation studies also vary with respect to the scope of the validated components. The architecture may be evaluated as a whole or selected modules or submodules may be evaluated. Table 3.1 in the 1998 NRC report on human behavior modeling (National Research Council, 1998, p. 104) provides a useful summary of validation studies performed prior to 1998. A word of caution is in order, however, since not all the validation studies use the same criteria; in other words, a fully validated model using a panel of experts does not reflect the same degree of validity as a partially validated model using actual human performance data.

To date, none of the existing cognitive architectures has been fully validated against generalized human performance. There are, however, a number of task-specific validation studies for many of the established architectures and a larger number of validation studies for single-process cognitive models (e.g., models of memory retrieval, visual attention models, GOMS-based models of user performance on specific tasks using a particular interface). The GOMS family of models has proved to be particularly useful in HCI, in which they have been used to evaluate and select from candidate designs, often saving large amounts of money (e.g., Gray, John, and Atwood, 1993; see also Olson and Olson, 1990). One of the earliest examples of a cognitive architecture validated against human performance is EPIC, which successfully predicted multitasking performance in telephone operators (Kieras, Wood, and Meyer, 1997). Validation against empirical data continues to be a focus of EPIC research.

As cognitive architectures proliferate in mission-critical contexts, more opportunities exist for their validation in complex task settings. For example, Purtee and colleagues (Purtee, Krusmark, Gluck, Kotte, and Lefebvre, 2003) validated an ACT-R model controlling unmanned aerial vehicle operation, using verbal human data and protocol analysis. Andre et al. (2000) discuss validation studies of ACT-R, Soar, COGNET, and MIDAS. In general, three factors hinder systematic cognitive architecture validation studies:

1. Lack of established validation metrics and associated methods, including benchmark problems, and an understanding of when to

apply which metric, using a particular method, in a specific task context. Different validation criteria are appropriate for different system objectives and operational characteristics. Currently, however, no systematic taxonomy exists of either the system objectives or the operational contexts.

2. Frequent confusion between verification—Does the system do what it was programmed to do?—and validation—Does the system accurately represent the modeled system? (Campbell and Bolton, 2005). Verification studies are often presented as proofs of model validity, with the architecture developers showing how the system generates behavior that is consistent with the behavior of human agents in some limited context. Such studies are almost meaningless, however, in establishing the model validity.

3. The extensive effort required to conduct studies comparing human and cognitive architecture performance on a given task. These studies require first the development of a simulation environment for the particular task (e.g., air traffic control), and the development of the human-task and cognitive-architecture-task interfaces, to enable both the humans and the architecture to perform the task. In addition, the system must support human subject performance tracking and data collection. Given the general lack of interoperability among cognitive architectures, establishing these interfaces is a labor-intensive endeavor, and only one case exists in which several architectures were systematically compared with a set of benchmark problems: the AMBR project. The AMBR project represents the more promising validation approach: the systematic comparison of the cognitive architecture performance on a particular task with human performance on the same task and under identical circumstances (Gluck and Pew, 2005).

Relevance, Limitations, and Future Directions

Relevance

Cognitive architectures are built both for basic research and for applied purposes. Research architectures aim to develop a model of some aspect of human information processing, to enhance understanding of these phenomena by identifying the mediating structures and mechanisms. Specific applications of cognitive architectures include the control of autonomous synthetic agents and robots in a variety of settings, including operational systems in hostile or adverse environments, control of synthetic characters and agents in virtual reality environments, stand-ins for humans to enhance realism and believability in simulation-based training and assessment environments, and as alternatives to human subjects in empirical studies sup-´ porting human factors analyses (e.g., user interface design and operation, task allocation between human and machine, risk assessment and reduction, personnel-task matching). Recent advances in gaming technologies and the proliferation of games into a variety of settings, including military training, enable the integration of interactive gaming, virtual environments, and cognitive architectures to create immersive environments with increasing levels of realism. Such environments are increasingly being used in training, assessment, rehabilitation, and human factors analyses.

Cognitive architectures can also be used for behavior prediction in a variety of settings, both individual and team, and across a range of task types and contexts. While some success has been achieved in predicting simple behavior in highly constrained task contexts, primarily HCI contexts (e.g., EPIC has generated accurate prediction of reaction times in simple dual-task contexts; Kieras et al., 1997), forecasting individual behavior is complex, under-constrained contexts is difficult, and often impossible. In spite of recent attempts (e.g., Silverman, Bharathy, and Nye, 2007), "it is currently not within the state of the art to develop a model of a particular person, or to predict the likelihood of a single-act at a particular point in time. Instead, the predictive value of cognitive architectures lies more in their ability to generate probabilistic distributions of a *range of possible behaviors* that a particular type of individual might exhibit in given circumstances, rather than to generate predictions of the likelihood of single acts by particular individuals" (Hudlicka, 2006b, p. 14).

The increasing emphasis on complex cognitive processes in military modeling is creating a broad range of applications for cognitive architectures modeling individual entities. Both the research and the applied cognitive architectures are relevant. Cognitive architectures are relevant for three of the core areas in military modeling: analysis and forecasting in planning, simulation for training and rehearsal, and design and evaluation for acquisition. These architectures are critical components of specific modeling and simulation applications: disruption of terrorist networks, prediction of adversaries responses to specific courses of action, prediction of societal reactions to specific events, crowd behavior modeling and crowd control training, and organizational design.

These modeling needs, along with the increasing transitions to teams and nontraditional warfare, also highlight the increasing importance of modeling individual motivation and behavior variability, via explicit focus on models of emotion and personality traits. Both of these are addressed in the emerging cognitive-affective architectures, discussed in the next major section.

Major Limitations

While there have been great theoretical, methodological, and technological advances in the development of cognitive architectures, many limitations remain.

The most critical one is in the area of validation. This includes a lack of established validation criteria and methodologies, frequent confusion between verification and validation, lack of methods for validating architecture memory and knowledge bases, and the lack of any fully validated, domain-independent cognitive architectures. Currently no validated cognitive architecture exists and systematic validation efforts, including validation methods and appropriate metrics, are just beginning to emerge (e.g., Gluck and Pew, 2005).

Another limitation is the time and effort required to develop a cognitive architecture and the associated bottleneck of knowledge engineering required for these models. As discussed above, the instantiation of an architecture in a new domain requires large amounts of human performance and task data, as well as information about the nature of internal problem solving and decision making. Whether obtained from empirical studies or from cognitive task analyses and knowledge elicitation interviews, the process of obtaining the necessary human data is highly labor-intensive and represents a major bottleneck in the development of cognitive architectures capable of emulating human problem solving, decision making, and performance. In addition, once built, the resulting long-term memories typically require extensive tuning to produce the desired behavior and match human performance data.

Even with the required tuning, cognitive architectures exhibit the "brittleness" problem that plagues expert systems—that is, a lack of graceful degradation when limits of the domain knowledge (the model's long-term memory) are reached. This is one of the factors that limit the scope and degree of realism, and it applies equally to nonlearning systems and architectures with limited learning capabilities, such as Soar.

Some researchers question whether the process of "manual" long-term memory construction can ever produce long-term memories capable of supporting robust performance, as is the case in biological agents. It is possible that long-term memories may need to be automatically constructed (learned) from ongoing, long-term interaction with the environment, as is the case with intelligent biological agents, including humans (Mathews, 2006), to produce robust knowledge bases capable of matching human performance and to enable the accurate representation of a range of behavior moderators, including emotions and personalities.

Regardless of a theoretical position on this matter, it is becoming apparent that automated construction of cognitive architecture memories or knowledge bases may be the most pragmatic solution to the difficult and labor-intensive task of knowledge base development.

A related challenge is posed by the differences in representational resolution between the cognitive architecture representational capabilities and needs on one hand, and the empirical methods available for knowledge extraction on the other. Computational models offer a higher degree of representational resolution for the internal processes than currently available human empirical data. In other words, while it is now possible to build detailed models of situation assessment, planning, learning, metacognition, and similarly complex cognitive processes, one cannot unequivocally identify the internal mechanisms and structures that mediate these functions in biological agents. This state of affairs has serious implications for model validation, discussed below.

While extensive human performance data exist at the periphery of human problem solving and performance—that is, sensory and motor data that define the model inputs and outputs—these are more suitable for black box input-output models. Cognitive architectures enable, and frequently require, the specification of the detailed nature of internal mental processes, at a level of resolution that is currently not matched by the ability to obtain the required data. The data required to represent the internal mental structures and processes (e.g., situations, expectations, goals, beliefs) can be obtained only via indirect inference from observable behavioral data or self-reports. It should also be noted here that the current enthusiasm for in vivo brain imaging techniques (such as fMRI or PET scans) being able to provide these data at the required level of resolution is considered premature by many neuroscientists.

A more pragmatic limitation is the lack of established domain ontologies, standardized modeling languages, and scenario and data repositories, which further hinder the architecture development process. Similarly, the lack of model standardization and the lack of interoperability limit the ability to exchange components across architectures and research groups. Both of these contribute to the fragmented state of affairs in architecture development, as well as the lack of established benchmark problems, against which different architectures could be compared, both to establish their validity and to facilitate systematic comparisons of the capabilities of different architectures.

Another factor limiting the realism and fidelity of cognitive architectures, as well as the believability of the associated agents, is the lack of models of many mental processes that influence human perception, cognition, and behavior and give rise to the type of variability and adaptability observed in humans. This is discussed further in the next section.

Performance can also be an issue, particularly in applied agent and robotic systems that require real-time responses. New hardware and non–von Neumann machine architectures are likely to contribute to solving this problem in the future (Martínez, Gomes, and Linderman, 2005).

Last is a limitation that is particularly appropriate in the context of this study: the relative lack of interactions and collaboration among the research communities centered on particular architectures. The two most established architecture communities, Soar and ACT-R, have until very recently dominated the market (as evidenced by the ICCM biennial conference). Newcomer architectures often have a difficult time getting established and recognized, and potentially productive interactions among architectures with complementary strengths are not exploited. Morrison highlights this issue when discussing the BRAHMS architecture (Sierhuis, 2001), noting that its focus on social interaction and the focus of Soar and ACT-R on detailed models of cognition would make for an ideal collaboration, which has not occurred (Morrison, 2003, p. 39). The "everyone in his or her own sandbox" phenomenon is a common social one. However, it is important to recognize to what extent this situation limits the continued development of these important models and the successful addressing of the limitations outlined above. The development of standardized problem sets for architecture comparison would go a long way toward addressing this situation, as would the development of shared memories and domain ontologies. A concerted effort to promote long-term collaborations among different research groups is probably the single most critical element in advancing the state of the art.

Future Directions

Expanding on the earlier discussions, we briefly list the main points and augment them with additional suggestions resulting from a recent workshop that brought together researchers from the cognitive science and architecture-development communities (Martínez et al., 2005).

- *Facilitate architecture development:* via the use of standardized domain representation languages (e.g., human modeling markup languages), interchangeable plug-and-play components of generic architectures, and construction of cognitive architecture development environments.
- *Facilitate architecture instantiation:* via shared domain ontologies and human performance data repositories.
- *Facilitate knowledge base development:* via the use of automatic knowledge acquisition methods and machine learning, to eliminate the need for labor-intensive knowledge engineering.
- *Enhance model explanation capabilities:* via the development of visualization and explanation tools that support the understanding of the complex processing in a cognitive architecture.

- *Address the brittleness problem:* via a combination of hybrid knowledge representation approaches (symbolic and connectionist), learning and automatic knowledge acquisition to develop architecture knowledge bases, and the representation of commonsense knowledge.
- *Enhance realism:* by integrating architectures with embodied agents, either synthetic agents in virtual environments or robots, and by including emotion, personality, and cultural factors to produce the type of behavioral patterns and variabilities characteristic of human behavior.
- *Validation:* develop validation methods, metrics, accreditation procedures, and environments facilitating the comparison of model performance with human data and with other architectures in a set of well-defined benchmark problems. Support the development of such validation suites, in terms of shared simulation environments and benchmark test suites, broadly available to researchers and model developers. Support validation of the system as a whole, but also component validation, such as function-based or module-based validation.
- *Explore new modeling formalisms:* explore the applicability of additional representational and inferencing mechanisms to enhance cognitive architecture performance, including nonsymbolic approaches such as chaos theory, and learning methods, such as genetic algorithms.
- *Models of groups and teams:* apply cognitive architectures to models of groups and teams, in which the decision-making processes of the entity of interest can be sufficiently abstracted to enable the development of a cognitive architecture model representing the group as a whole.
- *Context and task models:* enhance the understanding of model limitations by specifying the range of tasks and operational contexts for which a particular model is applicable and defining task and context taxonomies. Identify situations in which behavior can or cannot be predicted with varying degrees of specificity and accuracy.

AFFECTIVE MODELS AND COGNITIVE-AFFECTIVE ARCHITECTURES

Computational models of emotion represent a relatively recent development in computational models of mental phenomena. This development follows a rapid growth in emotion research in both psychology and neuroscience over the past 15 years. Although computational approach to emotion research represents a recent development, the recognition of the importance of emotion in decision making and individual and social behavior is not new (e.g., Simon, 1967), nor is the recognition that understanding emotion is critical for understanding cognition and adaptive behavior in general (Norman, 1981).

Like architectures focused on cognition, cognitive-affective architectures are simulation-based models of human information processing. In contrast to purely cognitive architectures, cognitive-affective architectures also include some aspects of affective processing. Like their purely cognitive counterparts, cognitive-affective architectures are used for both research and applied purposes. In addition to the objectives discussed for cognitive architectures, these models also serve to explore the nature of affective processes, the mechanisms of cognition-emotion interaction, and, in more applied contexts, to enhance the realism, believability, and effectiveness of synthetic agents and robots. Given the critical role of emotion in interpersonal communication, these architectures are thus particularly relevant for organizational modeling (Hudlicka and Zacharias, 2005).

In spite of their relatively recent appearance in cognitive science and AI research, significant progress has been made in computational emotion modeling and cognitive-affective architectures, particularly in the more applied areas of synthetic and believable agents (e.g., Dautenhahn, Bond, Cañamero, and Edmonds, 2002; de Rosis, Pelachaud, Poggi, Carofiglio, and De Carolis, 2003; Prada, 2005).

What Are Cognitive-Affective Architectures?

Cognitive-affective architectures are computational simulation models of particular affective phenomena (e.g., effects of emotions on behavior), some aspects of affective information processing (e.g., generation of emotion via cognitive appraisal of the current situation), and associated affective factors (i.e., specific emotions, moods, or affective personality traits). The process modeled most frequently is the generation of emotion via cognitive appraisal and the effects of emotion on behavior (e.g., Bates, Loyall, and Reilly, 1992; Gratch and Marsella, 2004b; Reilly, 2006). Less frequently, these architectures also include models of emotion effects on perception and cognition (Hudlicka, 1998, 2002a, 2002b, 2007b; Ritter, Avramides, and Councill, 2002).

The affective factors modeled in cognitive-affective architectures include both transient states and more permanent traits. The states include short-lasting emotions, such as joy, fear, anger, and sadness, as well as longer lasting moods (e.g., fearful, happy, sad). Traits include affective personality traits, such as emotional stability and extraversion of the five-factor personality model (Costa and McCrae, 1992). Some models also include mental states that have both cognitive and affective components, such as attitudes.

It is beyond the scope of this section to discuss the extensive literature in emotion research in psychology and neuroscience, both theoretical and empirical, which serves as the basis for computational emotion models. The reader is referred to the excellent recent handbooks on research in emotion and the affective sciences (Ekman and Davidson, 1995; Lewis and Haviland-Jones, 2000; Scherer, Schorr, and Johnstone, 2001; Davidson, Scherer, and Goldsmith, 2003).

Briefly, however, we define emotions at the most abstract level as mental states that involve evaluations of current situations (internal or external; past, present, or future) with respect to the agent's goals, beliefs, values, and standards. Note that this evaluation does not imply conscious, deliberative cognitive processes. A key aspect of emotions, and affective factors in general, is their multimodal nature. These complex phenomena involve physiological components associated with changes in the autonomic nervous system processes (e.g., heart rate, blood pressure, galvanic skin response); cognitive components (e.g., changes in attention and working memory properties); behavioral components associated with the expression of emotions, moods, and traits (e.g., facial expressions, effects on speech, gestures, posture, behavioral choices); and subjective components (e.g., idiosyncratic individual feelings associated with particular emotions and moods). It is critical to keep in mind this multimodal nature of emotions, since many misunderstandings of these complex phenomena can be traced to a focus on only a subset of these modalities—for example, misleading questions such as "Is emotion a thought or a feeling?" It is both and more. Izard (1993) provides a framework for integrating the multiple modalities of emotion, in the context of emotion generation.

Emotions play a number of critical roles in biological agents, both intrapsychic and interpersonal. Examples of the former include goal management, reallocation of resources, rapid activation of fixed behavior repertoires, all designed to enhance adaptive behavior (Hudlicka, 2003a). Examples of the latter include mediation of attachment behaviors and communicative and expressive functions of emotion (e.g., rapid communication of behavioral intent to facilitate coordination). See Hudlicka (2007a, 2007b) for a more in-depth discussion of emotion research background from a computational perspective.

Emotion research in psychology and neuroscience provides strong evidence that cognitive and affective processes function in parallel and in a closely coupled manner (e.g., LeDoux, 1998; Phelps and LeDoux, 2005). Most modern theories of emotion therefore consider cognition to be an important component of affective processing, and vice versa. This, along with a definition of cognition that includes both conscious/deliberative and unconscious/automatic processing, makes earlier debates regarding the primacy of cognition (Lazarus, 1984) versus primacy of emotion (Lazarus, 1984; Zajonc, 1984) in the generation of emotion a matter of semantics. The current consensus regarding this issue is that these debates were largely a result of terminological vagueness and misunderstanding regarding exactly what constitutes cognitive processes.

Cognitive-affective architectures share a number of features with the cognitive architectures discussed above. Like their cognitive counterparts, emotion models can be standalone models of particular aspects of emotions, particular affective processes, or affect-related phenomena. Cognitive-affective architectures are most frequently symbolic, but they can also contain connectionist components and thus be characterized as hybrid architectures. Purely connectionist approaches are used only for limited-scope models of single phenomena, rather than for entire architectures. The specific constructs and processes represented in a particular cognitive-affective architecture depend on its objective, level of resolution, the specific processes modeled and their theoretical underpinnings, and any particular application, as well as the particular implementation approaches. Like their purely cognitive counterparts, cognitive-affective architectures typically include modules and functions that correspond to specific functions identified in biological agents, for example, emotion generation via cognitive appraisal, and generation of facial expressions.

Given the broad range of proposed roles and characteristics of emotions, a systematic description of the variety of existing models addressing these phenomena can be challenging. Below we structure the description of existing models in terms of a categorization of core affective processes proposed by Hudlicka (2007b)—processes mediating emotion generation, and those mediating emotion effects on cognition and behavior. Hudlicka further suggests that "the mechanism mediating these two fundamental processes then enables the variety of emotion roles identified in biological agents, such as resource re-allocation, goal management, etc." (Hudlicka, 2007a).

The majority of existing cognitive-affective architectures focus on the generation of emotions, most frequently via cognitive interpretive processes, termed *cognitive appraisal*. The state-of-the-art section below discusses examples of these models and architectures. In the majority of these architectures the outcomes of the generated emotions, the emotion effects, are typically limited to influences on observable behavior. This includes specific behavioral choices by synthetic agents or robots, as well as "emotion expression" in terms of distinct facial expressions, speech, and gestures and movement (e.g., Andre et al., 2000; Paiva, 2000; de Rosis et al., 2003; Breazeal and Brooks, 2005). A few cognitive-affective architectures focus also, or instead, on modeling the effects of emotions on the perceptual and cognitive processes that mediate decision making and action selection, problem solving, and learning (e.g., the MAMID architecture—Ritter et al., 2002; Bach, 2007; Hudlicka, 2007a, 2003a, 1998). Figure 5-3 illustrates an example of a cognitive-affective architecture with a dedicated affect appraiser module for emotion generation, a number of cognitive modules for the cognitive and perceptual functions supporting the necessary interpretive processes, and a range of modulating parameters that implement the effects of emotions on cognitive processing.

Given the tight integration between cognitive and affective information processing, it follows that cognitive-affective architectures necessarily include purely cognitive processes, such as attention, planning, situation assessment, action selection, and different types of memories (working memory and long-term memories). These functions are necessary to provide the cognitive infrastructure in which the affective processes can be modeled. Thus, for example, cognitive appraisal necessarily requires representation of the actual current state of the world and self (referred to as "situation assessment" or sometimes "beliefs"), and the desired state of the world (referred to as "goals" or "desires"). Cognitive appraisal models also require knowledge about the mappings among specific stimuli (elicitors) and the resulting emotions (e.g., a large, rapidly approaching unknown object is likely to induce fear). More complex models of appraisal may also require the representation and generation of expectations and the agent's own abilities to cope with a particular situation. Depending on a particular research objective or application, specific cognitive processes of interest may need to be represented (e.g., learning, planning). Depending on the particular implementation approach, there may or may not be a one-to-one correspondence between the modeled process (e.g., appraisal) and an architecture module (e.g., appraisal module).

As with cognitive architectures, cognitive-affective architectures aim to be domain-independent, and their instantiation in a particular domain requires the specification and development of domain-specific long-term memories that contain the problem-solving knowledge required to perform a particular task.

Applications and Benefits of Cognitive-Affective Architectures

The applications and benefits of cognitive-affective architectures are similar to those of purely cognitive architectures, in both the theoretical and the applied realms. In addition, there are further categories of benefits, which follow from the primary roles of emotion in biological agents, as outlined above. The intrapsychic roles of emotion, such as goal management, rapid resource reallocation, and coordination across multiple cognitive functions, enable more robust and effective autonomous behavior by facili-

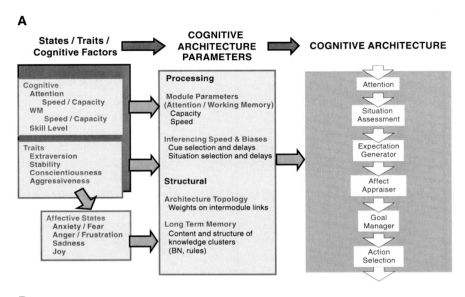

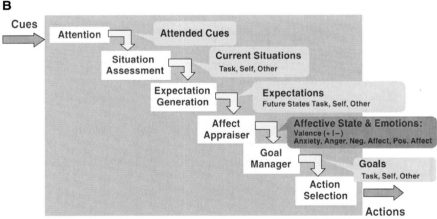

FIGURE 5-3 MAMID, a cognitive-affective architecture and its modulating parameters. Part A illustrates the modules, data flow, and mental constructs that mediate emotion generation via cognitive appraisal and decision making. Part B illustrates how the effects of emotions, personality traits and other individual differences are translated into architecture parameters that control processing in the individual modules.
SOURCE: Adapted from Hudlicka (2003a).

tating agent adaptive behavior in complex, uncertain environments (e.g., Velásquez, 1999; Scheutz, 2004; Scheutz and Schermerhorn, 2004; Scheutz, Schermerhorn, Kramer, and Middendorff, 2006; Bach, 2007). The rationale for using emotion to enhance agent autonomy rests on the assumption that since emotions mediate critical adaptive mechanisms in biological agents (e.g., goal monitoring and management, reward and punishment processes, resource reallocation), they are likely to enhance adaptive behavior in synthetic agents and robots.

The interpersonal roles of emotion, such as communication of internal mental states and behavioral intent, help improve human-machine interaction by enhancing the synthetic agents' realism and believability. The integration of emotions into purely cognitive architectures also enables affective expressiveness and behavioral variability that begins to resemble human behavior and thus enhances agent realism and believability, thereby promoting more engaging human-machine interactions. Examples of these applications include work in pedagogical applications (Prada, 2005; Prendinger and Ishizuka, 2005; Zoll, Enz, Schaub, Paiva, and Aylett, 2006), adviser and recommender systems (e.g., de Rosis et al., 2003), and training (Gratch and Marsella, 2004b). As mentioned above, models of the interpersonal role of emotions are particularly critical in organizational modeling, in which explicit models of social interactions must be represented. Augmenting purely cognitive architectures and models with emotion also enables more accurate and realistic modeling of users in a variety of training and tutoring applications.

Finally, since emotions play critical roles in biological agents, any computational model of biological information processing must necessarily take into consideration affective factors. This view reflects the current consensus in the neurosciences: to understand cognition one must also understand emotion (e.g., Phelps and LeDoux, 2005). Representation of emotion is thus necessary to develop realistic models of human information processing and behavior, whether for research or applied purposes.

The results of the theoretically motivated models of cognition-emotion interactions have a range of practical applications that include the following:

- Improved pedagogical strategies in education and training.
- Design of more effective and safer human-computer systems through improved human-machine function allocation, task design, and user interface design.
- Improved decision making and performance through the development of affect- and workload-adaptive decision support systems.
- More effective personnel selection for both team and individual tasks.

- More realistic models of social groups, teams, and larger organizations.
- Assessment and treatment for a range of affective and cognitive-affective disorders.

Like their purely cognitive counterparts, cognitive-affective architectures can also be used for behavior prediction in a variety of settings, both individual and team, and across a range of contexts, ranging from simple task behavior prediction to adversary modeling for a variety of purposes, including counterterrorism. Since they include affective factors, which are considered to be key sources of human behavioral variability and an essential component of motivation, it can be argued that these models are superior to purely cognitive architectures regarding behavior prediction. The caveat mentioned in the cognitive architecture section regarding the current limits in individual behavior prediction also applies to cognitive-affective architectures. Perhaps more so, since the behavioral variability and emotion-induced individual idiosyncrasies make accurate prediction of single acts virtually impossible. However, the addition of emotion to purely cognitive models does enable more realistic modeling and prediction of the possible ranges of behavior, due to varying individual personalities, emotions and moods, and consequent variabilities in interpretative processes, motivation, and behavioral expression (Hudlicka, 2007a).

State of the Art

Existing emotion models and cognitive-affective architectures are being used both as research platforms, to investigate the mechanisms and social roles of emotions, and in a wide range of applications to enhance agent and robot behavior and HCI. The latter are primarily in the form of cognitive-affective user models and cognitive-affective agents, used to enhance some aspect of HCI in training, education, and gaming environments. The majority of emotion models have been developed in academia, with some in industry research laboratories. The recent emergence of gaming and virtual environments has been a key factor in stimulating an interest in applied models of emotion and affective factors (e.g., personalities). No comprehensive review of emotion models and cognitive-affective architectures currently exists, analogous to the reviews of cognitive architectures (i.e., National Research Council, 1998; Morrison, 2003; Ritter et al., 2003). An earlier review by Hudlicka and Fellous (1996) provides descriptions of several older models, a more recent review of some cognitive-affective models can be found in Bach (2007), and Hudlicka (in preparation) will include an overview of existing emotion models and cognitive-affective architectures. Mellers, Schwartz and Cooke (1998) provide a review of some models of emotion effects on decision making but focus on more traditional, decision-theoretic models rather than cognitive architecture models.

This section provides a brief overview of the state of the art in emotion modeling, not an exhaustive catalogue of the large number of existing models. Cognitive-affective architecture structures are most frequently developed de novo (e.g., Velásquez, 1999; Breazeal and Brooks, 2005; Sloman, Chrisley, and Scheutz, 2005; Bach, 2007), although frequently following an established structure used for cognitive or agent architectures (e.g., the Belief-Desire-Intention agent architecture is often used as a starting point), or a particular model of information processing (e.g., RPD; Hudlicka, 2007b). In some cases, emotions are integrated into existing established architectures. For example, the Soar cognitive architecture has served as a framework for the implementation of several models of appraisal and emotion effects on behavior (e.g., Henninger, Jones, and Chown, 2003; Gratch and Marsella, 2004a). ACT-R has been used to model effects of emotion on cognition (Belavkin, 2001; Ritter et al., 2002).

Given the complexity of affective phenomena, the wide range of roles that emotions play in adaptive behavior and social interactions, and the lack of understanding of these processes, it is challenging to present the wide range of models in a systematic manner. Below we follow Hudlicka's approach (2006a, 2007a) and divide the discussion of existing models into two categories, based on the fundamental affective processes emphasized in the model: emotion generation via appraisal, and emotion effects on perception, cognition, and behavior (Hudlicka, 2008). We conclude the section with a brief discussion of two broadly scoped cognitive-affective architectures.

Models of Cognitive Appraisal

Cognitive appraisal is the dominant theory of emotion generation and the most frequently modeled aspect of emotion. A few architectures aim to incorporate additional modalities into the appraisal process (e.g., the "somatic marker" hypothesis: Damasio, 1994; Breazeal and Brooks, 2005; Stocco and Fum, 2005), and other noncognitive components (Velásquez, 1999). In computational terms, the objective of appraisal is to map the emotion elicitors (stimuli relevant for the generation of emotion) to the resulting emotion(s). This mapping may be either direct or via an intermediate stage of domain-independent appraisal dimensions (Scherer et al., 2001), which include novelty, valence, goal relevance and goal congruence, responsible agent, coping, and individual and social norms. The specific elicitors may also be mapped onto a set of two or three dimensions that can be used to characterize emotions; typically these are valence and arousal. These mappings are determined in the context of a specific set of the agent's goals and beliefs.

Different models of appraisal vary in the following: theoretical foundations used as basis for the computational model (different theories vary in the degree of elaboration of the processes involved, stages of processing, specific functions included); specific methods used to implement the elicitor-to-emotion mapping (e.g., rules, vector spaces, decision-theoretic formulations, belief nets); the degree to which goals and beliefs are represented explicitly by the model and the complexity of their representation and relationships; the capability of the model to generalize across ambiguous triggers, reason under uncertainty, and to perform approximate matches; whether domain-specific triggers are mapped directly onto the emotions or whether this mapping is performed via a domain-independent "layer" of appraisal dimensions (e.g., novelty, valence, goal congruence, etc.); the specific triggers, appraisal dimensions, emotions, or affective dimensions represented in the model; the ability to represent appraisal idiosyncrasies in terms of variability of the matching functions from elicitors to emotions; and whether the model exists in isolation or integrated in an overall architecture (Hudlicka, 2007a, 2007b).

The OCC model of appraisal (Ortony et al., 1988) remains the most widely used theoretical basis for computational appraisal models. The OCC model defines an elaborate taxonomy of emotion triggers and clusters them in terms of three broad categories: event-based emotions, reflecting desirability (or lack thereof) of an event with respect to the agent's current goals; attribution emotions, reflecting praiseworthiness (or lack thereof) of an event or situation with respect to the agent's values; and attraction emotions, reflecting the degree of like or dislike of an entity. Models of the appraisal using the OCC theory include the *Affective Reasoner* (Bates et al., 1992, the first implementation of the OCC theory), the EM (Reilly, 2006), the personality and emotion model of Andre et al. (2000), and the work of Paiva and colleagues (Martinho, Machado, and Paiva, 2000), all of which have been used to enhance the believability of synthetic agents.

Recently, the appraisal theories of Scherer (Sander, Grandjean, and Scherer, 2005; Scherer et al., 2001) and Smith and colleagues (Smith and Kirby, 2001) have begun to be used as theoretical bases for modeling. Scherer's theories provide an elaborate description of the domain-independent appraisal variables of novelty, valence, goal congruence, and coping potential, whose values are extracted from the domain-dependent stimuli. The theories of Smith and colleagues, based on the previous work of Arnold and Lazarus, are similar but emphasize the role and mechanisms of coping. Both theories reflect a trend toward more process-oriented theories, which lend themselves to computational implementations by providing more detailed descriptions of the mechanisms of the appraisal processes. These theories have recently served as basis for several computational models, including EMA (Gratch and Marsella, 2004a).

Increasingly, theoretical bases for particular appraisal models combine elements of multiple theories and approaches. Examples of these architectures include MAMID (see Figure 5-4) (Hudlicka and Canamero, 2004), which combines elements of the Scherer and Smith models of appraisal; the EMA architecture (Gratch and Marsella, 2004a), which combines elements of the Scherer, Smith and Lazarus, and OCC models of appraisal; the architecture for the robot KISMET (Breazeal and Brooks, 2005), which uses elements of somatic marker hypotheses and a three-dimensional model of the emotion space (arousal, valence, and dominance); and the robot Yuppy (Velásquez, 1999), which uses emotion as a core component of the robot control system and integrates both cognitive and noncognitive triggers in the emotion generation process.

Another promising trend in computational models of appraisal is the attempt to develop abstract formalisms, in which different theories can be compared. The work of Broekens and DeGroot represents an example of this trend (Broekens and DeGroot, 2006).

A number of appraisal models have been developed in the past decade and it is beyond the scope of this section to describe all of them. The interested reader is referred to the following recent publications which include descriptions of a number of cognitive-affective architectures and a variety of approaches to the implementation of emotion generation via appraisal (Dautenhahn et al., 2002; Trappl, Petta, and Payr, 2003; Hudlicka and Canamero, 2004; Fellous and Arbib, 2005).

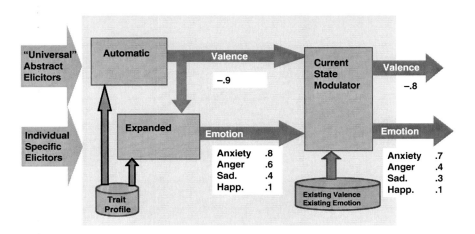

FIGURE 5-4 Affect appraiser module of the MAMID cognitive-affective architecture. SOURCE: Hudlicka (2005).

In general, several trends are evident in recent models of appraisal. First, there is increased complexity and fidelity (one hopes) of the emotion dynamics (i.e., the functions calculating emotion intensity and decay rates). Second, increased effort is made to integrate multiple emotions and to model appraisal as an evolving, dynamic process. Third, modelers are recognizing the need to differentiate among emotion states based on their duration and to model both emotions (lasting seconds and minutes) and longer lasting moods, as well as stable personality dispositions (traits). Fourth, increasing attempts are made by psychologists to develop more mechanism-oriented theories of appraisal. These so-called process models then provide more of the details necessary to develop computational versions, and can in turn benefit from the empirical hypotheses generated by computational models. Fifth, attempts are made to identify domain-independent appraisal dimensions as the intervening variables between domain-specific situations and the resulting emotions. While early models provided primarily domain-specific triggers and mapped these directly to specific emotions, more recent models interpose an intermediate step, whereby more abstract appraisal dimensions are first identified, such as relevance, novelty, unexpectedness, desirability, and ego involvement, which are then linked to specific emotions.

Models of Emotion Effects on Cognition and Cognitive-Affective Interactions

Architectures that focus on appraisal typically link the resulting emotion to specific behavioral results, most often to facial expressions, gestures, speech, or behavioral choices by the associated agents. The effective and realistic expression of emotion by synthetic agents represents a considerable technological challenge. Much progress has been made in this area in the social agent and robot research community. It is beyond the scope of this section to address the theoretical, methodological, and technical challenges. A recent book provides an overview of the methods and challenges (Prendinger and Ishizuka, 2003), and a brief overview of the state of the art is provided by Gratch and colleagues (Gratch, Rickel, Cassell, Petajan, and Badler, 2002). We focus here on an aspect of affective processing that remains underemphasized on cognitive-affective architectures: models of the effects of emotion on perception, cognition, and the appraisal processes themselves.

One of the earliest models in this category was the work of Araujo (1991, 1993), who implemented a connectionist (recurrent associative network) model of two phenomena in cognitive-affective interaction: the effect of emotional state on performance and the effect of emotional state on memory and recall, based on neuroscience data. The model represented two separate but interacting systems mediating cognitive and affective processing, each with different characteristics: fast processing of survival-related stimuli in the affective system, yielding approach/avoidance output, and slower, differentiated processing in the cognitive system.

MAMID (Hudlicka, 1998, 2002b, 2003a) represents an example of a cognitive-affective architecture whose primary focus is the modeling of the multiple, interacting effects of emotions and affective traits on perception, cognitions, and behavior. MAMID is a domain-independent architecture that implements a generic methodology for modeling a broad range of individual differences (also referred to as behavior moderators), in terms of a series of external parameters that control processing within the individual modules (see Figure 5-2). MAMID dynamically generates emotions via the affect appraiser module (see Figure 5-4). The resulting configuration of emotions (and prespecified personality traits) are translated into specific values of the architecture parameters, which then control aspects of fundamental processes within the architecture—speed, capacity, and specific content bias (e.g., bias for processing threatening information). MAMID's primary purpose is to elucidate the mechanisms mediating emotion-cognition interaction, with particular emphasis on the effects of emotions on the cognitive appraisal process itself and on emotion regulation.

Two other examples of parameter-based models of emotion effects are the work of Ritter and colleagues (Ritter and Avraamides, 2000; Ritter et al., 2002) and the MicroPsi architecture (Bach, 2007). Ritter follows the model proposed by Hudlicka and applies it to the modeling of emotion effects in the ACT-R architecture. The focus is on models of stress, and the parameters modeling these effects influence the ACT-R rule selection and conflict resolution algorithms (Ritter, Reifers, Klein, and Schoelles, 2007). In addition to the traits, states, and cognitive individual differences modeled in MAMID, Ritter also includes such factors as fatigue.

The MicroPsi architecture uses four parameters to model emotion effects: arousal, which determines degree of action readiness; resolution level, which influences the degree of elaboration of perceptual and memory processes; selection threshold, which influences the extent to which an agent persists in its current activity (versus changing its goals and behavior); and sampling rate/securing behavior, which controls the agent's orienting and novelty-seeking behavior. The MicroPsi architecture controls the behavior of simple agents in simulated environments, focusing on navigation and searching for objects of interest (e.g., food sources).

Several agent and robot cognitive-affective architectures also model some aspects of emotion-cognition interaction. For example, the Yuppy robot's attention and perceptual processes are influenced by emotions and display differences in orienting response and perceptual biases (Velásquez, 1999).

Several attempts have been made to model emotion effects on decision making in the context of decision-theoretic models, which need to be augmented to allow for variability of the utility functions as a function of the current emotion or mood. Busemeyer, Dimperio, and Jessup (2007) have developed an augmented decision-theoretic formalism to model the affective and motivational dynamics over time, termed "decision field theory" (DFT). Specifically, DFT models both the changing goals and differences in the time required to meet particular goals as a result of specific action. DFT currently models affective states in terms of valence (positive/negative). Behavioral alternatives are evaluated in terms of the anticipated valence that would be generated, and the alternative that generates the most positive valence is selected. The work of Lisetti and Gmytrasiewicz (2002) provides another example of augmenting older utility models with affective factors.

Work in modeling behavior moderators (termed "performance moderator functions" or PMFs) represents another attempt to model the effects of personalities on behavior (Silverman, Johns, Cornwell, and O'Brien, 2006; Silverman et al., 2007). The PMF-based models combine a variety of theoretical models, including the OCC appraisal model and decision-theoretic formalisms, and apply the resulting models to simulations of individual and group behavior.

Cognitive-Affective Architectures

Several cognitive-affective architectures have already been mentioned in the context of controlling agent or robot behavior and are described above in the context of either emotion generation or emotion effects on cognition and behavior (Velásquez, 1999; Breazeal and Brooks, 2005; Bach, 2007). Here we highlight two additional cognitive-affective architectures that aim to provide a broad model of intelligent behavior and integrate both cognitive and affective processing: the implemented Cog_Aff architecture (Sloman, 2003; Sloman et al., 2005) and a design for a cognitive-affective architecture proposed by Ortony, Norman, and Revelle (2005). Both architectures share a number of features, and it is interesting to note that while developed independently, there is a degree of convergence in the design.

Both models propose three levels of functioning, with a reactive stimulus-response layer mediating simple, hardwired behaviors; an intermediate level handling simple and routine, but learned, behavior (termed "deliberative" by Sloman and "routine" by Ortony); and a third level handling complex reasoning and problem solving (termed "meta-management" by Sloman and "reflective" by Ortony). Processing occurs in parallel at all three layers—complex feedback mechanisms among the layers coordinating the independent processes and influencing the final outcome. Both models also propose different degrees of complexity in the affective reactions arising at each level, with the reactive level generating rather undifferentiated affective states corresponding to positive and negative valence; the middle level generating simple, primary emotions such as fear, joy, sadness, and anger; and the top level generating both complex versions of the primary emotions, as well as complex emotions requiring explicit representations of the self and having a strong cognitive component (e.g., shame, pride, guilt). Existing agent and robot architectures typically implement a subset of these, usually just one level, although increasingly multilevel processing is being implemented; for example, the FearNot! agent implements both a reactive and deliberative level of processing in emotion generation (Paiva et al., 2005).

Relevance to Modeling Requirements

Cognitive-affective architectures are relevant for three core areas in military modeling: analysis and forecasting in planning, simulation for training and rehearsal, and design and evaluation for acquisition. In addition, the ability of cognitive-affective agents to enhance autonomous behavior is also critical for such applications as unmanned vehicle control. As mentioned above, cognitive-affective architectures are particularly relevant for modeling team and organization behavior, in which the emotion not only influences individual behavior, but also plays a key role in interpersonal interactions. The extensive existing work in social agents (e.g., Dautenhahn et al., 2002; de Rosis et al., 2003) is directly relevant here. One can envision integration of existing social network models with aspects of cognitive-affective architectures to improve the validity and utility of larger organizational models.

Of particular importance in the case of cognitive-affective architectures are training and assessment systems, in which the addition of affective factors increases the effectiveness of the training system by enhancing the realism of any social aspects of the training environment. A critical role is also played by affect-adaptive systems, capable of assessing the user's (trainee's) emotional state and adapting the pedagogical strategies accordingly. This is also a critical factor in operational decision support systems. One may wonder whether incidents such as the downing of an Iranian airliner by the U.S.S. *Vincennes* would have happened if an affect-adaptive decision-support system had been in place. The potential for reducing accidents via the use of such systems needs to be explored (e.g., see Hudlicka, 2002a).

Finally, the application of these models to behavior prediction, in both friendly and adversary situations, is also critical. Given the importance of emotion in motivation and behavior control, one can argue that any models attempting prediction must in fact include affective factors, while keeping in mind the general limitations of predictions of individual behavior already discussed. These applications include those outlined in Chapter 9: disruption of terrorist networks, prediction of adversaries to specific courses of action, prediction of societal reactions to specific events, crowd behavior modeling and crowd control training, and organizational design.

Major Limitations

Emotion models and cognitive-affective architectures have the same limitations as their cognitive counterparts, already discussed in the cognitive architecture section, exacerbated by the difficulties associated with modeling the transient, idiosyncratic, and poorly understood affective processes. These include lack of underlying theory to support model development, difficulties in obtaining required data, brittleness, the labor-intensive nature of model development, and lack of validation. The issue of data is particularly critical: while increasing amounts of empirical data are available about affective effects at the periphery (attention and behavior), the effects of emotions on the internal cognitive states (e.g., situation assessment, learning, goal management) are difficult to assess unequivocally. Furthermore, it is unlikely that the exact nature of these internal states can be identified to the extent required for process-level models in the near future.

As with cognitive architectures, the most critical limitation is architecture and model validation, although progress is being made in this area. This includes the same issues already discussed with respect to cognitive architectures: lack of established validation criteria and methodologies, frequent confusion between verification and validation, and the lack of a fully validated, domain-independent cognitive-affective architecture. These issues are discussed in more detail below.

Verification and Validation Issues

In spite of the challenges associated with validation of emotion models and cognitive-affective architectures, progress is being made in this area. A promising trend in emotion modeling is the increasing emphasis on including evaluation and validation studies in publications. As is the case with cognitive architectures, no existing emotion models or cognitive-affective architectures have been validated across multiple contexts or a broad range of metrics. However, some important evaluation and validation approaches and studies exist.

First, it is important to make the distinction between evaluation and validation. Given the increasing proliferation of cognitive-affective architectures in synthetic agents, there is increasing emphasis on evaluating the effectiveness of the resulting models in improving HCI. These evaluation studies do not necessarily address model validity or, if they do, they focus on limited black box validation approaches. They are nevertheless critical in establishing the need for, and effectiveness of, augmenting synthetic agents with affective processes, for particular purposes and applications—enhancing agent likeability, realism, believability, empathy, etc. Examples of these types of evaluation studies include the work of Prendinger and Ishizuka (2005) in evaluating the effectiveness of a synthetic agent capable of limited emotion recognition in reducing user frustration. Results of these studies indicate that users experience less stress and perceive the task as less difficult when provided with "empathic" feedback from the synthetic agent. Additional examples of this approach to agent evaluation include the work of de Rosis et al. (2003). Studies have also addressed the degree to which a social agent can improve human performance in a mixed human-robot team. Scheutz and colleagues (2006) have demonstrated improved effectiveness of human team members' performance as a robot team-member "expresses" emotions.

Some evaluation studies also focus on assessing the degree to which cognitive-affective agents are better able to negotiate complex, novel, and uncertain environments than purely cognitive agents. Examples of these studies include work by Hille (1999), cited in Bach (2007).

In addition to these evaluation studies, attempts are beginning to validate the underlying models themselves. As is the case with cognitive architectures, these validation studies are performed via a range of methods, including the weaker heuristic and qualitative evaluations and increasingly focusing on comparisons with human data. Examples of these efforts include evaluation of MAMID's parameter-based model of emotion effects, which used a heuristic evaluation approach to evaluate the model's ability to match human data at a qualitative level; establishing the validity of an augmented ACT-R architecture to model effects of stress on subtraction, using data from existing empirical studies (Ritter et al., 2002); and recent work by Gratch and Marsella (2004a) establishing a correspondence between aggregated empirical data from coping questionnaires and a model of emotion generation and coping implemented in the EMA architecture. The key challenge in these validation studies is the selection of the most appropriate dataset. This refers to selecting data from a comparable context, as well as selecting the appropriate method and degree of data aggregation. It is not clear to what extent comparison of performance at the aggregated level can be used to reflect model validity when such highly variable phenomena as emotions are considered.

The cognitive-affective architecture validation has not yet reached the stage of systematic comparisons that is beginning to be used for their cognitive counterparts, such as the AMBR project (Gluck and Pew, 2005). However, given the recent emphasis on validation in the computational emotion research community, such studies are likely to be taking place in the near future.

Future Research and Development Requirements

Future research and development requirements for cognitive-affective architectures are similar to those for cognitive architectures. Additional requirements reflect the challenges in building these architectures mentioned throughout the text, as well as the major limitations discussed—that is, issues related to a lack of underlying theory regarding emotion and emotion-cognition interactions to support model development, difficulties in obtaining required data for these transient and multimodal processes, brittleness, labor-intensive nature of model development, and lack of validation. In addition, there are technical (and theoretical) issues associated with accurate recognition of emotion in humans in affect-adaptive applications in training and gaming, as well as issues in realistic generation of affective behaviors (e.g., facial expressions, effects on natural language generation and speech). These represent important issues in the development of cognitive-affective agents and robots capable of social interaction. Some of these issues are discussed in a recent review of requirements for modeling synthetic agents (Gratch et al., 2002).

The very nature of emotion and affective processes as complex, multiple-modality phenomena makes modeling affective processes and cognitive-affective architecture more challenging than modeling purely cognitive architectures. It is not clear to what extent the types of abstractions typically made in these models (e.g., using sequential processes to model inherently parallel and distributed phenomena, abstracting an identified function as a single module within an architecture) hold when it comes to modeling the multimodal nature of affective processes. Cognitive-affective architecture development trends may also experience a more pronounced split between the research-oriented and the application-oriented architectures. Due to the increasing demand for more realistic and believable agents enabled by incorporating affective factors into agent architectures, the future developments in these models are likely to be driven by practical considerations for rapidly developing such agents for such applications as interactive gaming. This is likely to contribute to emerging standards for affective markup languages and other tools to facilitate rapid development of largely black box models of these phenomena.

EXPERT SYSTEMS

A key feature that differentiates expert systems (ESs) from more traditional software programs is the explicit representation of knowledge, stored in knowledge bases that are distinct from the inferencing mechanisms that control how the knowledge is used. This feature facilitates the editing of the knowledge base to accommodate additional or changing task knowledge and provides flexibility in how the knowledge embedded in the system can be used (e.g., to answer previously unanticipated questions about the problem).

ESs have increasingly become integrated with more traditional software development. Yet it would be a mistake to think of them as simply another programming paradigm, analogous, for example, to object-oriented programming, since a number of important factors distinguish ESs from these lower-level paradigms, including the architectures of these systems, the emphasis on explicit representation of knowledge and the associated knowledge representation formalisms, separation of knowledge and control, and the frequent use of human expertise and heuristics. These factors also make ESs well suited for modeling both individual and organizational behavior (see Hudlicka and Zacharias, 2005, for a discussion of how expert systems can be used in these contexts).

ESs should not be confused with cognitive architectures. ESs and cognitive architectures are different in both their objectives and their architectures. The objective of an ES is to solve a particular problem, frequently by simulating human expertise and the use of heuristics. The objective of cognitive architectures is to emulate human perceptual and decision-making capabilities, frequently in the context of basic research aimed at advancing understanding of these processes, or to control the behavior of synthetic agents or robots.[4] ES architectures are typically much simpler than cognitive architectures, the latter typically containing modules that correspond to functional components of the decision-making process (e.g., situation assessment, goal selection) or the mind (e.g., attention, long-term memory).

What Is an Expert System?

ESs are software programs that aim to simulate the decision making and problem solving of human experts on highly specialized tasks, such as medical diagnosis or mechanical system troubleshooting. ESs achieve their "expert" performance by applying large amounts of domain-specific knowledge to a particular problem. They are therefore also known as knowledge-based systems.

Three essential components define ESs:

1. *Knowledge base:* an explicit representation of domain and problem-solving knowledge for a particular task. This knowl-

---

[4]To the extent that some systems may contain elements of both ESs and cognitive architectures (e.g., knowledge bases, rule-based problem solving, characteristics of the working memory), they may be considered to partially fall within both categories (e.g., Soar; Hill et al., 1998).

edge is typically represented in a modular, symbolic format, such
as rules, frames (objects), logical propositions, semantic nets,
constraints, or cases, and includes factual knowledge as well as
heuristics used by human experts. For example, "IF (patient has
high fever) AND (patient is covered with red spots) AND (patient
is a child not vaccinated against chicken pox) THEN (patient has
chicken pox w/ probability 80%)." A typical rule base can contain thousands of rules.

2. *Working memory:* the component containing the specific data representing the current problem at hand (e.g., the current case), along
with particular goals to satisfy or specific constraints. The data
must be in a format that is compatible with the knowledge base
format (e.g., "Patient's fever is 104," "Patient is covered w/ red
spots," "Patient is 6 years old," "Patient has not had the chicken
pox vaccine").

3. *Inference engine:* an inferencing mechanism capable of combining
the existing knowledge with the current data to derive conclusions
of interest and thereby solve the problem at hand (e.g., derive a
diagnosis or interpretation of the data in the framework of the
knowledge provided). In the example above, a forward-chaining
rule interpretation mechanism would derive that there is a 80 percent chance of the patient's having chicken pox. Other inferencing
mechanisms include theorem proving for knowledge bases using
predicate calculus or case-based reasoning for cases.

ESs may also include one or more of the following components:

- *A (graphical) user interface and intelligent front end* to facilitate the
developers' and end users' interaction with the ES during development, refinement, and use.
- *Explanation capabilities* to explain the inferencing chains to the
end user, to ensure that the reasoning process is transparent and
that the final conclusions are accepted by the users.
- *Knowledge acquisition capabilities* to facilitate the acquisition
(from existing technical materials) or the elicitation (from human
experts) of the necessary knowledge and its modification during the
knowledge base refinement stage.
- *Learning capabilities* to help acquire additional knowledge from
patterns identified as the system performs its tasks.

ESs have been developed for a range of problem types (e.g., diagnosis,
design) across a variety of domains, including medicine, computer engineering, process control, banking, law enforcement, and others. ESs are useful

as decision aids, for training purposes, and to capture knowledge and preserve expertise in a particular area.

ESs can be developed using any computer programming language, typically a language that facilitates symbolic representation and inferencing, such as LISP. However, the use of ES shells is more common. Shells are development environments that facilitate ES development by providing system components and templates for structuring the necessary knowledge, thereby facilitating the knowledge engineering required to obtain the necessary knowledge from the expert(s) and encode it within a particular representational formalism. Shells also help maintain and modify the knowledge base and may provide a range of additional functionalities, such as a graphical user interface and explanation facilities.

Specific ESs differ along a number of dimensions. Most important are the domain represented and the type of problems the system can solve. Additional differences include the following:

- *Representational formalism* used to encode the task knowledge (e.g., rules, frames, procedural knowledge sources, predicate calculus).
- *Reasoning mechanisms* implemented within the inference engine and the type of control implemented by the inference engine (e.g., forward versus backward chaining, implemented in rule-based ESs; mixed or opportunistic, implemented in blackboard systems).
- *Type of knowledge represented* (e.g., deep versus shallow domain knowledge).
- *Source of the knowledge* (e.g., acquisition from existing technical materials or elicitation from human experts).
- *Type of problem-solving (control) knowledge* used to help determine which of several competing pieces of knowledge should be used at a given point in the inferencing.
- *Management of uncertainty* in both the knowledge representation and the reasoning (e.g., use of representational mechanisms inherently capable of representing uncertainty, such as Bayesian belief nets, explicitly representing uncertainty in terms of certainty factors, using fuzzy logic).
- *Knowledge about the structure of the ES itself* (meta-knowledge).
- *Degree to which intermediate results* are available for explanatory purposes (e.g., unstructured versus highly structured, allowing the tracing of the inferencing processes).
- *Ability to learn* additional knowledge or to acquire knowledge automatically.

State of the Art

ESs represent one of the more successful applications of AI and are used extensively in multiple types of industrial and government applications in the United States and abroad, particularly in Asia. ESs have been applied to a range of problem types and across a broad range of domains. The generic problem types (Chandrasekaran, 1986) include diagnosis and troubleshooting, data interpretation, design, and prediction and induction. Some domains spanning the time from early ESs to the present include

- *Medicine*, including diagnosis (e.g., web-based self-diagnosis programs), medication management systems (Hagland, 2003), medical emergency management, and toxicology (the DEREK system in England provides in silico testing of the adverse effects of chemicals and drugs, thereby avoiding live animal testing—Buckle, 2004).
- *Image interpretation,* such as the TriPath FocalPoint system, which screens about 10 percent of Pap smear slides in the United States).
- *Chemistry*, including interpretation of spectroscopic data.
- *Computer engineering*, including the early XCON system for configuring computers (Barker, O'Connor, Bachant, and Soloway, 1989) and software development and database design.
- *The oil industry*, including identification of promising wells for oil drilling (Cannon et al., 1989).
- *Agriculture and land management*, including interpretation of satellite images, hurricane damage assessment (Drake, 1996).
- *Real-time process control*, including system monitoring and performance optimization in power plants, such as a Japanese steel plant that uses an expert system SAFIA to control the operation of a blast furnace (Feigenbaum et al., 1993).
- *Manufacturing, troubleshooting, maintenance, and performance optimization* for a variety of electromechanical systems and telecommunication networks, for example, NASA's space shuttle engine diagnosis (Marsh, 1988).
- *Law enforcement and homeland security*, for example, PortBlue (http://www.portblue.com/pub/solutions-law-enforcement).
- *Training and tutoring in various subjects*, for example, the MITRE Corporation's F-16 Maintenance Skills Tutor used to train Air Force technicians (Marsh, 1999).
- *The automotive industry*, for example, diagnosis (Gelgele and Wang, 1998).
- *Financial advising and insurance underwriting analysis* (Pandey, Ng, and Lim, 2000).

- *Route planning and scheduling* (Nuortio et al., 2006; Sheng et al., 2006).
- *Contract administration and management* (Trimble, Allwood, and Harris, 2002).
- *Organizational design* (Burton and Obel, 2004).

Terms such as "knowledge technology," "hybrid intelligent systems," and "business-process reengineering" frequently indicate the use of ES technologies (Liebowitz, 1997). A number of recent advances in ES development contribute toward more rapid development, flexibility and extensibility, improved performance, enhanced interaction with human users, and more natural integration in the work flow. We discuss the most critical ones below.

Expert System Shells and Development Environments

A wide variety of shells are now available, which greatly speed up the development of ESs. The shells facilitate the knowledge engineering process required to build and maintain the knowledge bases by providing knowledge templates required for particular tasks. By enforcing consistency, these templates reduce common knowledge-base errors. The shells vary along a number of dimensions, including overall complexity, number and type of knowledge representation formalisms supported, number and types of problem-solving tasks supported, ease of knowledge base development and maintenance, degree of automatic knowledge engineering supported, and cost. A number of freeware shells are available, ranging from general rule-based languages, such as NASA's CLIPS, to specialized shells. The costs of commercial shells range from $50 to over $100,000. Increasingly, shells are tailored for a particular type of problem (e.g., diagnosis, design, scheduling, real-time control, planning) to support more efficient knowledge engineering and performance.

Automatic Knowledge Acquisition and Learning

Knowledge acquisition is the major bottleneck in building ESs. To help address this problem, a number of automatic knowledge engineering tools have been developed, some of which use established domain ontologies (Puerta et al., 1993), and researchers are exploring the application of machine learning methods to the automatic development of knowledge bases from training cases. In some cases, the learning methods may involve the use of additional representational and inferencing schemes, such as connectionist approaches or artificial neural nets.

Hybrid and Embedded Systems

Frequently, the most successfully deployed ESs are those that are integrated as components of larger, conventional systems. These embedded systems represent an important trend in which multiple methodologies or representations and inferencing mechanisms are applied to the solution of a particular problem. Examples of technologies that may augment an ES include fuzzy logic, neural networks, case-based reasoning, database management systems, genetic algorithms, chaos theory, statistical analysis, and data mining.

Representing and Reasoning Under Uncertainty

An essential aspect of expert reasoning is the ability to manage uncertainty. ESs must therefore be able to represent uncertainty in the facts and the knowledge and propagate uncertainties through the inferencing process. Early approaches included rather ad hoc "certainty factors," associated with rules. More recently, formalisms capable of integrating uncertainty representation and inferencing have become popular. These include multivalued fuzzy logic (Zadeh, 1965), and Bayesian belief nets (BBNs) (Pearl, 1986). BBNs especially have found extensive use in the development of "soft" ES-based decision-aiding systems in the DoD because of their intuitive graphical representation of causality and their ability to "reason" in the face of sometimes vague rules and often uncertain information.

Relevance, Limitations, and Future Directions

Relevance

From the list of current applications of ESs above, it is clear that those dealing with human individual or social behavior could be useful in many ways. ESs might be used with knowledge bases comprising profiles of individuals (e.g., political or military leaders) or groups to support what-if exercises estimating the probability of various behaviors, given different courses of action. They might be applicable to diagnosis of the intentions of adversaries, given knowledge of those adversaries' former behavior and current intelligence information. They might also be applicable to organizational design problems. Because of the ability to support the capture and direct representation of knowledgeable experts in DoD (e.g., strategic planners, counterintelligence specialists, psychological operations officers, etc.), ES-based assessment tools and decision aids are likely to continue to be developed for specialized DoD applications in all of these areas. This will be driven by the many benefits afforded by ESs already demonstrated in other domains, including:

- Improved quality and consistency of solutions, because of the ability to explicitly store and retain expertise over time and situation, ensuring permanence, the capturing and distribution of critical knowledge throughout an organization (Stylianou, Madey, and Smith, 1992).
- Increased availability of limited expertise, reduced down time, and increased reliability of human-system decision-making performance.
- Improved training via ES-based tutoring systems supporting situation assessment, planning, and decision making in understanding individual, group, and organizational behaviors.
- Extensibility and flexibility, the ability to explain its reasoning, and the ability to handle uncertainty in data and knowledge (Georgeff and Firschein, 1985; Giarratano and Riley, 1998).

Major Limitations

In spite of the successes and the potential for the future, some researchers have expressed the opinion that the idea of ESs is futile (Dreyfus and Dreyfus, 2004) and that such systems are doomed to perpetual mediocre performance simply by virtue of the fact that they are not human. This may well be true, but one must remember that their aim is to perform routine, well-established tasks, not to behave like Renaissance men. Perhaps the best solution to this problem is to have the system simply recognize the limits of its expertise and refer the problem to another ES. Nonetheless, several limitations contribute to this pessimistic view of ES potential.

One major limitation of ESs is the rapid degradation of their performance once the limits of their expertise (knowledge base) are reached. This is referred to as the brittleness problem. Unlike human experts, who display "graceful degradation" in their performance when faced with an unknown problem (by drawing on their large amounts of stored knowledge and general problem-solving methods), ESs can function well only within the very narrow scope of the specific task for which their knowledge base was developed. ESs thus resemble idiot savants: They may match or exceed the performance of human experts in a very narrow area of expertise, but they cannot perform simple tasks outside this area of expertise.

Another limitation is the extensive effort required to build the necessary knowledge bases and to maintain consistency when the knowledge base is modified. Ideally, the developer or user could add, delete, or modify the knowledge base as desired, taking advantage of its symbolic, modular structure, and the system would still derive the correct conclusions using the preexisting inference mechanism. In practice, this is not always the case. Frequently, when a particular piece of knowledge is added, deleted, or modified, the dependencies in the knowledge base cause unintended inferences, requiring further modification of the knowledge base (tweaking) and, less frequently, changes in the inference engine control algorithms. The main approaches addressing this problem are automatic knowledge engineering tools, shared ontologies, and standardized domain languages.

Finally, one of the major limitations is the difficulty in deciding whether an ES-based system is the most appropriate solution to the problem at hand, given the costs and effort often required for ES development. Depending on the task difficulty and problem stability, access to appropriate experts, and use of appropriate tools, the required time may range from weeks to many person-years. It is therefore critical that ES technology is applied appropriately. Several characteristics of the problem help determine whether an ES is the appropriate solution:

- *Stability or persistence of the problem:* Is the problem likely to exist long enough to justify the investment required to develop an ES?
- *Appropriate problem complexity:* Is the problem sufficiently difficult to warrant the development of an ES, yet sufficiently routine that the necessary knowledge and procedures can be obtained and encoded within the ES formalisms? It has been said that ESs are appropriate for tasks that would take an expert an hour or two (Bobrow, Mittal, and Stefik, 1986).
- *Appropriate problem familiarity:* Is the problem sufficiently familiar and can a sequence of steps be defined for solving it? ESs are not suitable for situations in which each problem is unique and novel methods must be developed to solve each problem. They are appropriate for automating tasks that are fairly routine and mundane, not exotic and rare (Bobrow et al., 1986).
- *Availability of the necessary knowledge:* Is the required knowledge available, either from technical materials or from human experts? Are the experts capable of articulating the necessary knowledge and are they available as necessary throughout the system development process, including evaluation and validation?
- *Availability of test cases:* Are sufficient test cases available to support a systematic evaluation and validation process?
- *Type of knowledge:* Is the knowledge highly task-specific or is a high degree of commonsense knowledge required? ESs are appropriate for problems that can be solved with highly domain-specific

knowledge, rather than the creative application of a broad range of commonsense knowledge.

It is critical to understand that ESs do not perform magic. ESs can solve only problems for which well-defined solutions already exist and the necessary knowledge can be obtained and encoded in the knowledge base.

Future Research and Development Requirements

To ensure continued use of ES technologies, the limitations outlined above need to be addressed.

To address the general issue of the narrow scope of applicability, effort needs to be devoted to developing technologies and systems that can recognize the limits of expertise and, when exceeded, refer the problem to another ES. This attempt at "self-awareness" is the underlying motivation of the emerging multiagent systems in ES research.

To address the issue of brittleness, one can pursue several strategies, including the development of:

- An ability of the ES to automatically acquire additional knowledge or problem-solving strategies by automatic knowledge acquisition and learning.
- An ability to represent large amounts of commonsense knowledge.
- An ability to draw on deep models of the domain and reason from first principles about an unfamiliar problem.

To address the issue of the extensive effort needed to build and maintain ESs, guidelines need to be developed to determine if an ES-based solution is appropriate to the problem at hand. In addition, effort needs to be put into the development of shared ontologies, standardized domain languages, and automatic knowledge engineering tools.

Finally, effort needs to be invested in developing methods for dealing with uncertainty and for addressing verification and validation to ensure consistency and correctness of the knowledge bases underlying ESs.

DECISION THEORY AND GAME THEORY

Overview

This section provides a brief overview of decision theory and game theory and their relevance to the individual and organizational modeling problem. In the earlier sections of this chapter we discussed many of the ongoing efforts in developing cognitive architectures and affective models to support the understanding of individual behavior within a psychological and situational context, and, in Chapter 3, the importance of culture as a means of providing social context and as a determinant of both individual and group behavior. And, as we have discussed, even these multidimensional approaches often prove too stark to capture the rich variety of individual human and collective group behavior that we observe out in the real world.

Hence, one cannot but be surprised when one looks at the formal modeling literature in economics and political science. Most of that literature ignores culture entirely and only recently have cognitive models become part of the mainstream in these areas (Camerer, 2003). Instead, the standard assumption in these disciplines is that people maximize their payoffs. Payoff maximizing behavior is not to be confused with self-interested behavior. A person can be both payoff maximizing and altruistic at the same time. Knowing payoffs requires an understanding of the motivations of the other players. That may not always be possible. Nevertheless, game theory and decision theory can handle this type of uncertainty as we discuss below.

Those outside economics and political science criticize the rational choice assumption, that is, the assumption of payoff maximizing behavior, on the grounds that it lacks descriptive accuracy. People don't make optimal decisions given a payoff function. Sometimes people make mistakes. Sometimes they don't have well-defined payoffs. That is true. Nevertheless, the assumption of optimizing behavior has several reasons to recommend it, at least as a baseline model. First, it is well defined, which means that analytically tractable models can be built. These models may not be 100 percent accurate, but they serve as gold standards against which models with relaxations in this assumption can be tested. Second, it enables prescriptive reasoning. Thus, using this model we can assess what people *should* do and then use this to generate hypotheses against which to compare actual behavior and identify the sources of deviation.[5] Third, some theoreticians argue that even though people do not optimize initially, they should head in that direction over time, particularly as the fallacy of their behavior is pointed out. In this way, the model becomes a forecaster of ultimate behavior. Fourth, under special circumstances, there is some empirical evidence that people may act *as if* they optimize. The empirical evidence is strongest when the stakes are large and when the situation is repeated or familiar.

---

[5] Often factors that are not typically thought of as rational, including religious and political beliefs, have major motivating effects on behavior. Ignorance—of the actual situation, the relative costs and payoffs of carrying out a decision, and other factors—may contribute to choices and behavior as well.

In general, to measure cognitive and cultural effects, a benchmark is needed for behavior (in the absence of those effects).[6] The two most widely used benchmarks are that people behave randomly and that people behave optimally. Myerson (1999) argues that rationality (i.e., acting optimally, given imperfect and/or incomplete information available to them) makes more sense. Many economists and game theorists use the optimal behavior assumption. However, for much of the social, statistical, and computer sciences and for the network models and link analysis models discussed later in Chapter 6, the random behavior assumption is used as the baseline.

We can distinguish between two types of models within the rational actor paradigm: *decision theory models* and *game theory models*. In a decision theory model, the payoff to a person or group's action does not depend on the actions of others (Raiffa, 1997). In a game theory model, payoffs depend both on the person or group's own action and on the actions of the other players (Bierman and Fernandez, 1998). We call the former insulated actions and the latter interdependent actions. This distinction creates a demarcation line between decision theory and game theory. Two highly simplified examples illustrate the difference. A military commander confronted with the problem of how to assign troops to responsibilities during peacetime faces a decision problem. That same commander allocating troops in the heat of battle often plays a game—the payoffs from the commander's action depend on the actions of the adversary.

What Are Decision Theory Models?

In decision theory models, the actor chooses from among a set of possible *actions* in order to satisfy some objective. In many situations that objective is to maximize a *payoff function*. Without uncertainty, decision theory models are not very interesting: the actor chooses the action with the highest payoff. The payoff depends on the action as well as on the *state*. The state literally means the state of the world—the set of factors that are payoff relevant. A country's oil reserves, its military strength, and its cash reserves would all be part of its state. Formally, we write the payoff as a function, $f(a|s)$, of the action, a, conditional on the state, s. In other situations, an actor's objective might be to minimize regret. The concept of regret can be formalized as the difference between what the agent receives and what the agent could have received with perfect information.

In a decision theory model, an agent has beliefs over the set of possible states. Formally, beliefs represent what someone thinks is likely to be true either at present or in the future. These beliefs are captured in a probability

---

[6]Many would say there is no such thing: all behavior occurs in a cognitive and cultural environment.

distribution over the possible outcomes. The expected payoff of an action equals the payoff of the action in each state multiplied by the probability of that state occurring as a result of the action taken. Consider a military commander who must decide whether or not to enter a hostile village.[7] The commander has three options: to enter with firepower, to enter with a small group and attempt to negotiate, or to enter the village with food and medical supplies. We define these as actions: attack, negotiate, and supply. The value of each of these actions depends on the hostility level of the village leaders. The village leadership might be hostile, moderate, or accepting. The leadership's attitude can be thought of as the state. We assume that payoffs to the military commander equal the number of lives lost, making lower payoffs better. We further assume that the military commander can make accurate assessments of the number of lives lost by following each action conditional on each state and that those are shown in Table 5-1.

This scenario illustrates why some consider decision theory a useful modeling tool and the reasons why decision theory ends up being not that useful in practice. First, the decision theorist must be able to specify the complete set of states and the consequences of the actions. Such information is generally not known by the military commander, and the time to gather such information may inhibit rapid response. For military actions, timeliness is often at least as important as accuracy. Second, the decision theorist needs to assume that the military commander knows the probability distribution over the attitudes of the village leadership—that the commander has accurate beliefs. In general, the commander does not have such information; that is, the commander and his staff do not have well-founded beliefs over all of the states.[8]

Finally, the decision theorist needs to assume that only first-order effects are critical; that is, the second-order effects of the actions are negligible. However, as most commanders will tell you, there are unintended consequences of actions (second-order effects) that are often more critical than the first-order effects. For example, a second-order effect of putting in to port in a city and enabling shore leave is an increase in money in the city and a consequent increase in corruption. To deal with this, the decision theorist has to make the model more complex so that it captures these second-order effects. The problem here is that these effects are not known a priori. There is simply insufficient understanding to predict the consequences of any action on any population or actor, especially given the influence of group think and social influence on behaviors.

---

[7]We consider an expanded version of this scenario later in our discussion of model verification and validation.

[8]In fact, assessing the attitudes of the population correctly is a key challenge facing today's military and requires a multidisciplinary approach not including decision theory.

TABLE 5-1 Number of Lives Lost Depending on State and Action

State/Action	Negotiate	Attack	Supply
Hostile leaders p = 0.25	20	16	6
Moderate leaders p = 0.25	10	8	6
Accepting leader p = 0.5	2	4	6

If, however, we were to assume that these many obstacles could be somehow overcome, then decision theory might still be a useful tool. For example, one can assume that the probability of an accepting leadership equals one-half and that the probability of each of the other types equals one-fourth. Given these assumptions, entering with food and medical supplies is the best action. It results in a loss of only six lives regardless of the leadership type, whereas both negotiating and attacking result in expected losses of eight and a half and eight lives, respectively.[9] Let a denote the probability of an accepting leadership, m denote the probability of a moderate leadership, and h the probability of a hostile leadership. These probabilities must sum to one. With a little effort, it can be shown that attacking is not optimal for any beliefs. Thus, the question is whether to supply or to negotiate with the village leader.

An important caveat is that, were it possible to overcome the obstacles to applying decision theory, the commander would still need computational support to correctly apply a decision theory model. That is, the scenario assumes that people think in Bayesian terms and that they do not make mistakes when computing probabilities. Ample evidence suggests that people are not Bayesian and that they're particularly bad at computing conditional probabilities and at estimating very low-probability events (Camerer, 2003). Thus, even if the commander could get the requisite information and knew all the probabilities, the commander would still not find the answer that decision theory would suggest. Many decision aids being developed currently are designed to overcome these two limitations and provide for the commander a "recommendation" based on decision theoretic reasoning. However, as noted, the recommendation will be faulty if the assumptions of knowing all states, all responses, and the probabilities are not met. At the current time, little is known about how to put confidence intervals around such recommendations.

Another potential role for decision theory is in determining the value of perfect or improved information. Suppose that the military commander

---

[9]We arrive at these numbers by taking the probability of each type of leader by the number of lives lost. In the case of the attack strategy, we multiply 16 by 0.25, 8 by 0.25, and 4 by 0.5 to get the expected value.

has beliefs about the village leadership based on incomplete information but that he can become perfectly informed at some cost. To be specific, suppose that at the cost of four lives the military commander can find out the attitude of the village leadership. Returning to the initial assumption about beliefs that the leadership was accepting half of the time, one can show that the cost of gathering the information exceeds its value. The information changes the military leader's action only when it reveals the leadership to be accepting. In that instance, the leader should negotiate rather than bring supplies. This action saves four lives. However, the possibility of saving four lives half of the time would not be worth the certain cost of four lives. Thus, the value of the information is less than the cost. In practice, however, there are several problems with this argument. First, it is often as difficult to determine the cost of information as it is to assess the probabilities. Second, the cost of information and the impact of the decisions may not be measurable in commensurate ways. That is, while an action may save lives, it may not cost lives to gather information; rather it may be an issue of time, purchasing surveillance cameras, etc. If this is the case, another limitation arises to the decision approach—that of converting all outcomes into the same currency.

In theory, decision theory can also be used to model the decisions of an adversary. However, to use decision theory effectively, one needs to know the adversary's beliefs over the relevant states of the world. One of the key difficulties in adversarial modeling in general is understanding the adversary's beliefs, capabilities, available resources, etc. If these were known, behavior modeling would not be as difficult as it is. Since the adversary tends to operate in a deceptive framework, hiding actions, and is adaptive with beliefs and attitudes that change in practice, decision theoretic models of the adversary tend not to be that valuable. In fact, historic models of this type tended to assume that the adversary had beliefs similar to those held by one's own forces, similar ways of engaging in battle, etc. Basing decisions on "mirror beliefs" can readily lead to disastrous consequences.

Even if the adversary's beliefs were known, the model is incomplete, because the adversary's attitude toward risk must also be known. Current models do not take into account the adversary's goals or strategies but take them as fixed. To model strategic adversaries, game theory is used, which we cover next. In our earlier discussion of cultural and cognitive models, we noted that people and cultures differ in how they respond in uncertain environments. Substantial evidence shows that people exhibit uncertainty aversion (Ellsberg, 1961). People prefer to take risks with known odds than risks with unknown odds. They will even take actions that appear unreasonable—from a rational choice perspective—in order to avoid uncertainty (Bewley, 1986). However, the extent to which they act in this way may be a function of the culture, and the relation of culture to risk-taking still needs exploration. A second aspect of risk that is typically overlooked in decision theory is the role of emotions. The emotional state of the actors can alter dramatically, and possibly in complex, nonlinear ways, their risk-taking behavior.

A second branch of decision theory, multiattribute decision theory, takes a more normative approach. It considers how to make good decisions when those decisions influence multiple dimensions. A military action has military, economic, political, and social implications. Rarely will one action be dominant, that is, lead to a better outcome on every outcome dimension. Thus, decision makers must come up with some process that either explicitly or implicitly assigns weights to the various outcome dimensions (Edwards and Barron, 1994). Multiattribute decision theory suffers from all the limitations already discussed. Nevertheless, a multiattribute approach is part of the most sophisticated of the agent-based models discussed in Chapter 6. In these multiattribute decision models, the actors in the models pursue multiple but simple and very well-defined goals. We must be careful not to be overconfident in our predictions, a point we discuss at length in our analysis of voting models in Chapter 6.

What Are Game Theory Models?

In a game theory model, as in decision theory, one assumes that each actor has a payoff function. And the same caveats apply as with decision theory: it may be impossible to know this function or take too much time to determine it. In game theory, an actor's payoff depends not only on his own action, a, but also on the action of other actors, and this action is called o. For this reason, the actors are referred to as *players,* and the payoff function is written formally as $f(a,o)$.

One can differentiate between games in several ways. Games can be *sequential,* like chess, or *simultaneous,* like rock, paper, scissors. This distinction is important because in some games advantages accrue to either the second mover or to the first mover. Games can also be *one shot* or *repeated.* In a standard repeated game, the same game is played in every period. Repeated games can be *finitely* or *infinitely repeated*; in the latter, cooperation is easier to sustain as the future always casts a shadow; that is, there are always more rounds to play that can create incentives. Games can be *zero-sum* or *nonzero-sum.* In a zero-sum game, for every winner there is a loser. In a nonzero-sum game, it's possible for the total payoff to all players to increase or decrease. Negotiation is often improperly seen as zero-sum, when in fact bargains can be reached that benefit both parties (win-win).[10] Below we discuss Colonel Blotto, a zero-sum game, and show how the competitive nature of such games makes predictions difficult.

Game theorists distinguish between two types of uncertainty. In games of *imperfect* information, the players do not know the actions of the other players, even if those actions have happened in the past. For example, a military might not know where an adversary has stored its weapons. In contrast, in games of *incomplete* information, players do not know the state of the world. For example, the military might not know how many weapons the adversary possesses. The state of the world most often influences payoffs, but it can also change the set of possible actions. The distinction between these two types of uncertainty is of more than academic interest. Problems due to incomplete information can be overcome by gathering data about preferences, capabilities, and relevant environmental features. Problems due to imperfect information require observation of the other players. Alternatively, such problems can be overcome by changing the rules of the game, for example, by creating new rules that reduce the amount of imperfect information. A problem for the commander, however, is that he may not know what he doesn't know. Adversaries may act deceptively and provide evidence that makes it appear that information is more complete or perfect than it is, thus limiting further the value of game theoretic models.

Game theory distinguishes between *actions* (what the players do), and *strategies* (the rules they use to decide what to do). For example, in a repeated prisoners' dilemma, in which players must either cooperate or defect in each period, a player's action is either to cooperate or to defect, but the player's strategy is the rule used to decide what action to take based on the past history of plays. A player might use the strategy *tit for tat*, mimicking the action of the other player in the previous play of the game. This distinction hides an important assumption of game theory—that the actors are assumed to always act strategically and not just to be reactive. However, heightened emotional states, exhaustion, religious fervor, a reduction in basic needs (no water or food or shelter), and countless other conditions can actually all lead the players to simply react rather than to behave strategically.

When one thinks of a game, be it football or chess, one thinks of sequences of moves, of ebbs and flows. Game theory focuses instead on equilibria. In what follows, we assume that the game has only two players, even though these models can handle any number of players. Normally, an equilibrium is written in terms of strategies. Here we simplify the definition

---

[10] A key limitation of game theory arises because of these different types of games: in practice, one must know what type of game is being played; however, the commander often does not have the information to make that determination, and assuming the wrong type of game can lead to suboptimal and indeed disastrous results.

and write it with respect to actions. In this formulation, an equilibrium (a^*,o^*) consists of an action for each player such that each action is optimal given the action of the other player. We write this as follows:

The actions (a^*,o^*) are an *equilibrium* if a^* optimizes $f(\cdot,o^*)$ and o^* optimizes $f(a^*,\cdot)$

Equilibria in which both players take single actions are called pure strategy equilibria. Often pure strategies fail to exist, and players must randomize their strategies across multiple actions. Each player still takes only a single action but chooses that action from a set of possible actions so as to confuse the other player. For example, an attack might be either by land or by sea. These equilibria are called mixed strategy equilibria. In general, pure strategy equilibria need not exist, but under some fairly mild conditions some type of equilibrium always exists. These mild conditions are mathematical assumptions about continuous payoff functions and the like.

Although one thinks of equilibria as places of rest, they can include finite punishment regimes, in which one player punishes the other for a fixed period of time, and bubbles, in which everyone is overly optimistic about the future (Blanchard and Watson, 1982; Green and Porter, 1984).

Game theorists use equilibrium as their solution concept. It's what they think the outcome of a game will be. However, equilibrium is a very strong assumption. Most social systems don't reach equilibrium. Natural disasters, scientific advances, belief changes, learning, external political coups, etc., all lead to adaptations that inhibit equilibrium from being reached. Understanding what behavior will be at equilibrium does not help the commander understand the ebbs, flows, and adaptations with which he is faced. Equilibrium as a solution concept for games has been justified on either of two grounds. First, optimizing players would locate equilibria. However, as we discussed, there is little evidence that the players actually optimize. Second, at least in some cases, agents who learn can locate equilibria. However, this tends to be true only for highly simplistic games, and generally those in which each individual game has only two players, even if many play in the overall tournament. Thus, a key difference between game theory models and nongame-theoretic agent-based models, which we discuss later, is what is assumed about behavior. Game theorists assume that behavior somehow gets the players to equilibrium, whereas, in most agent-based models, equilibrium is never reached and behavior is the result of various processes—cognitive, social, political, cultural, and so on.

Most game theory models assume two-person games. However, in most realistic situations, the adversary is not a single entity. In Iraq and Bosnia, for example, the adversary consists of sets of groups that come together and break apart, with varying strengths of alliances. Then there are coalition partners, nongovernment organizations, the population that might harbor ırgents or terrorists or not, and so on. In other words, there are multiple auʌors with ever-changing agendas.[11]

Relevance, Limitations, and Future Directions

Relevance

Decision theory and game theory can play four roles in constructing behavioral models:

1. They oblige us to define the actors, their possible actions and strategies, the states of the world, and payoffs.
2. They force us to think through what optimal behavior would be given our assumptions.
3. They enable us to gain a quick and powerful understanding of the primary incentives and their implications.
4. They provide simplistic, often mathematically tractable models against which deviations that engender greater realism can be assessed.

To see these four roles in an example, consider the Colonel Blotto game, which can be used to model military strategy and the actions of terrorist groups. Colonel Blotto is a simultaneous zero-sum game. Two players allocate fixed resources to a finite number of fronts. Whichever player allocates more resources to a front wins that front. A player's payoff equals the number of fronts it wins. In trench warfare, a front might literally represent a wall of troops. In using Colonel Blotto to model terrorism, one can think of a front as a potential target. This second, more modern application of Blotto is used here.

The two players would be the terrorist organization and the host country. Their possible actions would not necessarily be easy to characterize, but subject matter experts might be able to identify the set of potential targets. In the standard Blotto game, all fronts are of equal value. In a real-world scenario, that would not be true. Nevertheless, we might start by assuming that all targets take on equal value. If we assume that the host country can prevent terrorist acts on a target if they have sufficient resources, then the payoffs in the real world are approximated by Blotto. Already we see the value in using game theory in that it forced us to define the targets and their relative payoffs.

---

[11]Computational game theory *does* allow for more complex multiplayer games. However, the line between agent-based models and computational multiplayer game models is more a matter of theoretical intent than methodology.

Next, we can use game theory to solve for optimal behavior. In Blotto, a player tries to mismatch the actions of its opponent. Blotto can be thought of as a higher dimensional rock, paper, and scissors game. The equilibrium to both games is for both players to play mixed strategies: to randomly choose from among several actions. This solution provides a valuable insight: A smart player would randomize, making it impossible to know what action it will take. Formal analyses of Blotto reveal a second insight: If one player has more resources than the other, it does not necessarily win. To see why, suppose that the terrorist group has 5 units of the resource and that the host country has 20 units in an environment with 5 targets. If the host country evenly allocates its resources across the targets (as shown in Table 5-2), the terrorist group can win a front by putting all of its resources on one target.

Even though the terrorists win on only one front and the host country would officially be declared the winner of the game, most countries would not see this outcome as a win. They want to avoid any terrorist attacks. Blotto shows why that outcome is difficult to achieve.

Given that no pure strategy can win with certainty, the host country must still play a mixed strategy. In fact, if we assume that the host country wins ties, its optimal strategy in this case would be to assign five units to each of four targets and leave one target completely exposed. Thus, 20 percent of the time, the terrorist group succeeds and the host country has no resources allocated to the target despite having a four to one advantage in resources and behaving optimally. Counterintuitive results like this are a hallmark of game theory models. Often what we think is optimal won't be, when we think through all of the implications of actions. We might note that leaving a target uncovered might be politically infeasible. If so, that can be handled by changing the host country's payoff function.

In principle, we can use this model as a basis for a more elaborate and realistic model. Such a model could include more general payoffs, it could include externalities between the targets—perhaps resources at one target also partially guard another target—and it could make the game repeated. In that repeated game model, we might also restrict the ability of players to allocate resources. If so, we have a situation quite different from that

TABLE 5-2 Example of Allocation of Resources for the Host Country and Terrorist That Results in the Terrorist Winning on the Target 2 Front

Player	Target 1	Target 2	Target 3	Target 4	Target 5
Host country	4	4	4	4	4
Terrorist	0	5	0	0	0

described by Colonel Blotto but nevertheless informed by Colonel Blotto. However, as we have noted, the severe limitations of decision theory and game theory make this move to a more elaborate and realistic model impractical and not as trivial a step as the formal theorists might wish.

Game theory has been of moderate use in analyzing institutions. The game theoretic approach consists of four steps (Diermeier and Krehbiel, 2003):

1. Assume behavior.
2. Define the game generated by the institution.
3. Deduce the equilibria.
4. Compare the regularities to data.

If behavior is assumed to be optimizing, then equilibrium is achieved and institutions can be thought of as equivalent to equilibria. To compare two institutions, we need only compare their equilibria: the better the equilibrium (e.g., the greater utility to the relevant actors), the better the institution, and the more the actors will prefer it. The institutions as equilibrium approach proves powerful. If we want to compare a parliament with an open rule system, in which anyone can make a proposal, with a closed rule system, in which amendments are not allowed, or to compare a parliamentary system with a presidential system, we construct models of the two types of institution and compare their equilibria using game theory (Baron and Ferejohn, 1989). The institutions as equilibrium approach of game theory can be extended to include the game over institutions. In this game, the players first decide which institution to use. This meta-institutional game can explain not only how institutions perform but also why they may have been chosen in the first place. For example, we might use such a model to explain why a military leader chooses an open rule system even though that system allows greater voice to members of his cabinet. However, as noted, the assumptions that need to be made here are highly unrealistic, hence calling the entire approach into question.

When we expand game theory to include learning models, then we can capture some forms of cultural transference. Many game theorists think of culture as beliefs. That characterization provides some leverage, but it is far from adequate. More recent work considers cultural learning in which players learn from one another (Gintis, 2000). They can even learn from the other games that they play (Bednar and Page, 2007). Game theoretic models can also be expanded to include networks that can evolve over time. In sum, game theoretic models can include cultural forces, but those forces must be well defined and analytically tractable. The movement to expand game theory by taking networks and culture into account is promising. However, the research here is in its infancy.

Major Limitations

Decision theory models and game theory models tend to be overly simplistic, with few "moving parts" and with assumptions made with regard to the player behavioral characteristics that can be driven more by ease of solution criteria rather than fidelity of representation. Otherwise, the models become difficult or impossible to solve. For example, most game theory models assume either two players or an infinite number of players. The real world often takes place in the space in between, except for extremely artificial situations (e.g., chess games, two-candidate political races, etc.). Decision theory and game theory models require data about actors that often cannot be gathered with any reliability or within a reasonable amount of time determined by the decision window of the commander.

A further problem with game theory models is that they produce multiple equilibria. The Folk Theorem result states that, for repeated games, almost any outcome can be supported as an equilibrium. To overcome this problem of multiple equilibria, game theorists rely on refinements, such as symmetry. An equilibrium is symmetric if both players get the same payoff. Or they invoke Pareto efficiency: an equilibrium is Pareto efficient if no other equilibrium makes every player better off. Game theoretic models also often ignore the stability and attainability of the equilibria that they predict. Although recently game theorists have begun to study learning models, they tend to consider simple two-person games and not the more complex, multiplayer situations characteristic of the real world.

Future Research and Development Requirements

The potential for decision theory and game theory hinges on their ability to capture the complexities of real people and the real world. A concern with realism would seem to undercut the mathematical strength of these two approaches: their ability to cut to the heart of a situation. Nevertheless, the few degrees of freedom that these models allow can be tugged in the direction of greater realism with potentially large benefits. In decision theory, we can look to cultural and cognitive explanations to explain beliefs. We can also look to culture as a determinant of what is possible: some actions may be unlikely to occur in some cultures. Therefore, we can rule those actions out. However, as decision theory and game theoretic models become more nuanced to include cultural factors, they become less mathematically tractable, require increased data or more unrealistic assumptions, and require more effort for validation.

As already mentioned, game theorists have begun including culture in the form of beliefs, networks, and behaviors. This can also be accomplished less formally. For example, Calvert and Johnson (1999) argue for culture as a means of coordinating on an equilibrium. By coordination, they mean selection of one equilibrium from among many. In their approach, game theory becomes a preliminary tool: it defines the set of possible outcomes. Detailed historical and cultural knowledge from subject matter experts then selects from among those equilibria.

REFERENCES

Anderson, J.R. (1983). *The architecture of cognition*. Cambridge, MA: Harvard University Press.

Anderson, J.R. (1990). *The adaptive character of thought*. Hillsdale, NJ: Lawrence Erlbaum Associates.

Anderson, J.R. (1993). *Rules of the mind*. Hillsdale, NJ: Lawrence Erlbaum Associates.

Anderson, J.R., Bothell, D., Byrne, M.D., Douglass, S., Lebiere, C., and Qin, Y. (2004). An integrated theory of the mind. *Psychological Review, 111*(4), 1036–1060.

Andre, E., Klesen, M., Gebhard, P., Allen, S.A., and Rist, T. (2000). *Exploiting models of personality and emotions to control the behavior of animated interactive agents*. Paper presented at the International Workshop on Affective Interactions (IWAI), Siena, Italy.

Araujo, A.F.R. (1991). Cognitive-emotional interactions using the subsymbolic paradigm. In *Proceedings of Student Workshop on Emotions*, University of Southern California, Los Angeles.

Araujo, A.F.R. (1993). *Emotions influencing cognition: Effect of mood congruence and anxiety upon memory*. Presented at the Workshop on Architectures Underlying Motivation and Emotion (WAUME '93), University of Birmingham, England.

Bach, J. (2007). *Principles of synthetic intelligence: Building blocks for an architecture of motivated cognition*. Unpublished doctoral dissertation, Universität Osnabrück.

Barker, V. E., O'Connor, D.E., Bachant, J., and Soloway, E. (1989). Expert systems for configuration at Digital: XCON and beyond. *Communications of the ACM, 32*(3), 298–318.

Baron, D.P., and Ferejohn, J.A. (1989). Bargaining in legislatures. *American Political Science Review, 89*(4), 1181–1206.

Bates, J., Loyall, A.B., and Reilly, W.S. (1992). *Integrating reactivity, Goals, and emotion in a broad agent*. Presented at the Fourteenth Annual Conference of the Cognitive Science Society, July, Bloomington, IN.

Bednar, J., and Page, S.E. (2007). Can game(s) theory explain culture?: The emergence of cultural behavior within multiple games. *Rationality and Society, 19*(1), 65–97.

Belavkin, R.V. (2001). The role of emotion in problem solving. In *Proceedings of the AISB '01 symposium on emotion, cognition and affective computing* (pp. 49–57), Heslington, York, England.

Bewley, T.F. (1986). *Cowles Foundation discussion paper no. 807: Knightian decision theory: Part 1*. New Haven, CT: Cowles Foundation for Research in Economics at Yale University.

Bierman, H.S., and Fernandez, L.F. (1998). *Game theory with economic applications*. Reading, MA: Addison-Wesley.

Blanchard, O.J., and Watson, M.W. (1982). Bubbles, rational expectations, and speculative markets. In P. Watchel (Ed.), *Crisis in the economic and financial structure* (pp. 295–316). Lanham, MD: Lexington Books.

Bobrow, D.G., Mittal, S., and Stefik, M.J. (1986). Expert systems: Perils and promise. *Communications of the ACM, 29*(9), 880–894.

Breazeal, C., and Brooks, R. (2005). Robot emotions: A functional perspective. In J.-M. Fellous and M.A. Arbib (Eds.), *Who needs emotions? The brain meets the robot* (pp. 271–310). New York: Oxford University Press.

Broekens, J., and DeGroot, D. (2006). *Formalizing cognitive appraisal: From theory to computation.* Paper presented at Agent Construction and Emotions (ACE 2006): Modeling the Cognitive Antecedents and Consequences of Emotion Workshop, April, Vienna, Austria.

Buckle, G. (2004) *A different kind of laboratory mouse.* Available: http://digitaljournal.com/article/35501/A_Different_Kind_of_Laboratory_Mouse [accessed Feb. 2008].

Burton, R.M., and Obel, B. (2004). *Strategic organizational diagnosis and design: The dynamics of fit, third edition.* Boston: Kluwer Academic.

Busemeyer, J.R., Dimperio, E., and Jessup, R.K. (2007). Integrating emotional processes into decision making models. In W.D. Gray (Ed.), *Integrated models of cognitive systems* (pp. 213–229). New York: Oxford University Press.

Camerer, C. (2003). *Behavioral game theory: Experiments in strategic interaction.* Princeton, NJ: Princeton University Press.

Campbell, G.E., and Bolton, A.E. (2005). HBR validation: Integrating lessons learned from multiple academic disciplines, applied communities, and the AMBR project. In K.A. Gluck and R.W. Pew (Eds.), *Modeling human behavior with integrated cognitive architectures: Comparison, evaluation, and validation* (pp. 365–395). Mahwah, NJ: Lawrence Erlbaum Associates.

Cannon, R.L., Moore, P., Tansathein, D., Strobel, J., Kendall, C., Biswas, G., and Bezdek, J. (1989). An expert system as a component of an integrated system for oil exploration. In *Proceedings of Southeastcon 1989 Proceedings—Energy and Information Technologies in the Southeast* (volume 1, pp. 32–35). Los Alamitos, CA: IEEE Publications.

Card, S.K., Moran, T.P., and Newell, A. (1986). The model human processor: An engineering model of human performance. In K. Boff, L. Kaufman, and J. Thomas (Eds.), *Handbook of perception and human performance, volume II.* Hoboken, NJ: John Wiley & Sons.

Cooper, R.P. (2002). *Modeling high-level cognitive processes.* Mahwah, NJ: Lawrence Erlbaum Associates.

Cooper, R., Yule, P., and Sutton, D. (1998). COGENT: An environment for the development of cognitive models. In U. Schmid, J.F. Krems, and F. Wysotzki (Eds.), *A cognitive science approach to reasoning, learning, and discovery* (pp. 55–82). Lengerich, Germany: Pabst Science.

Corker, K.M., and Smith, B. (1992). *An architecture and model for cognitive engineering simulation analysis: Application to advanced aviation analysis.* Presented at American Institute of Aeronautics and Astronautics (AIAA) Conference on Computing in Aerospace, San Diego, CA.

Corker, K.M., Gore, B., Fleming, K., and Lane, J. (2000). Free flight and the context of control: Experiments and modeling to determine the impact of distributed air-ground air traffic management on safety and procedures. In *Proceedings of the 3rd FAA Eurocontrol International Symposium on Air Traffic Management,* Naples, Italy.

Corkill, D.D. (1991). Blackboard systems. *AI Expert, 6*(9), 40–47.

Costa, P.T., and McCrae, R.R. (1992). Four ways five factors are basic. *Personality and Individual Differences, 13,* 653–665.

Damasio, A. (1994). *Descartes' error: Emotion, reason, and the human brain.* New York: Avon Books.

Dautenhahn, K., Bond, A.H., Cañamero, L., and Edmonds, B. (Eds.). (2002). *Socially intelligent agents: Creating relationships with computers and robots.* Dordrecht, The Netherlands: Kluwer Academic.

Davidson, R.J., Scherer, K.R., and Goldsmith, H.H. (2003). *Handbook of affective sciences.* New York: Oxford University Press.

de Rosis, F., Pelachaud, C., Poggi, I., Carofiglio, V., and De Carolis, B. (2003). From Greta's mind to her face: Modelling the dynamics of affective states in a conversational embodied agent. *International Journal of Human-Computer Studies, 59*(1–2), 81–118.

Deutsch, S.E., and Pew, R.W. (2001). *Modeling human error in D-OMAR.* (Report No. 8328.) Cambridge, MA: BBN Technologies.

Deutsch, S.E., Cramer, N.L., Keith, G., and Freeman, B. (1999). *The distributed operator model architecture.* Available: http://stinet.dtic.mil/cgi-bin/GetTRDoc?AD=ADA364623 &Location=U2&doc=GetTRDoc.pdf [accessed Feb. 2008].

Diermeier, D., and Krehbiel, K. (2003). Institutionalism as a methodology. *Journal of Theoretical Politics, 15*(2), 123–144.

Drake, B.J. (1996). Expert system shell, multipurpose land information systems for rural. In *GIS/LIS '96 annual conference and exposition proceedings* (pp. 998–1005). Bethesda, MD: American Society for Photogrammetry and Remote Sensing.

Dreyfus, H.L., and Dreyfus, S.E. (2004). *From Socrates to expert systems: The limits and dangers of calculative rationality.* Available: http://socrates.berkeley.edu/~hdreyfus/html/ paper_socrates.html [accessed April 2008].

Edwards, W., and Barron, F.H. (1994). SMARTS and SMARTER: Improved simple methods for multiattribute utility measurement. *Organizational Behavior and Human Decision Processes, 60*(3), 306–325.

Eggleston, R.G., Young, M.J., and McCreight, K.L. (2000). Distributed cognition: A new type of human performance model. In *Proceedings of the 2000 AAAI Fall Symposium on Simulating Human Agents,* North Falmouth, MA. (AAAI Technical Report #FS-00-03.)

Ekman, P., and Davidson, R.J. (1995). *The nature of emotion: Fundamental questions.* Oxford, England: Oxford University Press.

Ellsberg, D. (1961). Risk, ambiguity, and the savage axioms. *Quarterly Journal of Economics, 75*(4), 643–669.

Feigenbaum, E., Friedland, P.E., Johnson, B.B., Nii, H.P., Schorr, H., and Shrobe, H. (1993). *Knowledge-based systems in Japan.* Baltimore, MD: World Technology Evaluation Center.

Fellous, J.-M., and Arbib, M.A. (2005). *Who needs emotions? The brain meets the robot.* New York: Oxford University Press.

Freed, M., Dahlman, E., Dalal, M., and Harris, R. (2002). *Apex reference manual for Apex version 2.2.* Moffett Field, CA: NASA Ames Research Center.

Gelgele, H.L., and Wang, K. (1998). An expert system for engine fault diagnosis: Development and application. *Journal of Intelligent Manufacturing, 9*(6), 539–545.

Georgeff, M.P., and Firschein, O. (1985). Expert systems for space station automation. *IEEE Control Systems Magazine, 5*(4), 3–8.

Getoor, L., and Diehl, C.P. (2005). Introduction: Special issue on link mining; Link mining: A survey. *SIGKDD Explorations Special Issue on Link Mining, 7*(2), 1–10. Available: http://www.sigkdd.org/explorations/issues/7-2-2005-12/1-Getoor.pdf [accessed Feb. 2008].

Giarratano, J.C., and Riley, G.D. (1998). *Expert systems: Principles and programming, third edition.* Boston, MA: PWS.

Gintis, H. (2000). *Game theory evolving: A problem-centered introduction to modeling strategic interaction.* Princeton, NJ: Princeton University Press.

Gluck, K.A., and Pew, R.W. (Eds.). (2005). *Modeling human behavior with integrated cognitive architectures: Comparison, evaluation, and validation.* Mahwah, NJ: Lawrence Erlbaum Associates.

Gratch, J., and Marsella, S. (2004a). A domain independent framework for modeling emotion. *Journal of Cognitive Systems Research, 5*(4), 269–306.

Gratch, J., and Marsella, S. (2004b). Evaluating a computational model of emotion. *Journal of Autonomous Agents and Multiagent Systems, Special Issue on the Best of AAMAS 2004.*

Gratch, J., Rickel, E.A., Cassell, J., Petajan, E., and Badler, N. (2002). Creating interactive virtual humans: Some assembly required. *IEEE Intelligent Systems, 17*(4), 54–63.

Gray, W.D., John, B.E., and Atwood, M.E. (1993). Project Ernestine: Validating a GOMS analysis for predicting and explaining real-world task performance. *Human-Computer Interaction, 8*(3), 237–309.

Green, E.J., and Porter, R.H. (1984). Noncooperative collusion under imperfect price information. *Econometrica, 52*(1), 87–100.

Grossberg, S. (1999). The link between brain learning, attention, and consciousness. *Consciousness and Cognition, 8,* 1–44.

Grossberg, S. (2000). Linking mind to brain: The mathematics of biological intelligence. *Notices of the American Mathematical Society, 47,* 1361–1372.

Hagland, M. (2003). Doctor's orders. *Healthcare Informatics, 39*(January).

Harper, K.A., Ton, N., Jacobs, K., Hess, J., and Zacharias, G.L. (2001). Graphical agent development environment for human behavior representation. In *Proceedings of the 10th Conference on Computer Generated Forces and Behavioral Representation,* Orlando, FL: Simulation Interoperability Standards Organization.

Henninger, A.E., Jones, R.M., and Chown, E. (2003). Behaviors that emerge from emotion and cognition: Implementation and evaluation of a symbolic-connectionist architecture. In *Proceedings of the Second International Joint Conference on Autonomous Agents and Multiagent Systems* (pp. 321–328), Melbourne, Australia.

Hill, R., Chen, J., Gratch, J., Rosenbloom, P., and Tambe, M. (1998). Soar-RWA: Planning, teamwork, and intelligent behavior for synthetic rotary-wing aircraft. In *Proceedings of the 7th Conference on Computer Generated Forces and Behavioral Representation,* Orlando, FL: Simulation Interoperability Standards Organization.

Hille, K. (1999). Artificial emotions: Angry and sad, happy and anxious behaviour. In *Proceedings of ICONIP/ANZIIS/ANNES Workshop and Expo: Future Directions for Intelligent Systems and Information Sciences,* Dunedin, New Zealand, November 22–23, University of Otago.

Hudlicka, E. (1998). Modeling emotion in symbolic cognitive architectures. In *Proceedings from AAAI Fall Symposium: Emotional and Intelligent: The Tangled Knot of Cognition.* (Technical Report #SS-98-02.) Menlo Park, CA: AAAI Press.

Hudlicka, E. (2002a). Increasing SIA architecture realism by modeling and adapting to affect and personality. In A.H. Dautenhahn, L. Bond, and B.E. Canamero (Eds.), *Multiagent systems, artificial societies, and simulated organizations.* Dordrecht, The Netherlands: Kluwer Academic.

Hudlicka, E. (2002b). This time with feeling: Integrated model of trait and state effects on cognition and behavior. *Applied AI, 16,* 1–31.

Hudlicka, E. (2003a). Modeling effects of behavior moderators on performance: Evaluation of the MAMID methodology and architecture. In *Proceedings of the 2003 Conference on Behavior Representation in Modeling and Simulation (BRIMS),* Scottsdale, AZ.

Hudlicka, E. (2003b). Personality and cultural factors in gaming environments. In *Proceedings of the Workshop on Cultural and Personality Factors in Military Gaming,* Alexandria, VA: Defense Modeling and Simulation Office.

Hudlicka, E. (2005). *The rationality of emotion . . . and the emotionality of reason.* Presented at the MICS Symposium, March 4–6, Saratoga Springs, NY. Available: http://www.cogsci.rpi.edu/cogworks/IMoCS/talks/Hudlicka.ppt#479,22,AffectAppraisal [accessed April 2008].

Hudlicka, E. (2006a). *Depth of feelings: Alternatives for modeling affect in user models.* Presented at the 9th International Conference, TSD 2006, September, Brno, Czech Republic.

Hudlicka, E. (2006b). *Summary of factors influencing decision-making and behavior. Psychometrix report #0612.* Blacksburg, VA: Psychometrix Associates.

Hudlicka, E. (2007a). *Guidelines for modeling affect in cognitive architectures. Psychometrix report #0706.* Blacksburg, VA: Psychometrix Associates.

Hudlicka, E. (2007b). Reasons for emotions. In W. Gray (Ed.), *Advances in cognitive models and cognitive architectures.* New York: Oxford University Press.

Hudlicka, E. (2008). What are we modeling when we model emotion? In *Proceedings of the AAAI Spring Symposium—Emotion, Personality, and Social Behavior.* (Technical report #SS-08-04.) Menlo Park, CA: AAAI Press.

Hudlicka, E. (in preparation). *Affective computing: Theory, methods, and applications.* Boca Raton, FL: Taylor and Francis/CRC Press.

Hudlicka, E., and Canamero, L. (2004). *Preface: Architectures for modeling emotion.* Presented at the AAAI Spring Symposium, Palo Alto, CA: AAAI Press, Stanford University.

Hudlicka, E., and Fellous, J.-M. (1996). *Review of computational models of emotion.* Arlington, MA: Psychometrix Associates.

Hudlicka, E., and Zacharias, G. (2005). Requirements and approaches for modeling individuals within organizational simulations. In W.B. Rouse and K.R. Boff (Eds.), *Organizational simulation* (pp. 79–138). Hoboken, NJ: John Wiley & Sons.

Hudlicka, E., Adams, M.J., and Feehrer, C.E. (1992). *Computational cognitive models: Phase I. BBN report 7752.* Cambridge, MA: BBN Technologies

Izard, C.E. (1993). Four systems for emotion activation: Cognitive and noncognitive processes. *Psychological Review, 100*(1), 68–90.

Jones, R.M., Laird, J.E., Nielsen, P.E., Coulter, K.J., Kenny, P., and Koss, F.V. (1999). Automated intelligent pilots for combat flight simulation. *AI Magazine, 20*(1), 27–41.

Kieras, D.E., Wood, S.D., and Meyer, D.E. (1997). Predictive engineering models based on the EPIC architecture for a multimodal high-performance human-computer interaction task. *Transactions on Computer-Human Interaction, 4*(3), 230–275.

Klein, G.A. (1997). The recognition-primed decision (RPD) model: Looking back, looking forward. In C. Zsambok and G. Klein (Eds.), *Naturalistic decision making.* Mahwah, NJ: Lawrence Erlbaum Associates.

Laird, J.E. (2000, March). *It knows what you're going to do: Adding anticipation to a quakebot.* (AAAI 2000 Spring Symposium Series: Artificial Intelligence and Interactive Entertainment, Technical Report #SS-00-02.) Palo Alto, CA: AAAI Press, Stanford University.

Langley, P., and Choi, D. (2006). A unified cognitive architecture for physical agents. In *Proceedings of the Twenty-First National Conference on Artificial Intelligence,* Boston: AAAI Press. Available: http://cll.stanford.edu/~langley/papers/icarus.aaai06.pdf [accessed April 2008].

Laughery, K.R., Jr., and Corker, K. (1997). Computer modeling and simulation of human/system performance. In G. Salvendy (Ed.), *Handbook of human factors and ergonomics, second edition* (pp. 1375–1408). Hoboken, NJ: John Wiley & Sons.

Lazarus, R.S. (1984). On the primacy of cognition. *American Psychologist, 39*(2), 124–129.

LeDoux, J. (1998). Fear and the brain: Where have we been, and where are we going? *Biological Psychiatry, 44*(12), 1229–1238.

Lewis, M., and Haviland-Jones, J.M. (2000). *Handbook of emotions, second edition.* New York: Guilford Press.

Liebowitz, J. (1997). Worldwide perspectives and trends in expert systems: An analysis based on the three world congresses on expert systems. *AI Magazine, 18*(2), 115–119.

Lisetti, C.L., and Gmytrasiewicz, P. (2002). Can a rational agent afford to be affectless? A formal approach. *Applied Artificial Intelligence, 16*, 577–609.

MacMillan, J. (2007). *Technical briefing: Modeling the group.* Available: http://www.tsjonline.com/story.php?F=2724207 [accessed Feb. 2008].

Marsh, C. (1988). *The ISA expert system: A prototype system for failure diagnosis on the space station.* In *Proceedings of the 1st International Conference on Industrial and Engineering Applications of Artificial Intelligence and Expert Systems, volume 1*, New York: ACM Press.

Marsh, C. (1999). The F-16 Maintenance skills tutor. *The Edge (MITRE Newsletter), 3*(1).

Martínez, J., Gomes, C., and Linderman, R. (2005). *Workshop on research directions in architectures and systems for cognitive processing.* Organized by Computer Systems Laboratory, Intelligent Information Systems Institute, on behalf of the Air Force Research Laboratory, July 14–15, Cornell University, Ithaca, NY.

Martinho, C., Machado, I., and Paiva, A. (2000). *Affective interactions: Towards a new generation of affective interfaces.* New York: Springer Verlag.

Mathews, R.B. (2006). The People and Landscape Model (PALM): An agent-based spatial model of livelihood generation and resource flows in rural households and their environment. *Ecological Modelling, 194*, 329–343.

Mellers, B.A., Schwartz, A., and Cooke, A.D.J. (1998). Judgment and decision making. *Annual Review of Psychology, 49*, 447–477.

Morrison, J.E. (2003). *A review of computer-based human behavior representations and their relation to military simulations.* (IDA Paper P-3845.) Alexandria, VA: Institute for Defense Analyses.

Myerson, R.B. (1999). Nash equilibrium and the history of economic theory. *Journal of Economic Literature, 37*(3), 1067–1082.

National Research Council. (1998). *Modeling human and organizational behavior: Application to military simulations.* Washington, DC: National Academy Press.

National Research Council. (1999). *Funding a revolution: Government support for computing research.* Committee on Innovations in Computing and Communications: Lessons from History. Computer Science and Telecommunications Board, Commission on Physical Sciences, Mathematics, and Applications. Washington, DC: National Academy Press.

National Research Council. (2003). *The role of experimentation in building future Naval forces.* Committee for the Role of Experimentation in Building Future Naval Forces. Naval Studies, Division on Engineering and Physical Sciences. Washington, DC: The National Academies Press.

Nawab, S., Hamid, Wotiz, R., and De Luca, C.J. (2004). Improved resolution of pulse superpositions in a knowledge-based system EMG decomposition. In *Engineering in Medicine and Biology Society, Proceedings of the 26th Annual International Conference of the EMBS '04 IEEE* (September, *1*, 69–71), San Francisco, CA. Available: http://ieeexplore.ieee.org/iel5/9639/30462/01403092.pdf?isNumber= [accessed Feb. 2008].

Newell, A. (1990). *Unified theories of cognition.* Cambridge, MA: Harvard University Press.

Norman, D.A. (1981). *Steps towards a cognitive engineering: System images, system friendliness, mental models.* Technical report for Program in Cognitive Science, University of California, San Diego.

Nuortio, T., Kytöjoki, Niska, H., and Bräysy, O. (2006). Improved route planning and scheduling of waste collection and transport. *Expert Systems with Applications, 30*(2), 223–232.

Olson, J.R., and Olson, G.M. (1990). The growth of cognitive modeling in human-computer interaction since GOMS. *Human-Computer Interaction, 5*(2–3), 221–265.

Ortony, A., Clore, G.L., and Collins, A. (1988). *The cognitive structure of emotions.* New York: Cambridge University Press.

Ortony, A., Norman, D.A., and Revelle, W. (2005). Affect and proto-affect in effective functioning. In J.-M. Fellous and M.A. Arbib (Eds.), *Who needs emotions?: The brain meets the machine* (pp. 173–202). New York: Oxford University Press.

Paiva, A. (2000). *Affective interactions, towards a new generation of computer interfaces.* New York: Springer.

Paiva, A., Dias, J., Sobral, D., Aylett, R., Woods, S., Hall, L., and Zoll, C. (2005). Learning by feeling: Evoking empathy with synthetic characters. *Applied Artificial Intelligence Journal, 19*(3–4), 235–266.

Pandey, V., Ng, W.-K., and Lim, E.-P. (2000). Financial advisor agent in a multi-agent financial trading system. In *Proceedings of the 11th International Workshop on Database and Expert Systems Applications* (pp. 482–486). Available: http://ieeexplore.ieee.org/iel5/7035/18943/00875070.pdf?isNumber= [accessed Feb. 2008].

Pearl, J. (1986). Fusion, propagation, and structuring in belief networks. *Artificial Intelligence, 29*(3), 215–288.

Phelps, E.A., and LeDoux, J.E. (2005). Contributions of the amygdala to emotion processing: From animal models to human behavior. *Neuron, 48*(2), 175–187.

Prada, R. (2005). *Teaming up humans and synthetic characters.* Unpublished doctoral dissertation, UTL-IS-Technical University of Lisbon, Portugal.

Prendinger, H., and Ishizuka, M. (2003). *Life-like characters: Tools, affective functions, and applications.* New York: Springer.

Prendinger, H., and Ishizuka, M. (2005). Human physiology as a basis for designing and evaluating affective communication with life-like characters. *IEEE Transactions on Information and Systems, E88-D*(11), 2453–2460.

Puerta, A.R., Neches, R., Eriksson, H., Szekely, P., Luo, P., and Musen, M.A. (1993). *Toward ontology-based frameworks for knowledge-acquisition tools.* Available: http://bmir.stanford.edu/publications/view.php/toward_ontology_based_frameworks_for_knowledge_acquisition_tools [accessed Feb. 2008].

Purtee, M.D., Krusmark, M.A., Gluck, K.A., Kotte, S.A., and Lefebvre, A.T. (2003). Verbal protocol analysis for validation of UAV operator model. In *Proceedings of the 25th Interservice/Industry Training, Simulation, and Education Conference* (pp. 1741–1750), Orlando, FL: National Defense Industrial Association.

Raiffa, H. (1997). *Decision analysis: Introductory lectures on choices under uncertainty.* New York: McGraw-Hill.

Reilly, W.S.N. (2006). *Modeling what happens between emotional antecedents and emotional consequents.* Paper presented at Agent Construction and Emotions (ACE 2006): Modeling the Cognitive Antecedents and Consequences of Emotion Workshop, April, Vienna, Austria.

Ritter, F.E., and Avraamides, M.N. (2000). *Steps towards including behavior moderators in human performance models in synthetic environments.* (Technical report #ACS-2000-1.) State College, PA: Pennsylvania State University.

Ritter, F., Avramides, M., and Councill, I. (2002). Validating changes to a cognitive architecture to more accurately model the effects of two example behavior moderators. In *Proceedings of 11th CGF Conference*, Orlando, FL.

Ritter, F.E., Reifers, A.L., Klein, L.C., and Schoelles, M. (2007). Lessons from defining theories of stress. In W.D. Gray (Ed.), *Integrated models of cognitive systems (IMoCS)* (pp. 254–262). New York: Oxford University Press.

Ritter, F.E., Shadbolt, N.R., Elliman, D., Young, R.M., Gobet, F., and Baxter, G.D. (2003). *Techniques for modeling human performance in synthetic environments: A supplementary review.* Wright-Patterson Air Force Base, OH: Human Systems Information Analysis Center.

Sander, D., Grandjean, D., and Scherer, K.R. (2005). A systems approach to appraisal mechanisms in emotion. *Neural Networks, 18*(4), 317–352.

Scherer, I.R., Schorr, A., and Johnstone, T. (2001). *Appraisal processes in emotion: Theory, methods, research.* New York: Oxford University Press.

Scheutz, M. (2004). Useful roles of emotions in artificial agents: A case study from artificial life. In *Proceedings of the Nineteenth National Conference on Artificial Intelligence, Sixteenth Conference on Innovative Applications of Artificial Intelligence.* Cambridge, MA: AAAI Press/MIT Press.

Scheutz, M., and Schermerhorn, P. (2004). *The more radical, the better: Investigating the utility of aggression in the competition among different agent kinds.* Available: http://citeseer.ist. psu.edu/cache/papers/cs/30796/http:zSzzSzwww.nd.eduzSzzCz7EairolabzSzpublicationsz Szscheutzschermerhorn04sab.pdf/the-more-radical-the.pdf [accessed Feb. 2008].

Scheutz, M., Schermerhorn, P., Kramer, J., and Middendorff, C. (2006). The utility of affect expression in natural language interactions in joining human-robot tasks. In *Proceedings of IEEE/ACM 1st Annual Conference on Human-Robot Interactions* (HRI2006) (pp. 226–233), Salt Lake City, UT.

Sheng, H.-M., Wang, J.-C., Huang, H.-H., and Yen, D.C. (2006). Fuzzy measure on vehicle routing problem of hospital materials. *Expert Systems with Applications, 30*(2), 367–377.

Sierhuis, M. (2001). *Modeling and simulating work practice. BRAHMS: A multiagent modeling and simulation language for work system analysis and design (SIKS Dissertation Series No. 2001-10).* Unpublished doctoral dissertation, The University of Amsterdam, Amsterdam. Available: http://www.agentisolutions.com/documentation/ papers/BrahmsWorkingPaper.pdf [accessed Feb. 2008].

Sierhuis, M., and Clancey, W.J. (1997). Knowledge, practice, activities and people. In *Proceedings of AAAI Spring Symposium on Artificial Intelligence in Knowledge Management,* Stanford University, CA. Available: http://ksi.cpsc.ucalgary.ca/AIKM97/AIKM97Proc. html [accessed Feb. 2008].

Silverman, B.G., Bharathy, G., and Nye, B. (2007). *Profiling as "politically correct" agent-based modeling of ethno-political conflict.* Paper presented at the Interservice Industry Training, Simulation and Education Conference, Orlando, FL.

Silverman, B., Johns, M., Cornwell, J., and O'Brien, K. (2006). Human behavior models for agents in simulators and games: Part I, enabling science with PMFserv. *PRESENCE, 15*(2), 139–162.

Simon, H.A. (1967). Motivational and emotional controls of cognition. *Psychological Review, 74,* 29–39.

Sloman, A. (2003). *How many separately evolved emotional beasties live within us?* Paper presented at the Workshop on Emotions in Humans and Artifacts, Vienna, Austria, August 1999 and to appear in Emotions in Humans and Artifacts, R. Trappl and P. Petta (Eds.), Cambridge, MA: MIT Press. Available: http://citeseer.ist.psu.edu/cache/papers/ cs/21550/http:zSzzSzwww.cs.bham.ac.ukzSzresearchzSzcogaffzSzsloman.vienna99.pdf/ sloman02how.pdf [accessed Feb. 2008].

Sloman, A., Chrisley, R., and Scheutz, M. (2005). The architectural basis of affective states and processes. In J.-M. Fellous and M.A. Arbib (Eds.), *Who needs emotions?: The brain meets the robot* (pp. 203–244). New York: Oxford University Press.

Smith, C.A., and Kirby, L.D. (2001). Toward delivering on the promise of appraisal theory. In K.R. Scherer, A. Schorr, and T. Johnstone (Eds.), *Appraisal processes in emotion: Theory, methods, research* (pp. 121–140). New York: Oxford University Press.

Stocco, A., and Fum, D. (2005). Somatic markers and memory for outcomes: Computational and experimental evidence. In *Proceedings of the XIV Annual Conference of the European Society for Cognitive Psychology (ESCoP 2005),* September, Leiden University.

Stylianou, A.C., Madey, G.R., and Smith, R.D. (1992). Selection criteria for expert system shells. *Communications of the ACM, 32*(10), 30–48.

Sun, R. (2003). *Tutorial on the Clarion 5.0 Architecture.* Technical Report, Cognitive Science Department, Rensselaer Polytechnic Institute.

Sun, R. (2005). *Cognition and multi-agent interaction.* New York: Cambridge University Press.

Tambe, M., Adibi, J., Al-Onaizan, Y., Erdem, A., Kaminka, G.A., Marsella, S.C., and Muslea, I. (1999). Building agent teams using an explicit teamwork model and learning. *Artificial Intelligence, 110*, 215–239.

Trappl, R., Petta, P., and Payr, S. (2003). *Emotions in humans and artifacts.* Cambridge, MA: MIT Press.

Trimble, E.G., Allwood, R.J., and Harris, F.C. (2002). *Expert systems in contract management: A pilot study.* Defense Technical Information OAI-PMH Repository (United States). (Accession# ADA149363.) Available: http://stinet.dtic.mil/oai/oai?verb=getRecord&metadataPrefix=html&identifier=ADA149363 [accessed April 2008].

Velásquez, J.D. (1999). An emotion-based approach to robotics. In *Proceedings of the IEEE/RSJ International Conference on Intelligent Robots and Systems,* Kyongiu, Korea.

Zacharias, G.L., Miao, A.X., Illgen, C., and Yara, J.M. (1995). SAMPLE: Situation awareness model for pilot-in-the-loop evaluation. In *Proceedings of the 1st Conference on Situation Awareness in the Tactical Air Environment,* Wright Patterson Air Force Base, OH. CSERIAC.

Zachary, W., Cannon-Bowers, J., Bilazarian, P., Drecker, D., Lardieri, P., and Burns, J. (1999). The Advanced Embedded Training System (AETS): An intelligent embedded tutoring system for tactical team training. *Journal of Artificial Intelligence in Education, 10,* 257–277.

Zachary, W., Jones, R.M., and Taylor, G. (2002). How to communicate to users what is inside a cognitive model. In *Proceedings of the Eleventh Conference on Computer-Generated Forces and Behavior Representation* (pp. 375–382), Orlando, FL: UCF Institute for Simulation and Training.

Zachary, W., Santarelli, T., Ryder, J., Stokes, J., and Scolaro, D. (2001). Developing a multitasking cognitive agent using the COGNET/iGEN interactive architecture. In *Proceedings of 10th Conference on Computer Generated Forces and Behavioral Representation* (pp. 79–90), Norfolk, VA: Simulation Interoperability Standards Organization (SISO).

Zadeh, L.A. (1965). Fuzzy sets. *Information and Control, 8,* 338–353.

Zajonc, R.B. (1984). On the primacy of affect. *American Psychologist, 39*(2), 117–123.

Zoll, C., Enz, S., Schaub, H., Paiva, A., and Aylett, R. (2006). *Fighting bullying with the help of autonomous agents in a virtual school environment.* Paper presented at 7th International Conference on Cognitive Modeling, Trieste, Italy.

6

Meso-Level Formal Models

In this chapter we describe and discuss formal models of human behavior at a level of aggregation and detail between the micro and macro levels. Such models are often referred to as meso-level models. Typically the models represent interactions and influences among individuals in groups and cover both individual and group phenomena and their interactions. These models include several voting and social decision models, social network models, link analysis, and agent-based modeling (ABM). The models have been developed in varied disciplines, including social psychology, sociology, anthropology, economics, and computer and communications sciences.

VOTING AND SOCIAL DECISION MODELS

Understanding and predicting social phenomena requires good models of individuals and groups. The behavior of a group can differ from that of the individuals that comprise it. A science of aggregation is needed to model the behavior and actions of collections of people. There is a need to know how individual beliefs, goals, and skills combine on various tasks, such as problem solving and decision making. This section covers voting models that assume people reveal their true preferences, game theory models that assume people vote strategically, and social psychological models that consider how individual preferences might change in a group setting.

What Are Voting Models?

The research and models from voting theory provide a natural place to begin an investigation into aggregation for both pragmatic and conceptual reasons.[1] Governments, terrorist groups, and alliances all make decisions by "voting." Some follow formal voting rules and procedures and others informally aggregate competing desires. Thus, our use of the term "voting" goes well beyond formal (e.g., electoral) registering of a preference to much less formal situations in which a preference is exercised or a decision is made with input from multiple individuals.

Conceptually, voting models are valuable for three reasons: (1) a substantial body of theory exists, (2) that theory shows no shortage of counterintuitive results, thus highlighting the challenges of aggregation, and (3) the theory highlights a key point: to model groups well, one must be able to model individuals and the interactions between them.

State of the Art in Social Decision Modeling

We first describe the basics of preference theory. We then discuss results from social choice theory that reveal the problems created by aggregation as well as briefly comment on game theoretic models of strategic voting. The distinction between social choice theory and game theory hinges on behavioral assumptions. Social choice theory assumes that people truthfully reveal their preferences. Game theory does not. It assumes that people act strategically, which may or may not lead them to reveal their true preferences. The game theory models also enable one to understand how and why various institutional rules matter. We also discuss research from psychology that addresses how choices are made in a group context.

Preference Theory

Preferences capture how much people value or desire things. They differ from choices, which are what people select. Modelers define preferences over a set of alternatives. These alternatives can be outcomes, or they can be policies that produce outcomes (Page, 2007). Preferences impose an ordering over the alternatives. It is customary to write the preferences of someone who prefers apples (A) to bananas (B) as follows: A > B. Most modelers make two assumptions about individual preferences: that a person can compare any two alternatives (completeness) and that a person does not exhibit any preference cycles or internal contradictions (transitivity).

---

[1] We might have alternatively considered models of riots or collective ecosystem maintenance, but the related literature is not as deep or well thought out.

If a person claimed to prefer apples (A) to bananas (B), and bananas to coconuts (C), and then claimed to prefer coconuts to apples, one might think that person was irrational. Formally, it would be said that the person exhibits a preference cycle in which A > B > C, but C > A. When individual preferences satisfy both completeness and transitivity (i.e., A > B > C and A > C), then they are called rational.

If a person has rational preferences and if the modeler rules out indifference, then that person's preferences can be written as an ordered list from the most to the least preferred alternative. Given a set of five alternatives, A, B, C, D, and E, one person's preferences might be written A > B > C > D > E, and another person's might be written E > D > C > B > A.

This construction does not represent strengths of preferences. One person might strongly prefer A to B and strongly prefer B to C. Another person might have the same preference ordering but strongly prefer A to B and only weakly prefer B to C. To capture these relative strengths, one can assign payoffs or utilities to each alternative. Payoffs are not considered here because comparing these utilities across people is considered a dubious practice.

Social Choice Theory

If the members of a group have identical preferences, then aggregating those preferences is straightforward. One can think of the group as one big individual—and for some groups that may not be a bad assumption. The aggregation of preferences becomes problematic when the group members' preferences are diverse. Preference diversity can be fundamental (people want different outcomes) or instrumental (people want the same outcomes but differ over the means to achieve them). In what follows, that distinction is ignored, but it becomes important when thinking about linking models. If voting models are to be linked with cognitive models, then the source of preference diversity is important to define because information can reduce instrumental preference diversity but has little effect on fundamental preference diversity.

A collection of individuals with rational preferences may fail to have rational preferences as a group. We give an example and then state a general theorem.

In this example, three military leaders have preferences over which city to use as a base of operations. The three candidate cities are Paris, London, and Berlin. The leaders are denoted L1, L2, and L3. Their preferences are as follows:

Leader L1: Paris > London > Berlin
Leader L2: London > Berlin > Paris
Leader L3: Berlin > Paris > London

Were these three leaders to vote on their choice between each pair of cities, London defeats Berlin two votes to one, Berlin defeats Paris two votes to one, and Paris defeats London two votes to one. Thus, the collective preferences exhibit a cycle. Although the collective consists of rational individuals, the collective is not rational. In theoretical terms, the property of rationality does not aggregate.

The possibility of a cycle is not an artifact of majority rule voting. Kenneth Arrow proved that any rule for aggregating preference orderings that is not a dictator produces cycles (Arrow, 1951). It requires only that preferences are diverse, rankings between two alternatives do not depend on a third irrelevant alternative, and rankings reflect unanimity—if everyone prefers A to B, then so does the collective.

Arrow's theorem does not imply that cycles are unavoidable, only that if one wants to avoid cycles, one has to sacrifice one of the other conditions of his claim—appoint a dictator, sacrifice unanimity, or violate independence of irrelevant alternatives. In general, as argued by Donald Saari, preference cycles are more a function of the voting system than the voter. He suggests that voting paradoxes arise when the voting system fails to respect the natural cancellations of votes and so generates preference cycles (Saari, 2001). For example, one such voting system or scoring rule, the Borda rule (Marchant, 2000), does not create cycles. Under the Borda rule with three alternatives, a person's top choice gets three points, her second gets two points, and her third gets only one point. Each alternative gets a score, making cycles impossible. Borda rule can, however, result in a tie, which is what would occur in the example of voting over cities. A tie isn't necessarily a bad thing. It reflects equal support for each alternative. Borda rule may thus seem to be better than majority rule, but we must keep Arrow's theorem in mind. Borda rule must violate one of his conditions, and, in fact, Borda does not satisfy independence of irrelevant alternatives. In the example above, a fourth, irrelevant city could be introduced and change the outcome under Borda rule. The fact that by introducing irrelevant alternatives someone could change the outcome argues against using Borda rule. The debate thus moves from a discussion of the voter to a discussion of the scoring rules (Saari, 2006).

Given that regardless of the voting rule individual agents may fail to reach a stable aggregation, organizational and institutional structures take on great importance. The rules for how a group makes decisions can have large effects on outcomes. For example, if someone has the power to set the agenda, then that person may have substantial power. Thus, even if an organization is democratic in principle, it may not be democratic in practice, especially if one person controls the agenda.

Strategic Voting

In aggregating preferences, it can be assumed either that people vote sincerely or that they vote strategically. Strategic voting occurs even in large groups in mature democracies; people vote for candidates who they think can win rather than the candidates whom they most prefer. Alan Gibbard and Mark Satterthwaite have shown this incentive to misrepresent to be universal (e.g., Satterthwaite, 1975).

Why does strategic voting further complicate matters? We have shown that rational individual preferences need not aggregate into a rational collective preference. Thus, it may not be possible to discern what a group would decide even if one knew the preferences of every member of that group. Given that people have incentives to misrepresent their preferences, they wouldn't reveal their true preferences anyway. Thus, one must discern how people's true preferences get mapped into their actions—in this case, their votes. And that requires a model of individual behavior in groups.

The possibility of coalitions further complicates the analysis of voting models. Subgroups may have an incentive to form a coalition to steer outcomes toward desired ends. This is seen in parliamentary systems, with Israel as an example. It may not be possible to predict which coalitions will form: politics makes strange bedfellows, and predicting those bedfellows can be difficult.

In a group setting, social influence dynamics can muddy the picture even further, as people may change their preferences to align with the real or the inferred preferences of others. Concern for the preferences of others and for one's own standing in a group creates more indeterminacy in collective decisions. A striking example of such social influence effects is provided by the Abilene paradox, in which each person privately prefers X but believes that others prefer Z. If all group members revealed their true preferences, the group would clearly choose X. However, the desire to conform to what is (incorrectly) perceived to be the normative opinion can lead a member to suggest Z and others to agree (Harvey, 1974).

While this counterintuitive outcome is probably rare in practice, it highlights the importance of realizing that social influence is not simply a matter of one person seeking to change the preference of another. People also actively seek to align their preferences with important others. Both computational models and empirical studies have demonstrated that the impact of others on individual preferences tends to create uniformity of preferences among people who are closely connected. Dynamic social impact theory (Latané, 1996) predicts that people will change their preferences to match those of others, with the impact based on both the strength (status) and the immediacy (social closeness) of others. The result is emerging pockets of uniform attitudes based on social network clusters. Research on minority influence (e.g., Nemeth, 1986) also shows that the views of a cohesive minority, when clearly and consistently stated, can also change the opinions of majority members. Hence, knowing what the majority of individuals prefer at time 1 may not allow one to predict confidently what a group will choose at time 2.

Relevance, Limitations, and Future Directions for Social Decision Models

While voting models per se, especially those that compare specific voting rules like Borda and majority rule, may seem more relevant to political science than to the situations that concern us here, the insights that can be drawn from these models are of critical importance. Two of the main recommendations of this report are that modelers should recognize diversity of background, activity, and preferences and that they should embrace uncertainty. Nowhere does that advice ring more clearly and loudly than in understanding the link from individual incentives to group behavior. Moreover, one can link diversity of background, activity, and preferences and uncertainty about all three into a general insight: the more diverse the members of a group in their general makeup (their background), their preferences, and their actions, the more uncertain one should be about their collective decisions and actions. For example, models that attempt to make predictions about the attitudes and behaviors of a group of noncombatant civilians must consider the diversity of that group in terms of the sociocultural-ethnographic-economic background, preferences, and available actions. The more diverse the group on any of these three dimensions, the less certain the predictions. At a very practical level, the implication of recognizing diversity is to make the models more complex. Another practical implication is that model results should often be characterized in terms of how the diversity of the population being modeled impacts the results (e.g., show entropy or diversity indices to characterize the initial population and show how outcomes change as the initial population varies on this metric).

Even if group models cannot be expected to make point predictions, they can provide a way to predict sets of possible outcomes. If one has even crude approximations of preferences, possible coalitions, and a set of possible voting rules, he can write game theoretic or agent-based models, and those models can provide some guidance for what might happen and, equally important, what probably will not happen. For a recent survey of these methods, see Kollman and Page (2006).

The ability to apply voting theory depends on data, knowledge, and theory. For many of the problems relevant to this study, one would not have information about individual-level preferences. And, even if one did have access, the theory tells us that it is not possible to predict outcomes with certainty from that data. Equally important, one may not have knowledge of the voting rules. And, as discussed above, the voting rule has a substantial influence on the outcome. Thus, even with information about preferences, one would also need to know something about the process of preference aggregation in the group of interest. Finally, to apply these models, one needs models of how people behave in groups. Are the group members strategic? Do coalitions form, and who belongs to those coalitions?

Imagine a model that includes the actions of a terrorist organization or of a nascent nation-state. One could make a black box assumption about how that organization or government acts. In other words, one could treat the group as an individual, presumably an individual who is the average of the group members.

The voting models reveal the problems with that approach. Groups do not make choices as though they were a single individual. The natural way to improve the model would be to make the black box transparent and to allow for multiple characterizations of the collective decision-making processes that produce those outcomes. This will require data, knowledge, and models of the group of interest, but the potential payoff is large, as it will provide a more accurate assessment of the likely distribution of behaviors over the set of possible actions.

Finally, empirical voting studies demonstrate that humans do not act in a strictly rational or strategic manner, hence calling into question the formal mathematical "rational" and "strategic" voting models. Summaries of the empirical literature point to the social rather than rational nature of voting behavior; for example, people vote primarily along ethnocultural lines rather than according to their economic interests and display widespread voter ignorance (Friedman, 2005). As another example, research on the "voter participation paradox"—in which it is asked why people vote at all, as each individual has virtually zero probability of affecting the outcome (Converse, 1964; Green and Shapiro, 1994)— both demonstrate this lack of rationality and suggests that there are huge variations in individual behavior. Turnout depends on a number of social factors, including the size of the electorate (as the size of the electorate grows, fewer voters turn out), the closeness of the competition (the closer it is, the higher the turnout), and the presence of an underdog (more turnout). This empirical work suggests both that the formal models and simplistic game theoretic models are inadequate and that the more detailed and nuanced behaviors possible in agent-based models (ABMs) are better at capturing the complexities of voting behavior.

SOCIAL NETWORK MODELS

Networks are ubiquitous, and many techniques have been developed for analyzing, predicting, and understanding the world in terms of the set of connections among entities—a network. As the focus here is on social and behavioral modeling, we limit our discussion of network modeling techniques to those that have been and are being used to address individual, social, organizational, political, or cultural issues, rather than, say, gene interaction networks or computer networks. For a review of the field of network analysis, see Freeman (2004), and for the methodology, see Freeman, White, and Romney (1991) and Wasserman and Faust (1994).

What Are Social Network Models?

Social network models view groups as consisting of a set of nodes (the members of the group) and a set of ties that connect them, which link together to form a network. The ties are often seen as pipes or roads along which various kinds of traffic flow, such as informational and material resources, as well as influences and coordination. Thus, a key aspect of network modeling is concerned with predicting (and controlling) what flows to whom at what time. Ties are also seen as providing a kind of underlying structure or topology that has effects on the performance of the group or individuals. A fundamental proposition of social network models is that a node's position in the network (in conjunction with its attributes) determines the opportunities for and constraints on action that it will encounter. A group-level corollary of this proposition is that the network structure of a group (together with other attributes of the group), determines the performance or outcomes of the group. Thus, network models differ from other models in placing less emphasis on characteristics of the nodes and more emphasis on the structure of connections between the nodes.

Social network analysis (SNA) has received a great deal of attention since the terrorist attacks of September 11, 2001 (Borgatti and Foster, 2003). Phrases for fighting terrorism, such as "disconnect the dots" and "it takes a network to fight a network," and for doing business, such as "it's not who you know but who or what who you know knows" and "are you networking?" have appealed to the imagination and raised awareness of this area. In addition, there have been successful applications of this approach. For example, social network information was used to locate Saddam Hussein, and several SNA tools have been used in various criminal investigations. Social network information is used in popular social networking web services, like Friendster, to help students vet their dates.

Traditionally, most SNA has focused on the analysis of relatively simple datasets involving a small number of social relations (often of just one kind) connecting a set of persons in some kind of group at a single point in time. Analysts in this area use computational techniques primarily to statistically analyze these networks. This area has a long tradition, predating World War II. It emerged from the social sciences, particularly from social psychology, anthropology, and sociology, and has now spread to organization science, economics, physics, and computer science.

More recent work has focused on more complex networks involving large numbers of nodes of differing types (see section on Multimode Networks below). For example, Carley (2003) has developed social network metrics that take into account not only relations among individuals, but also relations among tasks, relations among items of knowledge, assignments of tasks to individuals, relations of knowledge to individuals and other relationships.

In addition, a key research interest today is in understanding network dynamics, both in the sense of how networks change over time (especially in response to attacks) and in the sense of how things flow over the network links. Carley (2003) has used multiagent models in a network context to predict and reason about change in social and other networks.

State of the Art in Social Network Models

In this section, we lay out the key concepts of SNA, starting with a discussion of the nature of the data and followed by an outline of the key analytical constructs, namely cohesion, centrality, equivalence, and clustering. The section ends with a discussion of network evolution.

Nodes and Ties

The set of actors or agents that form the nodes of a network can consist of either individuals or collectives, such as organizations, cities, or countries. Nodes are assumed to possess characteristics that define their goals and affect their ability to achieve and exploit their network positions. These characteristics are modeled as a set of categorical and/or continuous attributes.

In general, relations among nodes are modeled as dyadic 2-tuples (called ties, links, or edges) that bind exactly two nodes to each other. Therefore, a conversation among three people A, B, and C is typically modeled as three separate dyadic interactions consisting of A with B, B with C, and A with C. For the most part, the ties modeled among nodes typically belong to a general class known as social relations. These include such things as acquaintance (e.g., knows), kinship (e.g., brother of, father of), other social roles (e.g., friend of, teacher of), and affective relations (e.g., likes, dislikes). Each type of tie can be further characterized by relevant characteristics or attributes. For example, a friendship tie can be characterized in terms of intensity, closeness, and duration.

In addition, network modelers often represent interactions over time—such as in-person meetings, communication, or fighting—as ties. Hence a tie is considered to exist between two nodes if at least one interaction between them is observed during a given period. The actual number of interactions may be recorded as an attribute of this tie. Interactions are inherently transitory and evanescent but are often seen as revealing the presence of underlying social relations.

Interactions, such as conversations, provide the mechanism by which things flow through social relations, as when an actor transmits information to a friend through communication or when a person infects another with a disease via personal contact. Thus flows represent a third category of tie that a network modeler can choose to model. Typical flows of interest have been information, ideas, infections, material goods (such as guns and money), and such intangibles as energy and motivation. These are often referred to by network analysts as "tokens."

Multimode Networks

When a categorical variable exists that distinguishes between different types of nodes, and, in addition, ties exist only between nodes of different types (and not within types), the resulting networks are referred as k-node networks or, in graph theory, as k-partite graphs, where k refers to the number of distinct types of nodes. These kinds of data typically arise in the context of recording affiliations between individuals and groups or events. For example, Davis, Gardner, and Gardner (1941) recorded which women attended which social events in a given season. Ties exist between women and events, but not among women and not among events. Similarly, it is common to record for each person in a group the organizations to which they belong(ed). And in organizational analysis, one can collect the number of hours that each person worked on various tasks or projects.

Multinode networks can be analyzed directly or converted into simple 1-node networks by deriving co-occurrence indices. For example, a 2-node women-by-events network can be converted into a 1-node women-by-women network in which a tie between each pair of women is characterized by the number of events they attended in common.

With multiple nodes, it is possible to represent the system as a whole as a meta-matrix (Carley, 2003). The meta-matrix is a conceptual device for identifying the set of networks within and among nodes of multiple classes. For example, given the three classes of nodes—people, knowledge, and activities—the set of subnetworks possible is shown in Table 6-1. The second key concept is the entity ontology—for network analysis, this is the set of categories that defines the node classes and the link classes among the nodes used in a particular study. The table illustrates a particular ontology; other ontologies are needed for other applications.

TABLE 6-1 Illustrative Meta-Matrix

	People	Knowledge	Activities
People	Social network	Knowledge network	Activity network
Knowledge		Information network	Needs network
Activities			Precedence network

Cohesion Models

A fundamental concept in network modeling is cohesion. Cohesion refers to the connectedness or structural integrity of a network, and it is often interpreted in terms of the network's potential for coordinating among its members or exploiting knowledge that is distributed across the network.

One aspect of network cohesion is density, which refers to the proportion of pairs of nodes that have a direct tie (i.e., are not dependent on an intermediary). A high density implies that, on average, each node is directly connected with many others. If the ties represent something like trust relations, this indicates a group in which information can flow quite freely.

Another aspect of cohesion is the average path distance, also known as characteristic path length. Path distance refers to the number of links in the shortest path between two nodes. A network with low average distance is one in which the lengths of the shortest paths between pairs of nodes are quite small, so that things flowing through the network can reach any or all nodes comparatively quickly. In the case of viruses or other infections, this is a measure of the vulnerability of the network to disease. In the case of the spread of best practices, it can be seen as a determinant of the potential performance of a continuously adapting system.

Centrality Models

A frequent analytical strategy in network modeling has been the identification of key players who are disproportionately important due to their structural position in the network (Borgatti and Everett, 2006). The structural importance of a node in a network is conceptualized as its centrality. One way to think about centrality is in terms of a node's direct or indirect contribution to the cohesion or structural integrity of the network. For example, degree centrality is defined as the number of ties that a node has. If the total number of ties in the network is a measure of the cohesion of the network, then clearly degree centrality can be seen as each node's "share" of the total cohesion. In this sense, the centrality measure implies a model of the sources of cohesion.

Other well-known aspects of centrality include closeness centrality, betweenness centrality, and eigenvector centrality. If the graph theoretic distance between two nodes in a network is defined as the length of the shortest path from one to the other, then closeness centrality is defined as the sum of distances from a node to all other nodes. To the extent that social ties among network members constitute pipes that transfer such traffic as information or influence, closeness centrality gives the average time until the arrival of something flowing along the shortest paths. Betweenness centrality is the share of all shortest paths in the network that pass through a given node. High betweenness nodes are a kind of glue holding the network together; deleting nodes with high betweenness from a network tends to disconnect the network or make all paths much longer. Eigenvector centrality can be described, in simplified terms, as the extent to which a node is connected to many nodes that are themselves well-connected—a kind of turbocharged version of degree centrality.

Another way to think about centrality is in terms of the exploitability of a node position. This is the perspective taken by social capital theorists, who see a node's position in the network as a kind of capital that the node can exploit for personal advancement or achievement. For example, a node with excellent closeness centrality is a short distance from other nodes in the network and is therefore well-positioned to hear information flowing through the network early, when it still confers a competitive advantage. A node with high betweenness centrality is in a position to make demands on others because these others need the central node in order to connect with others in an efficient manner.

Finally, centrality can also be thought of as providing expected values for certain node outcomes in a particular flow process. For example, the formula that defines betweenness centrality gives exact estimates of the expected number of times that something flows over a node in a process in which tokens travel exclusively along shortest paths. Similarly, closeness centrality gives the expected values of the time to first arrival of a token flowing through a network, again using exclusively shortest paths. Degree centrality gives the frequency of arrival of a token in a process in which tokens travel along unrestricted random walks through the network. Thus, definitions of centrality carry with them a model of how things flow in a network.

Equivalence Models

Equivalence modeling refers to the branch of network modeling concerned with detecting nodes that play similar structural roles in the network (Borgatti and Everett, 1992). The simplest equivalence model is that of structural equivalence. A pair of nodes is structurally equivalent to the extent that they are connected (and not connected) to precisely the same third parties, regardless of whether they are tied to each other. Structurally equivalent nodes are structurally indistinguishable and substitutable. A fundamental claim in this kind of modeling is that, by virtue of being structurally isomorphic, structurally equivalent nodes will tend to have similar outcomes. Structural equivalence can also be seen as providing a formal definition for concepts of node environment and niche.

Another equivalence model is called regular equivalence. This is a recursive model in which two nodes are regularly equivalent to the extent that they are connected to regularly equivalent third parties (but not necessarily the same third parties). Thus, two nodes do not have to have any contacts in common to be seen as equivalent, and indeed they can belong to entirely separate groups. As a result, the model can detect that both leaders of wholly unrelated organizations are playing the same role vis-à-vis their respective groups. Thus, it is a better model for the concept of social role than is structural equivalence. For example, given a network defined as the set of observed relationships among all people working in a hospital, regular equivalence can detect that two doctors of different patients are both playing the same role (i.e., doctor), whereas structural equivalence can detect only that two doctors of the same patients are playing the same role. The importance of regular equivalence is that it can discover latent or emergent social roles that have not been named and that the members of the network are themselves unaware of. However, the recursiveness of the definition, in which one needs to know the extent of regular equivalence between all other pairs of nodes in the network in order to calculate the regular equivalence of a given pair, makes this model computationally much more difficult than structural equivalence.

Cohesive Subgroup Models

An active area in network modeling is the identification of cohesive subsets—dense regions of a network that have more ties within than to the rest of the network—that operate as units. The fundamental assumption in this work is that members of a cohesive subset will have more in common with each other than with nodes outside the subset (Borgatti, Everett, and Shirey, 1990). This occurs both because nodes with common attributes will tend to seek each other out, forming the cohesive subsets in the first place, and because members of cohesive subsets have disproportionate influence on each other, creating homogeneity within the group. Thus the homogeneity of cohesive subsets results both from selection processes (similar nodes joining together) and from influence processes (interacting nodes becoming similar to each other).

228 BEHAVIORAL MODELING AND SIMULATION

Network Evolution

Because social network modeling is a relatively young field that until recently had to fight for legitimacy, it is natural that it has concentrated on the impact of network variables on non-network or "traditional" outcome variables, such as career success or team performance. However, as interest in networks has increased, so has research on the antecedents of network structure—in short, network evolution—and the antecedents of a node's position within network structures (e.g., why some nodes become more central than others) or of the emergent structural properties of the whole network (e.g., why some network structures are more robust than others).

Empirical research has demonstrated several key factors that determine who has ties with whom in a variety of networks. People who are physically near each other tend to communicate more, even in an age of asynchronous electronic communication. This effect of proximity (sometimes called propinquity) is a special case of a more general principle known as homophily—the tendency for individuals to have ties of various kinds with people who are like them on socially, culturally, or politically significant variables, such as geographic location, race, gender, age, social class, religion, culture, language, organizational affiliation, centrality, etc. Thus, these variables form the grist of most simulation models of network change. However, it should be noted that heterophilous mechanisms (in which opposites attract) also exist. Sexual relations, for example, are overwhelmingly heterophilous with respect to gender. In addition, most nonreciprocal relations such as "seeks advice from" or "gives orders to" are heterophilous, so that less knowledgeable people seek advice from more knowledgeable people rather than from those equally knowledgeable.

An important factor with elements of both homophily and heterophily is the activity focus. Common activities bring together people with similar interests, such as a bowling league or a political action group, creating homophilous linkages. However, as Alexis de Tocqueville noted as far back as 1835 (de Toqueville, 1835), these foci also tend to bring together people from different walks of life, creating heterophilous linkages across social boundaries.

Another important factor—not unrelated to homophily—is the transitivity induced by such mechanisms as cognitive dissonance (Festinger, 1957) or balance (Heider, 1988). For example, if node A likes node B, and node B likes node C, then in many circumstances node A experiences some pressure to at least not dislike node C, thus increasing the probability of a tie forming between A and C.

Finally, there are status-based mechanisms that are neither homophilous nor heterophilous in which nodes are sorted by status, and all nodes prefer to interact with high-status nodes. In such cases, the high-status nodes exhibit homophily, because they prefer each other, while the low-status nodes exhibit heterophily, because they prefer high-status nodes. The model of preferential attachment developed to explain the pattern of which websites link to which other websites is a kind of status model. In preferential attachment, new websites link to existing websites with a probability proportional to the number of links the existing website already has, creating a situation in which, in terms of incoming ties, the rich get richer.

In recent years, a number of researchers have modeled network change using simulation methods, typically variations of ABM (Zeggelink, 1994; Snijders, 2001; Carley, 2003). For example, in a stream of research she refers to as "dynamic network analysis," Carley (2003) defines a subclass of ABMs that have multiple agents who dynamically form a network that evolves as the agents themselves learn and adapt. Agents take action on the basis of what they know, whom they know, and their own internal cognitive architecture, possible actions, and other factors. These models have been used to explore information diffusion, the impact of new technology, the evolution of networks, and the impact of interventions (e.g., to analyze the relative impact of different courses of action on terrorist groups). Key features are that there can easily be thousands of actors, with the exact number of actors limited only by storage space on the computer. The cognitive/communicative complexity of the model is limited by available computational capacity and processing time.

Relevance, Limitations, and Future Directions

Social network models and dynamic network analysis models can be used to identify key actors or groups. They are useful in understanding terrorist networks and in analyzing the criticality of nodes in those networks. They are also useful in locating individuals: As mentioned, Saddam Hussein was located through an SNA of his contacts. Dynamic network analysis models can also be used to illustrate how the isolation of particular actors or groups will disrupt the flow of information or goods and services in both the short and long term dynamically. They can be used to show how groups or networks are likely to evolve under different conditions, technological environments, etc. For example, the Construct model (a combined dynamic network and ABM) was used to contrast the effect of removing the top leader of al-Qaeda (bin Laden) and of Hamas (then Yassin) (Carley, 2004) and suggested that, for Hamas, performance would improve temporarily and the next leader would be Rantissi; in contrast, for al-Qaeda, performance would decrease and the next leader was indeterminate. In addition, they can be used to examine the impact of changes in recruitment on organizational performance, the effects of policing policies on civil unrest, the effect of technology and information sharing on organizational performance, the effect of detection technologies on information flow, etc. Dynamic network analysis models can also be used to create dynamic war-gaming scenarios by predicting the effects of courses of action on enemy and noncombatant behavior.

Early work in social network modeling was mostly based on a branch of discrete mathematics known as graph theory. As a result, the models were fundamentally deterministic in character. These deterministic models do not lend themselves to prediction of populations (in which there tends to be a probabilistic distribution of behaviors and outcomes) nor of complex systems (see System Dynamics, Chapter 4). In these models, probabilistic thinking came into play only in relating deterministic variables to each other statistically. More recently, however, the field has begun to incorporate stochastic thinking at a more fundamental level. For example, the exponential random graph models (also known as P* models) seek to model networks in terms of their latent tendencies to form micro structures, such as transitive triples or starlike subgraphs (called "motifs" in the physics literature) (Milo et al., 2002). By estimating a parameter for each kind of micro subgraph, the models can achieve a parsimonious description of the network in terms of a string of estimated parameter values, together with standard errors. This begins to make it possible to compare networks statistically with each other or with theory.

Stochastic models also facilitate comparison of networks over time and indeed enable the estimation of rates of change in model parameters. In the long term, this line of work promises to yield continuous time models of network evolution, as opposed to current approaches to longitudinal analysis, which are limited to comparing snapshots of the network at discrete intervals in time.

Similarly, most network models that are based on graph theory (and most are) are designed for binary data (i.e., a tie exists or it doesn't). Validation of and extension of the metrics for nonbinary data is an ongoing research area that will eventually enable the capture of a wider range of social phenomena.

Finally, most standard social network tools available on the web, in practice, are limited in their ability to handle more than 100,000 nodes with the exception of ORA (Carley et al., 2007). Visualization routines in general tend to be underdeveloped and work best with small datasets; however, for most military users, the goal is not to be able to visualize millions of nodes, but to have good preprocessing systems that subselect just the small portion of the network to view. From a military standpoint, many of the existing tools, because they are oriented around metrics, are too complex for the average soldier to use and contain little guidance on when to use which metric. Finally, much military data are multimode and multilink, and as a result they are cumbersome to process with most social network tools, with the exception of ORA. Rarely in military applications is it the case that the social network exists separate from, and needs to be assessed separately from, other types of networks, such as the activity network.

All this being said, of the modeling tools described in this chapter, network models have had, to date, the biggest impact on military decision making. In particular, dynamic network tools that take into account the meta-matrix or that link to ABMs have been used to identify vulnerabilities in insurgent and terror networks, characterize political elites and track changes, identify local opinion leaders, and assess changes in beliefs and social influence. The most promising future directions involve linking these network approaches to other approaches, such as strategic reasoning à la game theory, or forecasting via ABMs, or geospatial identification by combining networks and map-based techniques. For example, placing network analysis in decision contexts enables reasoning about organizational change (Butts and Carley, 2006), while combining networks with spatial reasoning is facilitating analysis of the movement of terror groups to new locations of activity (Moon and Carley, 2007).

LINK ANALYSIS

Link analysis or link mining is related to SNA but has emerged as a distinct field centered on discovering patterns by looking at the relations among entities (see Getoor and Diehl, 2005, for a survey). Much of the work focuses on anomaly detection and link identification.

What Is Link Analysis?

Link analysis has emerged largely from computer science and forensics, with particular attention to work in machine learning. Historically, the term "link analysis" was used, particularly in the law enforcement area, to refer to approaches that let the analyst display and reason about the links between multiple types of nodes.

Modern link analysis is a new subfield largely centered in computer science and statistics. Researchers and analysts in this area use computational techniques to locate patterns and subgroups based on a given set of information about paths, in which a path consists of a series of links that may connect nodes of different types, such as Joe + hamburgers + McDonald's. Extraction of links often requires massive data preprocessing or restructuring of databases (Goldberg and Wong, 1998). Given a set of paths, advanced data-processing techniques are combined with machine learning to enable rapid database transformation and pattern extraction. Key questions often addressed are what paths are anomalies and what patterns can be inferred. Thus, much of the work in this area has focused on the identification and recognition of patterns, data mining, and node identification and deidentification. Inferred patterns are then used to infer the "cause" of the pattern or to make predictions about future links.

The main feature that distinguishes "social networks" from "link analysis" in general and from "link prediction" in particular is the richness of the phenomena that are being explored and modeled. In social networks, the focus is often on producing qualitative and quantitative assessment about various items, such as leadership or influence or performance, testing, and estimation, whereas in link analysis the focus is on predicting quantities, such as the number of nurses who work at the hospital and give blood. Therefore, in social networks, the goal is to produce realistic models based on believed and theoretically grounded assumptions, a good descriptive model is looked for, and the researcher worries about how to do inferences. Parameter values encode semantics of interest in a specific application, and the research asks what the estimated values are and how much can one believe such estimates. It is only at the last stage that the social network theorist worries about predictions. Good description is thought to yield good prediction.

In contrast, in link analysis, the richness of a model is often sacrificed to statistical or computational efficiency. The prediction task, not necessarily the link prediction task, is the key focus. Accurate predictions and a quick black box that produces them are often viewed favorably in the literature. The analysis and comparison of various link analysis approaches are typically weak; that is, little is done to compare the methods other than to compare speed. Furthermore, it is hard to make a case about why one should believe the guesses they produce other than on statistical grounds. In a sense there is no science that backs such predictions up, no theory for why these anomalies exist. Breiman (2001) discusses the differences between statistical and data mining approaches to analysis. These same differences apply to networks science (statistical) and link analysis (data mining).

State of the Art

Link analysis tools result in a mathematical representation of the relation of different entities to each other vis-à-vis some problem. This mathematical representation or "model" of the underlying social behavior is discovered from the data and can then be utilized in other types of models, such as multiagent systems, to characterize a type of behavior.

In modern link analysis, there are three fundamental concepts and two more general related concepts. First, there is the notion of similarity or distance among nodes. This distance is typically used to infer connectivity under the assumption that nodes that are similar or close will connect with other nodes in a similar fashion. Such a notion can derive from a formal (probabilistic) model or from other theoretical concerns and so is more deterministic. Link analysis algorithms can generally be categorized by their approach to similarity, or distance, and can be further divided by whether that distance is model based or algorithm based, and whether that distance is explicit or implicit.

The second core concept is that of groups or clusters of elements of a network (typically of nodes) and the way both single elements and groups of them interact with one another. The idea of groups is central to most other link analysis papers. The third key concept is the link function, which translates similarity into the presence, absence, or weight of a link. Key differences in link analysis algorithms are often expressed in terms of differences in the link function. Less central ideas include types of nodes and links and, of course, time.

In modern link analysis the analysis is done on the data itself, rather than on the network that has been inferred from the data. This avoids errors from the inference itself and from the relationship model that is being fitted. In addition, by assuming conditional independence of links, link analysts can leverage general statistical machinery. This provides an elegant way to deal with missing data—any data one has are just a tiny snapshot of a rich distribution. Link analysis also deals easily with "rich links," like multiparty links or multiple links between the same entities. In contrast, many of the social network tools have been developed with the "one dyad, one link" approach.

In link analyses, the paths typically include nodes of multiple types, such as people and events and resources. In contrast, in a typical social network, the nodes are generally all of the same type or at most of two types. Each of the paths in link analysis is a single observation, hence temporal information on when a path occurred is available. In link analysis, no effort is made to take the paths and form the implicit networks. No assumptions are made about the completeness of the underlying network. In contrast, the social network modeler starts with a network and typically does not preserve path information, in the sense of information about observed temporal trajectories. Furthermore, social networks assume that the links are not independent events, whereas much of the work in link analysis assumes that each instance of a link is an independent event but subject to parameters that can be deduced. Finally, link analysis as a theory of anomaly detection is agnostic about the types of links and nodes that form the paths, whereas social network modeling has historically focused on networks in which the nodes are information-processing entities, such as people, organizations, or groups, and the links are the various factors by which they are connected, such as friendship, mentoring, financial transactions, or marriage.

There are a growing number of link analysis tools, many of which are available on the web. Illustrative tools include GDA (Kubica et al., 2002;

Kubica, Moore, and Schneider, 2003), PROXIMITY (Jensen and Neville, 2002), and PRMs (an extension of Bayesian Nets) (Getoor, Friedman, Koller, and Taskar, 2001, 2002; Taskar, Abbeel, and Koller, 2002; Taskar, Wong, and Koller, 2003). Common tools exist for doing a variety of tasks, including extracting of links from databases (Goldberg and Senator, 1998) and texts (Lee, 1998) and analysis of the extracted links (Chen and Lynch, 1992; Hauck, Atabakhsh, Ongvasith, Gupta, and Chen, 2002).

Relevance, Limitations, and Future Directions

Link analysis has widespread military applications in the creation of actionable intelligence from large diverse data sources and in the development of network models—such as terrorist network models—from partial and incomplete transaction data. Modern link analysis has focused on anomaly detection in large datasets.

For many of the techniques that rely on machine learning, a key issue is having sufficient data with appropriate distributions so that the model can be "trained." In general, link analysis tools require a large quantity of labeled data that have been preextracted. A second key limitation of this approach is that many of the tools assume that the data exist in a file or database and cannot handle streaming data as they arrive or data that are "out of order" in terms of the entity classes. Third, many of the current nonproprietary methods do not scale well, greatly reducing the size of the datasets that can be handled. A fourth related limitation is that the models that are discovered when using a link analytic approach inherently assume that "tomorrow is like today." Hence using these models to predict future behavior may be limiting. Fifth, to be useful, these models need to be expanded to handle streaming data. This is work in progress and Bayesian updating rules are being developed for an increasing number of models. A final limitation of link analysis is that the models that result, although fitting strong mathematically based theory, may not be reasonable from a social or behavioral perspective. For example, knowledge discovery routines for finding groups will find groups that meet some predefined statistical requirement; however, these groups may not match the definition of a group in empirically grounded social theory or even the everyday sense of what constitutes a group.

While link analysis is generally useful for locating patterns and discerning structure, it is very limited in its ability to analyze downstream effects. For example, using link analysis to answer such questions as who has the most connections (degree centrality) can be done in only the most rudimentary way when links are viewed as independent. Consider the question, "Is Mustafa important because of the number of communications that go through him, or because his communications are the only ones connecting different sectors of a terrorist group?" Using SNA, this question can be addressed directly and easily with existing and well-validated metrics. When a link analysis approach is taken, in which links are viewed one at a time and treated as independent, a special-purpose and extremely complex model would need to be constructed.

In principle, link analytic tools can be used to locate and construct the networks, and then social network or dynamic network metrics can be applied for predictive purposes. This is a promising direction for creating actionable intelligence. However, current link analytic models always use customized representations of the underlying network, making it difficult to transfer their results to other tools designed for prediction. A generalized standard for representing the underlying network is needed. Advances in this direction include graphml and dynetml (Tsvetovat, Reminga, and Carley, 2004), which are XML languages for dealing with network data; however, both of these are insufficient to meet the needs for which link analysis results are used.

An important issue for network modeling is the robustness of the models in the face of errors in the data. This is particularly an issue for hidden or stigmatized populations (e.g., criminals, terrorists) and for illicit or private relations (covert operations, political influence, etc.). To date, very little work has been done to assess the robustness of different network models in the face of different kinds of errors in the data, such as missing ties, missing nodes, gratuitous ties, and gratuitous nodes (as when a person who uses two different names is mistakenly entered as two different nodes). Similarly, few network models provide standard errors or confidence intervals for their outputs. Thus, it is simply unknown how much error in the data can be tolerated or whether a network model of flawed data does more harm than good.

Another increasingly important issue for network modeling is the bounding of empirical networks—that is, determining which nodes to include and which ones to exclude. Part of the conceptual base of network modeling is the interdependence of nodes. This creates a problem for artificially bounding the networks that one wishes to model. One can arbitrarily choose to model the members of an organization or the residents of a village, but this does not stop the nodes from having ties with people outside the sample frame. To the extent that these unobserved ties affect what happens to the nodes, the models will fail to predict outcomes of interest. This problem cannot be eliminated, but it can be ameliorated by including larger chunks of the human network in the analysis, particularly chunks that correspond to natural boundaries. For example, if the computational and data collection issues can be overcome, modeling an entire village or other geopolitical unit is clearly preferable to arbitrarily modeling half of the village because of practical limitations. What is needed is investigation into the consequences of the different ways of bounding networks and into alternative ways of framing research issues to get around the boundary specification problem.

A major area for future research is the study of models and algorithms to recover and/or discover link connectivity patterns, rather than node connectivity patterns (in the sense of Milo et al., 2002). This has potential for application, for example, to network privacy (reidentification, deidentification), to subgraph matching, and to motif discovery. Of primary importance for real-world applications is the development of fast approximation algorithms that replicate the solution of successful algorithms for solving various problems, thus addressing the scalability issue that typically burdens algorithms that involve counting links in various ways. Another area that requires investigation is how to connect models of static and dynamic networks to observations and measurements. This is a general issue for all modeling techniques, and its implication for the broader impact of research is far-reaching and would include as a subtopic the integration of information from multiple sources, à la metamatrix, to support the discovery of interesting patterns.

In addition, any simultaneous advances in automated data collection and computational algorithms for very large networks would significantly improve the usefulness of link analysis for the problems at hand. It is already possible to construct communication networks based on telephone logs, e-mails, etc. However, the degree to which one can infer different kinds of social relations—such as trust, kinship, aid, conflict, etc.—from these data is still unknown, nor are alternative data currently available. Much of social network research has been based on survey research methodology, which is not applicable in the case of unwilling actors, such as enemies.

AGENT-BASED MODELING OF SOCIAL SYSTEMS

The social and organizational sciences seek to understand not only how individuals behave but also how interactions among individuals generate macro-level outcomes. Understanding a social system requires more than understanding the individuals in it. It also requires understanding how the individuals interact with each other and how the results can be more than the sum of the parts. Agent-based modeling is well suited for this objective.[2]

[2]Several research communities are currently exploring methodological approaches closely related to agent-based modeling under a variety of other names. Examples include multiagent-based systems, agent-based computational economics, agent-based social simulation, multi-agent systems, and individual-based modeling. A sample of introductory readings from these various research communities can be accessed online at http://www.econ.iastate.edu/tesfatsi/aintro.htm.

What Is Agent-Based Modeling?

Agent-based modeling is the computational study of systems that are complex in the following sense: (1) the systems are composed of multiple interacting entities and (2) the systems exhibit emergent properties—that is, properties arising from entity interactions that cannot be deduced simply by averaging or summing the properties of the entities themselves.

What distinguishes agent-based modeling from general complex systems modeling, however, is the form of the entities that make up the system. A system can be complex even if its constituent entities are homogeneous units, such as CO_2 molecules. In contrast, the constituent entities of an ABM are heterogeneous "agents" with internal states that can vary over time in response to internal deliberations as well as external forces, thus admitting the exploration of systems of heterogeneous agents with a range of social and learning capabilities.

More precisely, the agents in an ABM can represent people (e.g., consumers, sellers, voters). They can also represent social groupings (e.g., families, firms, communities, government agencies, nations), biological entities (e.g., livestock, crops, forests), and even physical systems (e.g., weather, geography, transmission grids). When the interaction network formed by agents is contingent on past experience, and especially when the behaviors of agents in this interaction network continually adapt to past experiences, standard mathematical and statistical tools typically have only limited ability to derive the dynamic consequences. In this case, agent-based modeling might be the only practical method of analysis.

Agent-based modeling is a general-purpose technology. On one hand, the only constraints are the modeler's purpose, imagination, and ability to encode. A modeler is free to make assumptions believed to be most relevant and realistic for an issue of interest. On the other hand, the realism of the resulting model will depend strongly on the extent to which the modeler's assumptions are driven by data. In general, the more tightly a model has been constrained by real-world data, the smaller the space of possible outcomes.

As detailed by Axelrod (1997, pp. 206–221), simulation in general, and agent-based modeling in particular, is a third way of doing science in addition to deduction and induction. Scientists use deduction to derive theorems from assumptions and induction to find patterns in empirical data. Simulation, like deduction, starts with a rigorously specified set of assumptions regarding an actual or proposed system of interest, but, unlike deduction, simulation does not prove theorems with generality. Instead, simulation generates data suitable for analysis by induction. In contrast to typical induction, however, the simulated data come from controlled experiments rather than from direct measurements of the real world. Consequently, simulation differs from standard deduction and induction in both its implementation and its goals. Simulation permits increased understanding of systems through controlled computational experiments. In particular, agent-based modeling can be used to investigate how macro-level effects and social behaviors arise from the micro processes of interactions among many agents.

A general phenomenon exhibited by ABMs is large events. The logic of the central limit theorem states that the sum of a collection of random events produces a bell curve. In such cases, deviations from the mean, large or small, are rare. In ABMs, random effects can accumulate. These accumulations can be more than additive or even multiplicative. The result can be huge cascades: forest fires, riots, stock market crashes, epidemics, and even the collapse of governments. Moreover, ABMs can be used to estimate the probability of such extreme events (Gladwell, 2000).

In summary, agent-based modeling applied to social, cultural, and organizational processes uses concepts and tools from social science and computer science. It represents a methodological approach that could ultimately permit three important developments: (1) the rigorous testing, refinement, and extension of existing theories that have proved to be difficult to formulate and evaluate using standard mathematical and statistical tools; (2) a deeper, more integrated understanding of fundamental causal mechanisms in multiagent systems, whose study is currently hampered by artificial disciplinary boundaries; and (3) a tool for exploration and evaluation of the potential impact of course of action and policy alternatives.

State of the Art

The goals pursued by ABM researchers take six general forms: empirical description, empirical prediction, normative analysis, behavioral understanding, heuristic understanding, and methodological advancement.

Researchers pursuing empirical description ask: Why have particular macro-level structures and social behaviors evolved and persisted, even when there is little top-down control? Examples include trade networks, socially accepted monies, mutual cooperation based on reciprocity, and social norms. Agent-based modelers seek causal explanations grounded in the repeated interactions of agents operating in specified environments. In particular, they ask whether particular types of observed macro-level regularities can be reliably generated from particular types of ABMs.

ABM researchers interested in empirical prediction ask: If this history of events were to take place, what would be the likely future consequences? These types of questions can be pursued in the context of ABM frameworks in which the modeler builds in scenarios of interest, introduces agents with realistic degrees of adaptability, and then tests to see how the agents react over time as the scenarios unfold.

A third goal is normative analysis: How can ABMs be used as laboratories for the discovery of good rules of operation? ABM researchers pursuing this objective are interested in evaluating whether policies and institutional arrangements proposed for various types of social systems result in desirable system performance over time. Examples include the design of auction systems, voting rules, and law enforcement practices.

A fourth goal is the understanding of diverse behaviors. The performance of markets, democracies, and even traffic laws varies around the globe. We chalk up these differences to cultural or behavioral differences, but we lack a calculus of culture. We do not know whether or not slight variations in behavioral rules will result over time in widely divergent outcomes. ABM can help to illuminate the accumulation of effects from diverse behavioral rules and the extent to which slight variations in behavioral rules have substantial effects.

This goal overlaps with both the empirical goal of prediction and the normative goal of good operational design, yet it is distinct from each. It necessitates a fundamental shift in how one looks at social systems. Standard models typically focus on the means of variables—the average expected outcomes. Yet often in agent-based modeling, the tail of the distribution wags the dog, so to speak. For example, to predict the likelihood of a riot, what matters most is not the average level of civil unrest among a population but the percentage of people enraged enough to trigger a riot through disruptive behavior that others will then mimic.

A fifth goal is heuristic understanding: How can greater insight be attained about the fundamental causal mechanisms in social systems? Even if the assumptions used to model a social system are simple, the consequences can be far from obvious if the system is composed of many interacting agents. The macro-level effects of interacting agents are often surprising because it can be hard to anticipate the full consequences of even simple forms of interaction. For example, one of the earliest and most elegant ABMs—the city segregation (or "tipping") model developed by Nobel laureate Thomas Schelling (1978, pp. 147–155)—demonstrates how residential segregation can emerge from individual choices even when everyone is fairly tolerant.

A sixth goal is methodological advancement: How best can ABM researchers be provided with the methods and tools they need to undertake the rigorous study of social systems through controlled computational experiments? How best can they examine the compatibility of experimentally generated theories with real-world data? ABM researchers are exploring a variety of ways to address these issues, ranging from careful consideration of methodological principles to the practical development of programming, visualization, and validation tools.

Perhaps the most provocative consequence of these methodological advancements is in the area of nonequilibrium science. Much of existing social science research, particularly research relating to organizations and institutions, is predicated on an assumption that systems are in equilibrium. This allows one to compare modeled systems by the equilibria they implement. In contrast, the real world routinely exhibits a wide variety of nonequilibrium phenomena, such as abrupt transitions, crashes, and path dependencies. Agent-based modeling permits researchers to study out-of-equilibrium behaviors, hence it should ultimately help them to understand, evaluate, and characterize these phenomena (Arthur, 2006; Page, 2008).

ABM Structural Properties

ABMs can be structurally specified in widely diverse ways. Five distinguishing structural properties of particular interest are as follows: the number of agents, the basic manner in which agents are represented, the cognitive sophistication of the agents, the social sophistication of the agents, and whether or not the agents are situated in a relational or spatial grid.

Table 6-2 illustrates how these five structural properties differ across four classes of models currently used by ABM researchers: cognitive ABMs; dynamic network ABMs; cellular automaton ABMs; and rule-based ABMs.

The table must be interpreted with some care. One caveat is that, in each model class, the actual level of realism depends on the degree to which agent attributes are based on actual data and the degree to which agent behavioral rules faithfully represent real-world processes. Another caveat is that, in principle, ABMs are ubiquitously applicable to problems that involve two or more agents whose behavior depends, at least in part,

TABLE 6-2 Structural Differences Commonly Exhibited by Agent-Based Models

Model	Number of Agents	Agent Representation	Cognitive Sophistication	Social Sophistication	Grid Based
Cognitive	Few	Rules	High	Low	No
Dynamic-network	Many	Equations + rules	Moderate	High	No
Cellular automata	Few to many	Equations or rules	Low	Low	Yes
Rule-based	Few to many	Rules	Low	Low	Often

on each other. Thus, differences commonly exhibited in current use do not necessarily reflect fundamental differences in capabilities. For example, the fact that cognitive ABMs currently tend to comprise relatively few highly sophisticated cognitive agents is due to processing power limitations and not to modeling or coding limitations per se.

We now consider the ABM structural properties in greater depth.

Number of Agents and Cognitive Sophistication

As a general rule, the cognitive sophistication of the agents in an ABM is inversely proportional to the number of agents. On one hand, a model could comprise from 2 to 10 very cognitively sophisticated agents doing very in-depth knowledge-intensive tasks. In such a model, interactions among agents would typically be prescribed by protocols for interaction and by hierarchical precedents regarding who does what. Such models are more common in computer science and engineering; illustrative models are those involving BRAHMS, Soar, ACT-R, or Neural Networks (see Chapter 5). Models of this type are valuable for studying aspects of small team behavior, including modeling small adversarial teams. However, they are generally not appropriate for more societal or cultural issues, such as state failure, crowd control, or adaptation in terrorist networks.

On the other hand, an ABM could comprise tens of thousands or millions of cognitively simplistic agents doing relatively simple tasks. In this case, interactions among agents would be the result of the agents meeting and greeting each other, trying to occupy the same space, or exchanging or consuming resources. Such models are more common in biology, physics, and the social and organizational sciences; illustrative models are those involving SWARM, REPAST, or MASON. Such models are often used to examine whether the complexity of real-world social processes can arise from agent interactions rather than from the complexity of individual agents. Models of this type are valuable for academic research, suggesting possible scenarios, providing very high-level guidance, and studying migration and crowd control.

Mid-range models often are comprised of 10 to 10,000 agents with moderately sophisticated learning capabilities. Such models are often written directly in high-level languages like C++ for reasons of processing speed. In this case, interactions among agents are the result of deliberate decision-making and learning processes that are strongly informed by empirical data. Agent behavior can be quite detailed, such as a detailed mapping of activities taken in a day and the influence of a bioattack on those activities. Such models are increasingly used in such application areas as epidemiology, state failure assessment, crowd control, organizational design, adversarial modeling, and counterterrorism. Models of this type, particularly when they are strongly tied to data and employ socially sophisticated agents, can provide actionable intelligence in the areas listed.

Social Sophistication

As a general rule, the social sophistication of an ABM varies with the number of agents, with the level of sophistication being highest for mid-size populations and lowest for models with only a few agents or with millions of agents. Realistic social behavior requires a certain level of cognitive sophistication (Carley and Newell, 1994). However, many social issues do not emerge as relevant until intermediate-sized social groupings are considered.

Typically, models with either a few agents or with millions of agents impose assumptions that limit the use of such models for examining social issues. For example, in ABMs with only a few cognitively sophisticated agents, social factors are typically either ignored or prescribed in terms of a communication and command hierarchy, implying that the structure governing social behavior is time invariant. For example, in models comprising millions of cognitively simplistic agents, real social networks are typically not modeled. Instead, agents are differentiated using from two to five sociodemographic dimensions and "network" links are characterized by nearness in a grid. As such, these models are insufficient for modeling terrorist networks. This high level of simplification means that such models rarely generate actionable intelligence.[3]

Agents in Grids

In many models, particularly those comprising large numbers of cognitively simplistic agents, the agents are generally constrained to interact within some form of grid structure. There are two main ways in which agents are laid out in grids: relational and spatial. In a relational approach, each grid cell represents an agent, and the attributes and actions of this agent are determined in part by the attributes and actions of agents in nearby cells. In contrast, in a spatial approach, each grid cell is a location that agents move through (right-left, up-down). Agents consume or leave resources in the cells they occupy, and they interact with the agents they meet in the same or neighboring cells.

The classic example of an ABM using a grid is John Conway's Game of Life (see Gardner, 1970). Today, many grid-based ABMs are barely more

---

[3] See http://www.econ.iastate.edu/tesfatsi/anetwork for annotated pointers to ABM research on the formation and evolution of social networks.

complex than the original Game of Life, although modern systems use a doughnut-shaped grid (torus) rather than a rectangular grid to avoid edge effects. Grid-based modeling facilitates rapid model development and is supported by agent-based modeling toolkits such as SWARM, REPAST, and Netlogo. However, it is not adequate to support the realistic modeling of four-dimensional social behavior (in space and time) or to capture social network effects in any great depth.

In particular, then, ABM frameworks with grid layouts are currently of limited utility for modeling military situations requiring high levels of realism. Space-time action sequences and sophisticated social network effects are important factors that need to be carefully accounted for in a variety of military models. For example, they are needed if one is to build a model of adversarial behavior in an urban setting in which an adversary can move from subways to rooftops and can receive shelter from friends.

ABM and Learning

In ABMs, the agents learn. A major issue is how to model the minds of the cognitive agents who populate ABM frameworks.[4] Should these minds be viewed as logic machines for planning and reasoning with appended data filing cabinets, the traditional artificial intelligence view (Franklin, 1995), or should these minds be viewed as controllers for embodied activity in keeping with the artificial life view (Clark, 1997)?

On one hand, as with any simulation system, if the purpose of an ABM framework is to determine an optimal design for a fully automated process, there is no particular reason why agent cognition should mimic that of real people. Indeed, this could be positively detrimental to good process performance. On the other hand, if the purpose is to replicate and forecast human social behavior, then mimicry of real human behavior might be essential to ensure predictive content.

As detailed in Brenner (2006), ABM researchers are increasingly moving away from the unconsidered adoption of off-the-shelf machine learning representations, such as conventionally specified genetic algorithms and reinforcement learning algorithms. Some ABM researchers are systematically investigating the performance of alternative learning representations in various multiagent decision contexts. Others are attempting to calibrate their learning representations to empirical decision-making data and human subject experimental data.

---

[4] See http://www.econ.iastate.edu/tesfatsi/aemind.htm for annotated pointers to ABM research on agent learning representation.

ABM and Social Networks

Social networks comprise one of the more active research areas in agent-based modeling.[5] One critical issue is the manner in which social networks are determined through deliberative choice of partners as well as by chance and necessity. For example, in economics a key concern has been the emergence of trade networks among collections of buyers and sellers who determine their trade partners adaptively, on the basis of past experiences with these partners (Tesfatsion, 1997).

A second critical issue concerns the management of a social network for a common (team) goal when participant agents have different motivations for when and how to interact. An example would be the optimal organization of a corporate enterprise comprising multiple divisions.

A third critical issue concerns the disruption of harmful social networks. For example, research on terrorist networks suggests that they are difficult to destabilize when they have a cellular organization, with participant agents in communication only on a need-to-know or similarity basis.

For each of these issues, it is important to consider the extent to which social networks affect the ability to predict social and cultural outcomes with accuracy based on observable structural conditions and institutional arrangements. More precisely, to what extent and with what fidelity does a modeler need to capture social network effects together with structural and institutional effects in order to achieve satisfactory predictive power?

For illustration, consider the case of markets. Some types of markets can be expected to display only weak social interaction effects, for example, pool-based wholesale electric power markets under the strong control of a system operator. In this case, the structural aspects of the market (e.g., numbers of buyers and sellers, costs, capacities) and the institutional aspects of the market (e.g., the legal contractual arrangements governing market participation) will presumably be the primary determinants of market outcomes. Other types of markets can be expected to display strong social interaction effects. This is true for labor markets, in which work contracts are highly incomplete and outcomes are strongly dependent on work site interactions between workers and employers. For such a market, given any single structural and institutional starting point, there will presumably be a wide variety of possible outcomes based partly on random social interaction effects.

As another example, consider modeling of state failure. A model that examines only the social network among the various stakeholders will not be able to predict state failure with accuracy, nor will a model that

---

[5]See http://www.econ.iastate.edu/tesfatsi/anetwork.htm for annotated pointers to ABM research on interaction networks. See, also, the volume of readings edited by Breiger et al. (National Research Council, 2003) and the surveys by Vriend (2006) and Wilhite (2006).

examines only the resources or actions available to the different participant actors. However, by combining these considerations into a single model in which agents are encouraged or discouraged from taking actions by those to whom they are linked, state failure can be better predicted.

In summary, applications that require the generation of actionable intelligence in social situations will generally require careful consideration of social network effects along with structural and institutional effects.

ABM Development Issues

A computational laboratory (CL) is a framework that permits the study of complex systems by means of controlled, replicable, computational experiments using an integrated array of specialized software tools.[6] In particular, CLs providing a variety of agent-based tools facilitate the integrated development of ABMs.

A number of critical issues arise regarding the development of CLs for ABM applications. For example, should a separate CL be constructed for each application, or should researchers strive for general multifaceted platforms? How can experimental findings be effectively communicated to other researchers by means of descriptive statistics and graphical visualizations without information overload? How might these findings be verified and validated by comparisons with output and data obtained from other sources? How might they tell researchers to look at existing data in different, more dynamic ways? A particularly important unresolved issue is the need to ensure that findings from CL experiments reflect fundamental aspects of a considered problem and not simply the peculiarities of the particular hardware or software platform used to implement the experiments.

CLs clearly ease the entry barrier for researchers wishing to use ABM in various problem applications. However, it is important to keep in mind that the use of such integrated development environments permits even novice simulators to build seemingly powerful ABMs in the course of a few months. As a result, we are now seeing thousands of small systems being built by individuals or small teams with little or no training in simulation, and the models are being used to inform critical decision making and policy. In the absence of any accepted criteria for validation (see discussion below and in Chapter 8), it is impossible to judge whether these models are adequate for their intended purposes.

On the positive side, the use of ABMs enables the analyst to systematically consider the interaction among more factors and so base decisions on

---

[6]See http://www.econ.iastate.edu/tesfatsi/acomplab.htm for annotated pointers to ABM research on CLs. See also Dibble (2006) for a detailed discussion of CL use for spatial agent-based modeling.

a more thorough analysis. On the negative side, the development of ABMs by those not trained in simulation means that the results of the models are often misinterpreted and classic mistakes are often made, which cause the results from the models to reflect incorrect simulation practices rather than interactions among the modeled factors.

In summary, great care must be taken in the development of ABM frameworks. Although CLs permit rapid individual development of ABM frameworks, detailed, sophisticated ABM frameworks that produce actionable results often need to be developed by a team working collectively for three to five years. It makes sense to use separate teams for data gathering, validation, and usability testing, as each of these areas requires different types of scientific skills. In addition, the team building the model often needs to employ many of the same techniques for development that are used in system engineering.

Relevance, Limitations, and Future Directions

Military operations intrinsically involve military engagements with rival forces, and forces intrinsically involve equipment and human participants in dynamic motion over geographic terrains. Stated more abstractly, military operations are complex dynamic processes involving multiple, heterogeneous, strategically interacting agents operating through time over spatial landscapes.

Framed in this way, the modeling of military operations is seen to be precisely the type of modeling challenge that agent-based modeling is designed to address. What, specifically, are its key advantages for military applications?

First and foremost, agent-based modeling provides flexibility. Agents can be modeled as autonomously driven entities operating on their own time scales in fulfillment of individual or group goals. Their methods of operation can be constrained by idiosyncratic personal and cultural considerations. They can be equipped with social communication capabilities permitting adaptive information acquisition and transmission. They can survive or not depending on their ability both to secure life-sustaining resources and to manage or prevent life-threatening situations.

In particular, agents in ABM virtual worlds can be designed to live in their world with the same degree of flexibility as their real-world counterparts. System behaviors emerge from the bottom up, through the decentralized actions of autonomous agents situated in space and time. This contrasts with a command and control approach to modeling in which outcomes are enforced from the top down. A top-down approach requires that every contingency be anticipated. A bottom-up approach need not anticipate all contingencies, but it must have a sufficiently rich behavioral repertoire at the individual level so that the system can respond to whatever situation arises. This is precisely the type of modeling flexibility that agent-based modeling provides.

Also, agent-based modeling is particularly well suited for studying information diffusion and the evolution of norms, trust, and reputation. The classical game theory approach to these issues seeks to explain behavior on the basis of individual rationality considerations, such as explaining the evolution of norms in terms of anticipations of future reciprocity (Gintis, 2000). In contrast, the ABM approach tends to place equal or greater stress on peer emulation, parental mimicry, and other socialization forces thought to underlie the transmission of culture.[7]

ABMs have been used to evaluate the likelihood that the general attitudes of the population would become more pro or con regarding the United States in the face of elections and changes in leadership. ABMs have also been used to forecast state failure, regime change, and the emergence of corruption in various nation-states (Popp et al., 2006).

In summary, the issue is not whether agent-based modeling is relevant for modern military operations: it clearly is. The issue is whether it has reached a sufficient stage of development to provide practical support for military operations.

Major Limitations

ABM frameworks as currently constructed have limitations that could affect their ability to meet critical military needs. This section discusses some of these limitations.

Degree of Realism

The value of any simulation, including any ABM simulation, is partly tied to the level of realism in the model. Any simulation system is a model and so should be less complex than the real world. However, oversimplification results in models so high level or so incorrect that the results can be misinterpreted and so should not be used for policy setting or decision making. The rule of thumb is to make the model only as complicated as it needs to be to address the issue of concern and to achieve the necessary level of fidelity.

Adding more rules or equations that increase the realism of the resulting model should presumably increase its usefulness for decision making. Yet opponents often argue that the more equations or rules, the worse the

---

[7]See http://www.econ.iastate.edu/tesfatsi/asocnorm.htm for annotated pointers to ABM research on the evolution of social norms. See also Young (2006) for a proposed ABM methodology for studying the long-run evolution of social norms.

model. Arguments include appeals to parsimony, Occam's razor, under-standability, and so on. A typical argument is that, as the model increases in complexity (number of variables and rules/equations), it becomes increasingly likely that the model can be made to fit any possible outcome. ("Over-fitting" is discussed further in Chapter 9.)

This argument derives from econometrics, in which, as the ratio of parameters to data increases, ultimately the data can be completely and perfectly modeled. This argument, however, is not directly applicable to ABMs. In them the addition of new rules and equations serves to increase the number of outcomes or dependent variables (data) that can be generated; the data are not given a priori as in econometrics. Moreover, the addition of empirically based rules and equations can increase the plausibility of these generated outcomes by reducing the possibility of implausible results.

The realism of ABMs can be increased, and their military value increased, as they are linked to real data. Most groups that build them have contrasted, at best, the results of one dependent variable with real data. Only a few agent-based models, such as some recently created for the Defense Advanced Research Projects Agency or the BioWar system, use massive amounts of real data to set the input specifications of the models and other data to validate the system. In general, this requires the linking of the models to database systems. The key technical challenge here is that, as the ontology in the database changes, the model needs to be augmented. There are currently no tools to facilitate such changes. A second challenge is that, for validation, it is important to have the model produce data in the same form as the real data, that is, to create a comparable database. There are currently no standardized tools for doing statistical comparison of data in two identically structured databases.

Model Trade-Offs

ABMs using cognitively sophisticated agents tend to require the use of knowledge engineering techniques. Such models tend to be special purpose and permit minimal reuse. The key value of such models is to take the place of human teams in war-gaming situations, equipment testing, and design situations and to evaluate processes that facilitate team behavior. In general, these models use various cognitive architectures with multiagent components added and so are often limited to only a small number of agents. Their strength is looking at detailed task-related behavior. As previously noted, such models tend to use predefined social interactions. This limits their use in war games because the ABMs do not exhibit a full range of adaptive interaction but model only limited task-based communications and actions.

The typical grid-based ABMs, with millions of cognitively unsophisticated agents, are generally useful only for high-level explorations of general concepts. They are valuable for starting groups to think outside the box and for provoking discussions. These models are rarely sophisticated enough to be used as an adaptive adversary in war-gaming or for evaluating task-based behavior. The strength of these models is their ability to look at population-level trends resulting from local action. As such, they show promise in such areas as marketing, impact of psychological operations, information diffusion studies, and disease transmission studies. Rarely do such models generate actionable intelligence.

Now consider dynamic network ABMs tied to empirical data. Such models utilize agents with moderate levels of cognitive sophistication and high levels of social sophistication. This results in models that can be used for war-gaming to look at adaptive adversaries. Given current technology, this combination results in models that can handle more agents but that run more slowly. The strength of these models lies in representing and reasoning about fairly large-scale units, such as the army's unit of action, cities at 20 percent population, or terrorist networks. The added cognitive and social sophistication inherent in these models makes it possible to produce actionable results. However, getting a model to the point of producing actionable results takes a multiperson, multiyear data collection effort on top of a multiyear model development effort.

Modeling of Actions

One of the key factors limiting ABMs from a military perspective is the modeling of actions. Currently actions can be modeled at a very high level (pro-con, hostile, friendly, or neutral) or at a very detailed level (fire a particular weapon). There is neither a middle ground nor a hierarchy relating actions at one level to another. Therefore, efforts to model actions tend to be either very generic or single use. A basic ontology of actions is needed for the state of the art to advance.

Research and Development Requirements

Several requirements can be identified for the further development of ABMs that could be of use in military settings. The next section discusses tool development, data farming, linkages of agent-based modeling to other modeling efforts, and the development of the human resources and expertise needed to support ABM development.

Tool Development

Key advances and applicability to military modeling require agent-based modeling and network analysis techniques to be integrated into tool chains. For example, pattern discovery techniques can be used to derive equations from historical data that can then be used in ABMs to evolve future systems. ABM techniques can be used to evaluate courses of action and to suggest areas for further data collection. Combining these techniques will enable new types of problems to be solved; for example, combining social network metrics with pattern discovery techniques is the key to building an understanding of how networks grow and evolve.

This is not to suggest that the military should move to large integrated behavioral models—quite the contrary. What is needed is increased interoperability of the tools. The development of ABM CLs and the explosion of network analytic tools are putting social behavioral modeling into the hands of the masses. Moreover, these trends are leading to the development of many small, single-purpose tools. This should be taken advantage of by encouraging interoperability (this is also discussed further in Chapter 8).

It is important to note that it would not be feasible to require all tools to be written in a single language or to require the use of a single framework; rather, the solution needs to enable the integration of models not only from diverse domains but also in diverse languages. Multiple models, visualization tools, and the like should be available to address diverse problems, but in such a way that data (real and virtual) can be shared easily among the various tools.

There are a variety of things needed to support such interoperability. Standards for the interchange of relational data need to be developed. Behavioral modeling tools need to be web enabled, and XML input-output (IO) languages need to be developed. A uniform vocabulary for describing relational data also needs to be developed; this is particularly critical because the tools and metrics are coming out of at least 20 different scientific fields.[8]

For defense and intelligence applications, common platforms and data sharing standards need to be explored and developed so that tools written in the unclassified realm can be rapidly moved, without complete redesign, to the classified realm. Enabling interoperability and providing a platform and common ontologies for these tools will enable novel problems to be more rapidly addressed by regrouping existing models. It will also enable various subject matter experts to interact through the interaction of their models. In turn, this will enable a broader approach to problems, reduce the likelihood of biased solutions, and facilitate rapid development and deployment.

---

[8]These fields include anthropology, sociology, psychology, organization science, marketing, physics, electrical engineering, geology, ecology, economics, biology, bioinformatics, health services, forensics, artificial intelligence, robotics, computer science, mathematics, statistics, information systems, medicine, civil engineering, communication, and rhetoric.

Current tools are either very data-greedy or become more valuable as they are linked to real data. However, there is a dearth of relevant data currently available in clean preprocessed form. Thus, to reduce the time analysts spend on data collection and to increase the time they spend on analysis, automated and semiautomated tools for data gathering, cleaning, and sharing are needed. Such tools should include natural language processing tools for extracting relational data from audio and text sources, "web-scraping" tools, automatic ontology generators, and visual interpretation tools to extract network data from photographs and visual images.

Appropriate subtools for node identification, entity extraction, thesaurus creation, and other functions are also needed. The development and availability of these tools in an interoperable environment are critical for providing masses of data that can be used for model tuning and validation. Moreover, these tools reduce time spent on data collection and thereby free the analysts' time for analysis. More rapid data collection would also mean the availability of more datasets for doing meta-analyses, thereby enabling improvements in the theoretical foundations of the field and in the understanding of social behaviors. Finally, these tools are essential for providing the wealth of data needed by social behavioral models to make reasonable forecasts or to provide reasonably accurate analyses of situations and organizations.

Improved speed for many of the algorithms could be provided by computer architectures designed for relational data or by the use of special integrated circuits with embedded versions of the less scalable algorithms. Note this would enable a speed savings beyond that afforded by the use of current vector technology. Such technology would facilitate faster processing and enable more real-time solutions, particularly for large-scale networks.

To reduce the "art" aspect of interpretation in this field, a living archive of collected network data is needed, replete with information on metrics for the nodes in each dataset. Such an archive could be used to set context information. For example, such information could be used to evaluate whether the density of particular networks is exceptionally high or low or to identify exceptional values of connectedness of individuals. Such an archive would facilitate meta-analysis and comparative analysis. This is critical for improving the theoretical foundations of the field as well as for the understanding of social behavior.

Forecasting and Possibility Analysis

Of the models described here, those that have shown the most promise in terms of forecasting are the voting models, the dynamic network models (that combine agent-based technology and meta-matrix of relations), and the social influence models. These models have had limited success in forecasting voting outcomes, changes in beliefs and attitudes at the macro level, and identifying emergent new leaders. For other modeling techniques, including the ABM and system dynamic techniques for complexity modeling, the models are best at providing insight into the space of possibilities, that is, demonstrating what possible futures might exist and their relative likelihood. However, for these models to provide an adequate map of the possibilities (a reasonable response surface), the models need to be run a vast number of times under diverse scenarios; hence, as is discussed in the next section, there is a need for placing these models in a data farming environment.

One question that arises is: How can these models be made more predictive? This topic, in and of itself, is quite complex and a full treatment is beyond the scope of this study. However, several factors are worth noting. As more of these models are placed in data farming environments, statistical tools are developed for mining the vast data so generated, and repositories of meta-matrices are developed and shared with scientists for testing and validating, one can expect that many of these models will become more reliable in their forecasts. However, there will still be many classes of social phenomena for which prediction, of the form used in engineering and physics, will simply not be possible due to the lack of stationarity in the underlying social processes, the paucity of data, and the lack of continuity in key variables.

A second question often arises regarding the concern that, if the models are truly predictive, the mere act of making a prediction public will cause actors to change their behaviors and so alter the outcome. While this issue is addressed in other sections of this report, several key factors directly related to the nature of the models described here are worth mentioning. For most of the models described here, other than the simple voting models, making the models transparent to the public (so that others can infer the predictions) or making the predictions themselves public is not likely to invalidate the predictions. There are three basic reasons for this: lack of temporal forecasting, level of specificity, and hyper-confluence. Temporal forecasting tends to be weak and predictions are often vague in terms of *when* something will occur; rather than point predictions, most predictions are of the form "A will likely occur after B" or "at some time in the future more than two weeks but less than two years from now." Most models produce rather general results, such as that a state will fail, civil violence is likely to erupt, or corruption will increase, rather than the more specific "the state will fail due to a regime change where General X takes over" or "civil violence will take the form of riots in these five cities" or "corruption will increase the most in the area of infrastructure development in county X." Finally, most models generate a prediction due to hyper-confluence, that is, the strongest predictions are those for which there are a large number of interconnected causes that weave together in complex ways. But single actors can best counter a specific event that is likely to occur at a specific time with only one or two actions or activities. Even with sufficient research funding, improved theory, and available data to overcome the issues of vague temporal forecasting and lack of specificity; the problem of hyper-confluence will remain. That is one of the key reasons why social and behavioral models need to be driven by the science of the possible, rather than the traditional science of point predictions involved in traditional physical science and engineering models.

Data Farming

ABMs designed for applied settings need to be placed in data farming environments. These environments need to be augmented with special-purpose tools for running massive virtual experiments. These tools should enable improved visualization and analysis and facilitate the development of semiautomated response surface generators. Current data farming tools often are cumbersome to use, require code modification of the ABM, and are limited by the processor speed and storage capabilities of the machines that they run on.

In order for ABM frameworks to run routinely in data farming environments, more flexible environments need to be developed and made easily available to researchers. Moreover, ABM frameworks need to be developed with wrappers,[9] so that they can be placed in these environments. Standardized IO formats need to be developed. By routinely placing ABM frameworks in a data farming environment, a better understanding of the space of possibilities predicted by the frameworks will be derived. This will enable ABM frameworks to better support policy and decision making.

Currently, when ABM frameworks are used to inform policy and critical decisions, they are typically run only a few times in carefully controlled computational experiments. While this approach enables the analyst to explore more possibilities more systematically than not using a simulation, it still leaves open the possibility that errors might be made if the results are generalized beyond the scope of the experiment. By placing ABM frameworks in a data farming environment, the number of computational experiments conducted, the space of possibilities examined, and the scope of

---

[9]"A wrapper is a software layer used to change the interface of a component or to give new properties, such as fault tolerance or security, to the interaction between components. Software wrappers are often used to glue existing subsystems into a larger system with new properties and functions. The wrappers know the protocols needed to make the subsystems work together, even if they were not originally designed for a common purpose" (Webber, 1997, p. 1).

analyzed conditions can be expanded, often by several orders of magnitude, thus providing a stronger basis for decision making. Furthermore, once an ABM framework has been validated, the response surface equivalent can be used as a "rapid" model in training situations in which the users do not have time to wait for an ABM experiment to finish running.

Cross-Disciplinary Initiatives

Another avenue that may promote major breakthroughs is the linkage of ABM social behavioral modeling to gaming environments, particularly online multiplayer games such as Everquest and America's Army (see Chapter 7). Research initiatives that explore the link of ABM social behavioral modeling to gaming tools may be valuable. Possible research areas include using agent-based modeling to explore the realism of the social behaviors exhibited in gaming models; using it to provide flexible opponents or to make the apparent number of game players larger and so force players to think about group scale issues; and using agent-based modeling to track and analyze game behaviors using dynamic network analysis techniques. Key benefits here would be improved training tools and visual what-if scenario evaluation.

As previously noted, additional ABM development needs to be done in a number of areas. These include attachment of ABM frameworks to data streams, improved ABM visualization, metric ABM robustness studies, and so on. Moving ahead in these areas will require linking social networks to other types of data, such as location and event information, and linking diffusion theory to other forms of theory, such as action and cultural theory. This will require the funding of both basic and applied research. It will also require an increased recognition for, and acceptance of, applied social science research in universities.

Currently there are a number of funded research efforts in the areas of cultural modeling, geospatial link analysis, and adversarial modeling, all of which are supporting work along these lines. A key to much of this work is that it combines dynamic network analysis with geospatial reasoning or anthropological data-gathering techniques. Much of this work is applied, directed at providing usable systems in several years. This is a positive development, particularly when such modeling efforts are based on strong empirical and theoretical foundations. However, there is still a huge amount of basic research to be done in such areas as the development of an ontology for tasks, a unified model of culture, or even a shared definition of culture. Relatively little research funding is being directed to the basic research questions in this area.

The key here is not simply to invest in the social sciences but to invest in the mathematical and computational social sciences to engender the development of work that will support defense needs. The benefit will be an improved understanding of basic social and cultural phenomena. Another benefit will be a decrease in the development of misleading models that appear to be social but that are not theoretically or empirically sound.

At the same time, most of the research community, particularly in the social sciences, is not focusing on strongly applied problems. The mere idea of hard deliverables, while accepted as common practice in engineering and computer science, is contrary to the basic culture of most social science departments. Thus, while there is a strong need for quantitative social science modeling on defense issues, there is a dearth of social scientists involved in and trained to do applied work.

Building Expertise

The lack of highly trained professionals is a key difficulty in this area. Universities need to expand their undergraduate social science curricula to include more of the mathematical and computational social sciences. In particular, undergraduate courses should be routinely taught that cover SNA and agent-based modeling, and that permit the mastery of ABM programming tools. Universities need to encourage and facilitate applied research. New curricula are needed that have an engineering style but that are focused on social and policy applications. Master's programs that combine social and computational science need to be developed. Military universities, such as West Point and the Naval Postgraduate School, should also offer social network courses and possibly ABM courses, particularly those for evolving networks, and they should integrate dynamic network measures of shared situation awareness, leadership, and power into the standard curriculum.

The development of these curricula and degree programs is vital to the nation's intellectual strength in order to remain at the forefront in this area. The clear benefit of these programs will be a stronger workforce of computational social analysts capable of developing and using social behavioral models.

Analysts engaging in ABM but trained in computer science, engineering, or physics should work in teams with social scientists to avoid duplicating work already done or making commonsense assumptions about social processes that have no empirical bases. Corporations need to provide time and resources for selected personnel to become jointly trained in computer and social science, either by increasing the number of personnel sent to master's programs, bringing in relevant faculty to teach short courses, or engaging in more joint research with universities as equal partners (in which the university provides the missing skill, social or computational). The key advantage of teaming is that it will enable improved model development and will serve as a stop-gap until more computational social analysts are trained.

Expected Outcomes

Across the board, success in the activities outlined above would facilitate the rapid development and deployment of agent-based modeling. The advantage is that it enables systematic reasoning about various courses of action in a wide range of complex environments. More courses of action could be evaluated in less time and more systematically than is done with conventional table-top war-gaming or current non-computer-assisted analysis of relational data. The dynamic social network and ABM tools outlined above reduce the time spent on data processing and increase time spent on analysis and interpretation. They would facilitate what-if analysis and could ultimately support near-real-time what-if analysis in the field. This would be a valuable force multiplier.

In summary, the activities listed above would increase the maturity of the modeling field, improve scientific theory, facilitate rapid linking of computational models to empirical data, particularly network data in a unified reasoning framework to solve novel problems, and encourage new discoveries. These activities would also promote the development of a new science that combines computation and society, just as the previous joining of computer science, design, and psychology led to the new science of human-computer interaction.

REFERENCES

Arrow, K.J. (1951). *Social choice and individual values.* Hoboken, NJ: John Wiley & Sons.

Arthur, W.B. (2006). Out-of-equilibrium economics and agent-based modeling. In L. Tesfatsion and K.L. Judd (Eds.), *Handbook of computational economics, volume 2: Agent-based computational economics.* Amsterdam, The Netherlands: Holland/Elsevier.

Axelrod, R. (1997). Advancing the art of simulation in the social sciences. In R. Conte, R. Hegselmann, and P. Terna (Eds.), *Simulating social phenomena* (pp. 21–40). Berlin: Springer.

Borgatti, S.P., and Everett, M.G. (1992). Notions of position in social network analysis. *Sociological Methodology, 22,* 1–35.

Borgatti, S.P., and Everett, M.G. (2006). A graph-theoretic framework for classifying centrality measures. *Social Networks, 28*(4), 466–484.

Borgatti, S.P., and Foster, P. (2003). The network paradigm in organizational research: A review and typology. *Journal of Management, 29*(6), 991–1013.

Borgatti, S.P., Everett, M.G., and Shirey, P. (1990). LS sets, lambda sets and other cohesive subsets. *Social Networks, 12*(4), 337–357.

Breiman, L. (2001). Statistical modeling—The two cultures. *Statistical Science, 16,* 199–231.

Brenner, T. (2006). Agent-learning representation: Advice on modeling economic learning. In L. Tesfatsion and K.L. Judd (Eds.), *Handbook of computational economics, volume 2: Agent-based computational economics.* Amsterdam, The Netherlands: Holland/Elsevier.

Butts, C.T., and Carley, K.M. (in press). Structural change and homeostasis in organizations: A decision-theoretic approach. *Journal of Mathematical Sociology.*

Carley, K.M. (2003). Dynamic network analysis. In National Research Council, *Dynamic social network modeling and analysis: Workshop summary and papers* (pp. 133–145). R. Breiger, K.M. Carley, and P. Pattison (Eds.), Committee on Human Factors. Board on Behavioral, Cognitive, and Sensory Sciences, Division of Behavioral and Social Sciences and Education. Washington, DC: The National Academies Press.

Carley, K.M. (2004). Estimating vulnerabilities in large covert networks using multilevel data. In *Proceedings of the NAACSOS 2004 Conference*, Pittsburgh, PA. Available: http://www.casos.cs.cmu.edu/events/conferences/2004/2004_proceedings/Carley,Kathleen.doc [accessed Feb. 2008].

Carley, K.M., and Newell, A. (1994). The nature of the social agent. *Journal of Mathematical Sociology, 19*(4), 221–262.

Carley, K.M., Columbus, D., DeReno, M., Reminga, J., and Moon, I.-C. (2007). *ORA user's guide.* (Report No. CMU-ISRI-07-115). Carnegie Mellon University School of Computer Science, Institute for Software Research.

Chen, H., and Lynch, K.J. (1992). Automatic construction of networks of concepts characterizing document databases. *IEEE Transactions on Systems, Man, and Cybernetics, 22*(5), 885–902.

Clark, A. (1997). *Being there: Putting brain, body, and world together again.* Cambridge, MA: MIT Press.

Converse, P.E. (1964). The nature of belief systems in mass publics. In D.E. Apter (Ed.), *Ideology and discontent* (pp. 206–261). London, England: Free Press of Glencoe.

Davis, A., Gardner, B.B., and Gardner, M.R. (1941). *Deep south: A social anthropological study of caste and class.* Chicago, IL: University of Chicago Press.

de Toqueville, A. (1835). *Democracy in America.* London, England: Saunders and Otley.

Dibble, C. (2006). Computational laboratories for spatial agent-based models. In L. Tesfatsion, and K.L. Judd (Eds.), *Handbook of computational economics, volume 2: Agent-based computational economics.* Amsterdam, The Netherlands: Holland/Elsevier.

Festinger, L. (1957). *A theory of cognitive dissonance.* Palo Alto, CA: Stanford University Press.

Franklin, S. (1995). *Artificial minds.* Cambridge, MA: MIT Press.

Freeman, L. (2004*). The development of social network analysis: A study in the sociology of science.* Vancouver, British Columbia: Empirical Press.

Freeman, L.C., White, D.R., and Romney, A.K. (Eds.). (1991). *Research methods in social network analysis* (revised edition). Piscataway, NJ: Transaction Press.

Friedman, J. (2005). Popper, Weber, and Hayek: The epistemology and politics of ignorance. *Critical Review, 17*(1–2). Available: http://www.criticalreview.com/2004/pdfs/ignorance_article.pdf [accessed March 2008].

Gardner, M. (1970). The fantastic combinations of John Conway's new solitaire game "Life." *Scientific American, 223,* 120–123.

Getoor, L., and C. Diehl (2005). Link mining: A survey. *SIGKDD Explorations, 7*(2). Available: http://www.cs.umd.edu/~getoor/Publications/getoor-kddexp05.pdf [accessed April 2008].

Getoor, L., Friedman, N., Koller, D., and Taskar, B. (2001). Probabilistic models of relational structure. In *Proceedings of the International Conference on Machine Learning.* Available: http://www.cs.umd.edu/~getoor/Publications/jmlr02.pdf [accessed April 2008].

Getoor, L., Friedman, N., Koller, D., and Taskar, B. (2002). Learning probabilistic models of link structure. *Journal of Machine Learning Research, 3,* 679–707. Available: http://www.jmlr.org/papers/volume3/getoor02a/getoor02a.pdf [accessed Feb. 2008].

Gibbard, A. (1973). Manipulation of voting schemes: A general result. *Econometrica, 41*(4), 587–601.

Gintis, H. (2000). *Game theory evolving: A problem-centered introduction to modeling strategic interaction.* Princeton, NJ: Princeton University Press.

Gladwell, M. (2000). *The tipping point: How little things can make a big difference.* Boston, MA: Little, Brown, and Company.

Goldberg, H.G., and Senator, T.E. (1998). Restructuring databases for knowledge discovery by consolidation and link formation. In *Proceedings of the 1st International Conference on Knowledge Discovery and Data Mining*, Quebec, Montreal. Available: http://citeseer. ist.psu.edu/cache/papers/cs/3649/http:zSzzSzeksl-www.cs.umass.eduzSzailazSzgoldberg-senator.pdf/goldberg95restructuring.pdf [accessed Feb. 2008].

Goldberg, H.G., and Wong, R.W.H. (1998). Restructuring transactional data from link analysis in the FinCEN AI system. In *Proceedings of the 1998 AAAI Fall Symposium on Artificial Intelligence and Link Analysis*. Available: http://kdl.cs.umass.edu/events/aila1998/goldberg-wong.pdf [accessed April 2008].

Green, D., and Shapiro, I. (1994). *Pathologies of rational choice theory.* New Haven, CT: Yale University Press.

Harvey, J.B. (1974). The Abilene paradox and other meditations on management. *Organizational Dynamics, 3*(1).

Hauck, R.V., Atabakhsh, H., Ongvasith, P., Gupta, H., and Chen, H. (2002). COPLINK concept space: An application for criminal intelligence analysis. *IEEE Computer Digital Government Special Issue, 35*(3), 30–37.

Heider, F. (1988). *The notebooks: Balance theory, volume 4.* New York: Springer Verlag.

Jensen, D., and Neville, J. (2002). Linkage and autocorrelation cause feature selection bias in relational learning. In *Proceedings of the Nineteenth International Conference on Machine Learning* (pp. 259–266), Sydney, Australia. Available: http://citeseer.ist.psu.edu/cache/papers/cs/30391/http:zSzzSzkdl.cs.umass.eduzSzpaperszSzjensen-neville-icml2002.pdf/jensen02linkage.pdf [accessed Feb. 2008].

Kollman, K., and Page, S.E. (2006). Computational methods and models of politics. In L. Tesfatsion and K.L. Judd (Eds.), *Handbook of computational economics, volume 2: Agent-based computational economics.* Amsterdam, The Netherlands: Holland/Elsevier.

Kubica, J., Moore, A., Schneider, J., and Yang, Y. (2002). Stochastic link and group detection. In *Proceedings of the Eighteenth National Conference on Artificial Intelligence* (pp. 798–804), Menlo Park, CA: AAAI Press/MIT Press. Available: http://www.cs.cmu.edu/~schneide/AAAI02_GDA.pdf [accessed April 2008].

Kubica, J., Moore, A., and Schneider, J. (2003). Tractable group detection on large link datasets. In X. Wu, A. Tuzhilin, and J. Shavlik (Eds.), *The Third IEEE International Conference on Data Mining* (pp. 573–576). Washington, DC: IEEE Computer Society.

Latané, B. (1996). Dynamic social impact: The creation of culture by communication. *Journal of Communication, 46*, 13–25.

Lee, R. (1998). Automatic information extraction from documents: A tool for intelligence and law enforcement analysts. In *Proceedings of the AAAI Fall Symposium on Artificial Intelligence and Link Analysis* (pp. 63–67). Available: http://citeseer.ist.psu.edu/cache/papers/cs/15373/http:zSzzSzeksl-www.cs.umass.eduzSzailazSzlee.pdf/richard98automatic.pdf [accessed Feb. 2008].

Marchant, T. (2000). Does the Borda rule provide more than a ranking? *Social Choice and Welfare, 17*(3), 381–391.

Milo, R., Shen-Orr, S., Itzkovitz, S., Kashtan, N., Chklovskii, D., and Alon, U. (2002). Network motifs: Simple building blocks of complex networks. *Science, 298*(5594), 824–827.

Moon, I.-C., and Carley, K.M. (2007). Modeling and simulation of terrorist networks in social and geospatial dimensions. *IEEE Intelligent Systems, Special Issue on Social Computing*, 22(5), 40–49.

National Research Council. (2003). *Dynamic social network modeling and analysis: Workshop summary and papers*. R. Breiger, K. Carley, and P. Pattison (Eds.), Committee on Human Factors. Board on Behavioral, Cognitive, and Sensory Sciences, Division of Behavioral and Social Sciences and Education. Washington, DC: The National Academies Press.

Nemeth, C.J. (1986). Differential contributions of majority and minority influence. *Psychological Review, 9*(1), 23–32.

Page, S.E. (2007). *The difference: How the power of diversity creates better groups, firms, schools, and societies*. Princeton, NJ: Princeton University Press.

Page, S.E. (2008). Agent-based models. In L. Blume and S. Durlauf (Eds.), *The new Palgrave dictionary of economics, second edition*. Hampshire, England: Palgrave Macmillan Ltd.

Popp, R., Kaisler, S.H., Allen, D., Cioffi-Revilla, C., Carley, K.M., Azam, M., Russell, A., Choucri, N., and Kugler, J. (2006). *Assessing nation-state instability and failure*. Paper presented at the Aerospace Conference IEEE 2006, March, Big Sky, MT. Available: http://ieeexplore.ieee.org/iel5/11012/34697/01656054.pdf?tp=&isnumber=&arnumber=1656054 [accessed April 2008].

Saari, D.G. (2001). *Decisions and elections: Explaining the unexpected*. Cambridge, England: Cambridge University Press.

Saari, D.G. (2006). Which is better, the Condorcet or Borda winner? *Social Choice and Welfare, 26*(1), 107.

Satterthwaite, M. (1975). Strategy-proofness and Arrow's conditions: Existence and correspondence theorems for voting procedures and social welfare functions. *Journal of Economic Theory, 10*, 187–217.

Schelling, T.C. (1978). *Micromotives and macrobehavior*. New York: W. W. Norton.

Snijders, T. (2001). The statistical evaluation of social network dynamics. In M. Sobel and M. Becker (Eds.), *Social methodology dynamics* (pp. 361–395). Boston and London: Basil Blackwell.

Taskar, B., Abbeel, P., and Koller, D. (2002). *Discriminative probabilistics models for relational data*. Paper presented at the 18th International Conference on Uncertainty in Artificial Intelligence. Available: http://www.biostat.wisc.edu/~page/rmn.pdf [accessed Feb. 2008].

Taskar, B., Wong, M.F., and Koller, D. (2003). *Learning on the test data: Leveraging "unseen" features*. Paper presented at the 20th International Conference on Machine Learning, August, Washington, DC. Available: http://ai.stanford.edu/~koller/Papers/Taskar+al:ICML03.pdf [accessed Feb. 2008].

Tesfatsion, L. (1997). A trade network game with endogenous partner selection. In H. Amman, B. Rustem, and A.B. Whinston (Eds.), *Computational approaches to economic problems* (pp. 249–269). Dordrecht, The Netherlands: Kluwer Academic.

Tsvetovat, M., Reminga, J., and Carley, K.M. (2004). *DYNETML: Interchange format for rich social network data*. (CASOS Technical Report #CMU-ISRI-04-105). Pittsburgh, PA: Carnegie Mellon University, School of Computer Science, Institute for Software Research International.

Vriend, N. (2006). ACE models of edogenous interactions. In L. Tesfatsion and K.L. Judd (Eds.), *Handbook of computational economics, volume 2: Agent-based computational economics*. Amsterdam, The Netherlands: Holland/Elsevier.

Wasserman, S., and Faust, K. (1994). *Social network analysis: Methods and applications*. New York: Cambridge University Press.

Webber, F. (1997). *Software wrappers to support nonstop computing*. Paper presented at the 20th National Information Systems Security Conference, October, Baltimore, MD. Available: http://csrc.nist.gov/nissc/1997/proceedings/730.pdf [accessed Feb. 2008].

Wilhite, A.W. (2006). Economic activity on fixed networks. In L. Tesfatsion and K.L. Judd (Eds.), *Handbook of computational economics, volume 2: Agent-based computational economics*. Amsterdam, The Netherlands: Holland/Elsevier.

Young, H.P. (2006). Social dynamics: Theory and applications. In L. Tesfatsion and K.L. Judd (Eds.), *Handbook of computational economics, volume 2: Agent-based computational economics*. Amsterdam, The Netherlands: Holland/Elsevier.

Zeggelink, E. (1994). Dynamics of structure: An individual oriented approach. *Social Networks, 16*(4), 295–333.

7

Games

This chapter deals with massively multiplayer online games (MMOGs) as a tool for social and organizational modeling. An MMOG is a type of computer game that enables hundreds or thousands of players to simultaneously interact in a game world to which they are connected via the Internet. Typically this kind of game is played in an online, multiplayer-only persistent world (Wikipedia, 2007).[1] These games are a different kind of animal from the models and modeling approaches previously discussed.[2] MMOGs are simultaneously tools that allow players to interact with behavioral models, frameworks for building such models, and laboratories in which these models can be tested.

WHAT ARE MASSIVELY MULTIPLAYER ONLINE GAMES?

Games, particularly videogames, are a recent addition to the modeling and simulation (M&S) tool suite. A videogame is defined as a mental contest,

---

[1]In this chapter, the committee makes references to online communities and game manufacturers. Because of the nature of the games world, scholarly references are not often available, nor are they the most up-to-date or accurate sources of information.

[2]They are also not to be confused with "game theory," a mature research area with strong mathematical foundations established by von Neumann and Morgenstern (1944), but, as noted in Chapter 5, having significant constraints for application to real-world problems, the most notable for this context being the massiveness of the multiplayer community-inhabiting MMOGs; few game theory studies consider more than a handful of independent players or "agents" (Moss, 2001). There are a number of other limitations to the game theory approach, discussed extensively in Chapter 5.

according to certain rules, played with a computer, for entertainment. In the Department of Defense (DoD), the term used is "serious game," which we define in this report as a mental contest, according to certain rules, played with a computer, that uses entertainment to further government or corporate training, education, health, public policy, and strategic communication objectives. An early examination of the potential for using games for modeling, simulation, and analysis originates in the report *Modeling and Simulation: Linking Entertainment and Defense* (National Research Council, 1997).

Games are an interaction medium, a set of engaging and immersive models, and an interactive laboratory with which models and simulations can engage. As an interaction medium, games provide a way for humans to provide input and receive feedback in real time, participating in a running simulation. If the game is immersive enough, this running simulation will fully engage the attention of the game player, and that player will focus on game play to the neglect of the external world. The most commercially successful interactive games cyclically increase the adrenaline levels of the player, while demanding little in the way of mental focus. Games that demand great mental focus do poorly in the market and typically lose player interest. M&S systems that are embedded in such an interaction paradigm need to take this into account if the expectation is to make the M&S system as engaging as a commercial game. The desired outcome for this paradigm is that the M&S system will be so engaging that soldiers will continue to work with the simulation during personal time. Many training simulations derived from the game America's Army (described in more detail below) and similar games belong to this category (Zyda, Mayberry, McCree, and Davis, 2005).

Games also contain a set of engaging and immersive models, models that look very interesting from the perspective of DoD. For example, a large number of meetings start out with the phrase "if only we could build an engaging game like SimCity"—that is, SimNavy for the Navy, SimAir for the Air Force, etc. (Zyda et al., 1998). There are many problems with such statements. The purpose for which the personal computer (PC) game SimCity [3] and The Sims,[4] its direct descendent, was written to entertain

---

[3]SimCity is a PC game in which the user controls several elements of managing a city, such as allocation of funding, distribution of community resources (police and fire stations, schools, etc.), and community layout. Maxis (now Electronic Arts Inc.) released the first version of SimCity in 1989. SimCity was the first game in the Sims franchise, and was the inspiration for other nonviolent open-ended games, such as Sid Meier's Civilization (Electronic Arts Inc., 2006b). SimCity is partly based on Jay Forrester's urban planning model, which is described in Chapter 4 (Electronic Arts Inc., 2007).

[4]The Sims is a PC game in which the user controls individual characters (Sims) in a "virtual dollhouse." The user is responsible for managing day-to-day needs of the Sims, such as their need for fun, hygiene, food, rest, and social activity.

and engage the game player. No real attempt was made in that game or its
successors to accurately model the real world, nor was there any attempt
to verify, validate, or accredit (VV&A) the results of that game—it is pure
entertainment. It does, however, suggest a way in which one can develop
potential outcomes or possibility spaces that can then be considered for
further analysis. Probably the most interesting aspect of games like SimCity
(http://simcity3000unlimited.ea.com/us/guide/), The Sims (http://thesims.
ea.com/), and Civilization IV (http://www.2kgames.com/civ4/home.htm) is
that these games were built for relatively small amounts of money and on
schedule, and they still perform as extremely successful entertainment. DoD
M&S programs with budgets two orders of magnitude larger have failed to
deliver even a tenth of the capability to create a space of potential outcomes
for consideration (Bennington, 1995).

Games are also an interactive laboratory with which models and simulations can engage. They can play the role in social and organizational
modeling that linear accelerators play in particle physics—testbeds built and
used to perform experiments and analyze results (Carley, Moon, Schneider,
and Shigiltchoff, 2005). Like linear accelerators, MMOGs are expensive
to build. The costs of successful immersive game development run from
$8 million for the first two years of game development for a Spartan effort
like America's Army to more than $100 million to develop an MMOG and
its infrastructure.

To use a game as an interactive laboratory, it must be built or acquired
before experiments can be performed with it. If the intention is to connect
a social model to an MMOG for validation or improvement of the social
model, money is needed either to build the MMOG or to acquire the use of
it and the tools and permissions that allow its modification from a willing
game development partner. Note, however, that the entire FY 2008 estimated budget of the Defense Advanced Research Projects Agency, $3.085
billion (see http://www.darpa.mil/body/budg.html [accessed July 2007]),
is comparable to the current cash assets of a gaming giant like Electronic
Arts (see http://finance.yahoo.com/q/bs?s=ERTSandannual [accessed July
2007]), plus the expected revenue of $1 to $1.5 billion from an operating
MMOG.[5] The size of the financial stakes for MMOG game companies
means that getting the attention of a game development partner may rely
more on personal connections or a fully funded joint basic research agenda
than on any financial incentives that DoD could offer.

[5]Consider the single MMOG World of Warcraft with an estimated 8 million+ players paying
$12.99 a month, for annual revenues of approximately $1.2 billion.

STATE OF THE ART

Our review of the state of the art in MMOGs considers the three roles of games separately—games as an interaction medium, games as a set of models, and games as interactive laboratories.

Games as an Interaction Medium

Games as an interactive medium are always changing and improving. The drivers for innovation in the game industry are new technology for making the games ever more immersive and interactive, as well as industry competition and emulation. The driver for many years has been the increasing graphics speeds for PCs and consoles. That drive has made photorealism one of the major pushes for interactive games. Soundscape complexity has also made games more immersive as PCs and consoles have improved their sound support capabilities. The first Dolby 5.1 certified game, America's Army, was developed in 2002, and now this feature is included in almost all games.

We are near the point of diminishing return for graphics improvements, and people are now focusing on "fully interactive worlds." The best example of the fully interactive world style is Rockstar's Grand Theft Auto: San Andreas (GTA-SA). While the story line may lack redeeming social value, the game is so popular because the game player can interact with everything in the game's world in a nonlinear fashion. This means that the player can navigate the world and do whatever comes to mind, without being constrained to a single story path, as in many games. The fact that there are missions to complete in GTA-SA is perhaps unimportant. It is the journey and the accompanying interaction that immerse and retain the player. If one wanted to have one game as representative of state of the art, then GTA-SA is that game with its fully interactive world paradigm, but the number of games attempting to copy that paradigm is quite large; the most notable is the Godfather game of Electronic Arts.

Games as a Set of Engaging and Immersive Models

The games Sims 2, by Electronic Arts Inc., and Civilization IV, by Firaxis, represent the state of the art with respect to games as a set of engaging and immersive models. Sims 2 is a game that allows the player to create virtual characters, or Sims, and then direct them over a virtual lifetime. Settable parameters include gene mix across generations, life goals, popularity, fortune, family, romance, knowledge, financial status, and lifestyle. Sims can be pushed to extremes "from getting busted to seeing a ghost, from marrying an alien to writing a great novel" (Electronic Arts

Inc., 2006a). The game allows the player to fulfill dreams, to try extremes, and to basically explore potential outcomes and possibility spaces.

Civilization IV is a game that allows the player to create a civilization from its inception to its pinnacle and eventual demise. Players can choose peace and growth or choose a war footing, all from an easy-to-use interface. Civilization IV comes with a stream of easy-to-use modification tools that allow players to create and integrate their own interests into the game. As in Sims 2, Civilization IV allows the player to explore possibility spaces and potential outcomes. It is that capability that makes the modeling and simulation of these games very interesting to DoD and to the Department of Homeland Security. The question is often asked, "How do we connect these games, with near-zero modification, to real news feeds so that we can compare their 'predictions' against what subsequently happens in the real world?" Of course, these games are written to explore potential outcomes and not to be predictive, but there is a continual quest to achieve predictions, as in the film "Minority Report."

Games as an Interactive Laboratory

MMOGs as interactive laboratories provide a state-of-the-art capability with respect to DoD goals. The idea is, for example, that if it were possible to test models of what causes insurgencies against large groups of real online people, one could then understand and run the models backward to change the conditions so that the insurgencies do not happen. This is a tall order, built on several premises. The first premise is that models exist of the cause of insurgencies, and there is no way to test those models in real life. An additional premise is that if one could test and prove those models against real people in MMOGS, one would then have greater confidence in deploying the ideas embodied in the models in real life. The interesting point of such discussions is the desire to test social models using existing MMOGs rather than having DoD create its own testbed, thereby saving $100 million of testbed development costs that could be used to create models.

To consider the top 10 MMOGs, how relevant or close to the problem are they? World of Warcraft, City of Heroes, City of Villains, Final Fantasy XI, Eve Online, Guild Wars, RuneScape, Everquest 2, Maple Story, Dark Age of Camelot, and Lineage 2 were the top 10 MMOGs listed by one site as of July 2007 (see http://www.the-top-tens.com/lists/top-ten-mmorpg-games. asp [accessed July 2007]). Although their visuals are far from realistic, and their stories are mostly about worlds that don't exist and quests not linked to real life, the stories are all about the fights between good and evil, not unlike today's global war on terrorism. So the thought is to take one or more of these MMOGs, modify the story a bit, put in links to the predictive models to be tested, and then see if one can begin the process of predicting player behavior in the MMOG with connected systems. If that can be done, then perhaps it will be possible to run the models backward to stop insurgencies before they form or interdict them earlier before they gain strength.

RELEVANCE, LIMITATIONS, AND FUTURE DIRECTIONS

This section explores how MMOGs can be used to address DoD problems, the limitations on that use, and the next steps needed to address those limitations. The discussion is organized around the three major capabilities offered by MMOGs: an interaction medium, a set of models, and an interactive laboratory.

Games as an Interaction Medium

Interactive games are great interfaces to models and simulations, because designers have created an interface typically more intuitive than that in comparable DoD-developed M&S systems. Interactive games typically require no reading of manuals and have the player up and running in three minutes or less. The corresponding time is typically months for comparable DoD M&S systems. So if the goal is to put models and simulations into the largest number of hands possible, then an interactive game interface is the right way to go. An additional advantage of interactive games is that their development and modification tools are easier to use than the simulation setup tools used by DoD. If defense simulations were as easy to set up as games, modelers' ability to explore possibility spaces and potential outcomes would be dramatically increased.

Games as an interaction medium are limited, at the moment, to games designed and implemented by the game development industry for entertainment purposes. For DoD use, those games must either be used as they are or modified with available tools. An additional limiting factor is that DoD does not typically have access to personnel skilled in game development.

Interactive games, their supporting hardware infrastructures, their supporting software, and their input devices are under constant pressure to innovate and evolve. The biggest change coming in the next few years will be in the underlying models of human and organizational behavior, particularly with respect to the modeling, display, and input of human emotion into the interactive game. Think of this as adding to the communication modalities already employed in games: visual display, auditory display, haptic display, and (coming soon) two-way emotional communication and display. Low-cost sensors that read parts of the human emotional state have already been designed for use as game input devices. These sensors provide virtual sensors indicating mental focus, adrenaline, surprise/response, and relaxation, along with physiological measures of heart rate, blink rate, breathing rate, and oxygen level in the blood. Software using these measures is already under development for use in evaluating games before they are shipped to determine what does or does not work in the produced game. Experiments are under way to determine how to use emotional inputs in games, including how to display appropriate emotions back to the player based on his/her personal state.

Games as a Set of Engaging and Immersive Models

The set of models inside engaging and immersive commercial games are proprietary and somewhat of a black box. We (DoD and its modeling researchers) cannot look at those models or modify them, other than the parameters exposed from the game's interface or provided via modification tools. We cannot VV&A those models—but we probably haven't really been able to achieve real VV&A with defense models and simulations, either (see Validation, Chapter 8). What we do know is that games like The Sims 2 and Civilization IV look quite capable for use in defense problems, if only we could modify them, even slightly, for defense purposes. It would be interesting to know whether one could explore more of the space of potential insurgency outcomes with Civilization IV, developed at a cost of some $20 million, than with JSIMS, developed at a cost of $1.8 billion. Could modelers do that exploration with just the available Civilization IV modification tools?

Likewise it should be possible to run experiments in virtual worlds similar to Second Life (http://www.secondlife.com), perhaps with a somewhat less benign set of rules, which would have military and strategic applications. For example, imagine Second Life with sovereign state entities, some of which were motivated to expand and dominate other regions of the game space. What would be the behavioral/organizational reactions of the other players? It is likely that genuine social experiments could be undertaken in settings like this, at a cost far below JSIMS. As another example, the popular board game Diplomacy is already available for online play (see http://www.diplom.org/index.py); it ought to be possible to modify it to bring it up to date in terms of state actors and allow for multiplayer states with their own internal decision-making processes, political parties, cultures, etc. Of course, there would be issues that would need to be thought through, such as access to the online games by hostiles, the potential for abuse of human subjects by traumatizing their avatars, and how to make the costs and benefits "real" (so that the players are not casual about starting virtual wars, for example). However, it seems clear that the potential gains are large enough to warrant some real effort devoted to overcoming these obstacles.

Since the models inside games are typically proprietary and not precisely what DoD requires, this mismatch makes it hard for DoD to accept and utilize such models. If DoD were to establish its own serious game development studio, this limitation could be overcome.

Also, since games are typically designed for entertainment, they often provide a misrepresentation of reality—for example, compressed time, inaccurate social networks, and missing cultural factors. If games are to be used for training, then greater attention needs to be paid to which aspects of reality need to be more carefully characterized in the games. This requires basic research on what factors are needed for what purpose, the inclusion of facts about social and cultural behavior, and the inclusion of social and organizational scientists as members of the game development team.

Games as an Interactive Laboratory

MMOGs are proprietary and written for a particular entertainment purpose, with rules very much unlike those found in real life. Players get to be heroes, villains, and superheroes in MMOGs and are often able to transport their virtual characters across large terrains without apparent cost in time or physics. So while it looks as if modelers might be able to do some experiments with MMOGs relevant to DoD concerns, there are definitely issues in the details.

MMOGs as interactive laboratories are limited in their use for DoD purposes because they were built for entertainment. For MMOGs to become widely used in DoD, DoD may need to establish its own studio to build such an MMOG, using a mix of game industry veterans and defense M&S personnel. A vision for what this might look like is the collection of art resources and animations from the America's Army game ported to a larger, more open platform (U.S. Army, 2007). Right now, America's Army is built on the Epicgames Unreal-2 game engine, an engine limited to small squad-on-squad play (32 players total) and small areas of terrain (1 km × 1 km). In addition, access to the art resources and code from America's Army has been restricted to DoD due to proprietary game engine license issues and close control of those resources by the Army project management team. Good small training systems have been built using the America's Army material (see http://info.americasarmy.com), but in general, the close hold of the source code and game resources has made it difficult for DoD scientists desiring to use that material to be able to build the additions and extensions necessary to carry out their research. As DoD moves to the larger realm of MMOGs for social model experimentation and to a game engine capable of handling much larger terrains of concern, the openness and accessibility challenge needs to be solved for the greater DoD good.

Funding a specialized MMOG game studio outside the government would be one approach to the challenge, perhaps within the environment afforded by a specialized cross-departmental university center, such as the Entertainment Technology Center at Carnegie Mellon University (see http://www.etc.cmu.edu/) or as a university-affiliated research center but with more internal development capability than seen at labs like the highly successful Institute for Creative Technologies at the University of Southern California. There are many applications for which this MMOG will be of value. A particular MMOG that would have great applicability is one that implemented the various alternative futures as described in the report *Mapping the Global Future* (National Intelligence Council, 2004). Understanding those potential outcomes and being able to roll back to a state in which the potential outcome does not happen would be a great tool for designing better national policies.

REFERENCES

Bennington, R.W. (1995, May). Joint simulation system (JSIMS): An overview. In *Proceedings of the IEEE 1995 National Aerospace and Electronics Conference (NAECON 1995)* (pp. 804–809). Available: http://ieeexplore.ieee.org/iel3/3912/11342/00522029.pdf [accessed Feb. 2008].

Carley, K.M., Moon, I.-C., Schneider, M., and Shigiltchoff, O. (2005). *Detailed analysis of factors affecting team success and failure in the America's Army game: CASOS technical report.* (Report #CMU-ISRI-05-120.) Pittsburgh, PA: Carnegie Mellon University. Available: http://stinet.dtic.mil/cgi-bin/GetTRDoc?AD=ADA456079&Location=U2&doc=GetTRDoc.pdf [accessed Feb. 2008].

Electronic Arts Inc. (2006a) *About the Sims 2.* Available: http://thesims2.ea.com/ [accessed Feb. 2008].

Electronic Arts Inc. (2006b). *Maxis: A timeline.* Available: http://www.maxis.com/about/about_timeline1.php [accessed April 2007].

Electronic Arts Inc. (2007). *Inside scoop: The history of SimCity.* Available: http://www.maxis.com/about/about_timeline1.php [accessed April 2007].

Moss, S. (2001). Game theory: Limitations and an alternative. *Journal of Artificial Societies and Social Simulation, 4*(2). Available: http://jasss.soc.surrey.ac.uk/4/2/2.html [accessed April 2008].

National Intelligence Council. (2004). *Mapping the global future: Report of the National Intelligence Council's 2020 project.* Washington, DC: U.S. Government Printing Office.

National Research Council. (1997). *Modeling and simulation: Linking entertainment and defense.* Committee on Modeling and Simulation: Opportunities for Collaboration Between the Defense and Entertainment Research Communities, Computer Science and Telecommunications Board. Commission on Physical Sciences, Mathematics, and Applications. Washington, DC: National Academy Press.

U.S. Army. (2007). *America's Army: The official U.S. Army game.* Available: http://www.americasarmy.com/ [accessed Feb. 2008].

von Neumann, J., and Morgenstern, O. (1944). *Theory of games and economic behavior.* Princeton, NJ: Princeton University Press.

Wikipedia. *Massively multiplayer online game.* Available: http://en.wikipedia.org/wiki/MMOG [accessed April 2007].

Zyda, M., Hiles, J., Rosenbaum, R., Roddy, K., Gagnon, T., and Boyd, M. (1998). SimNavy-Phase 0 building an enterprise model of the U.S. Navy. In *Proceedings of the 1998 Technology Initiatives Game*. Available: http://gamepipe.usc.edu/~zyda/pubs/SimNavyPaper98. pdf [accessed April 2008].

Zyda, M., Mayberry, A., McCree, J., and Davis, M. (2005). From Viz-Sim to VR to games: How we built a hit game-based simulation. In W.B. Rouse and K.R. Boff (Eds.), *Organizational simulation* (pp. 553–590). Hoboken, NJ: John Wiley & Sons.

8

Common Challenges in IOS Modeling

This chapter discusses broad issues and challenges that are encountered across the range of individual, organizational, and societal (IOS) modeling approaches and methods, highlighting problems that need to be solved for these modeling approaches to be most useful for the military's needs. We first describe issues of integration and interoperability, the challenges that confront modelers and simulation developers when they attempt to integrate multiple models and simulations, with the goal of making them interoperable—that is, able to use output from one model as input for another. Next we describe some of the challenges (and potential benefits) of developing and using modeling frameworks and tools that facilitate the development of IOS models. We then describe issues of model verification, validation, and accreditation (VV&A), issues that are especially challenging for the modeling of human behavior. Finally, we discuss some of the challenges posed by the data requirements of IOS models in light of the realities of the data and information available to model developers and users. In each section we note some potential solutions to the challenges.

INTEGRATION AND INTEROPERABILITY

In this section, we discuss the issues that confront modelers attempting to integrate models developed with different internal structures, at different levels of granularity, or with inconsistent inputs and outputs. The nature of the challenges requires that the discussion be quite technically sophisticated and use terminology and concepts that may be unfamiliar to many readers.

We have tried to define some of the terms in footnotes, but a simplified discussion would not do justice to the subject matter.

Model Interoperability: Incompatibilities and Functionality Gaps[1]

There are several fundamental issues (and associated hard problems) that need to be addressed in undertaking the development of an interoperable framework of IOS models. First and foremost is the problem of making existing or even new models interoperable, as these are developed independently (i.e., with no coordination) by different software design and development teams, in consultation with domain experts having various levels of skills and expertise. A very common approach is to build a wrapper around an existing model, thus converting it to an input-output (I-O) black box, or to provide an intelligent agent operating autonomously, which communicates with other models in the network. But this approach is likely to introduce other types of gaps and incompatibilities between models, some of which are identified in Table 8-1 and illustrated in Figure 8-1. We discuss here the need to identify an overall methodology to fill these gaps, including various intelligent automated techniques, processes, and guidelines, as well as aid from human subject matter experts and analysts whenever needed.

Interface Incompatibility

The first problem shown in Figure 8-1 (in the top row) concerns interface incompatibility between two models that either already exist or are being developed independently. If we intend to feed output from model A about a certain object X as input to model B, then some mismatch between the output and input may occur in terms of the assumptions about the numbers and types of X's attributes. This is often straightforward but tedious to deal with, often merely involving translation from one descriptive framework to another (e.g., from numerical values—1, 2, 3, . . . —to "fuzzy" values—low, medium, high, . . .). A bigger problem ensues when different levels of resolution are used to represent the same object in two different models. If model A provides a high-resolution object representation of X (e.g., a map, enemy force estimates) for model B, and model B needs a low-resolution representation (e.g., latitude/longitude of enemy center of gravity), then some aggregation process must be conducted, usu-

---

[1]Much of the work described in this section on model integration and interoperability was performed by John Langton and Subrata Das at Charles River Analytics with support from the Air Force Research Laboratory, Information Directorate (AFRL/IF) under contract FA8750-06-C-0076, and adapted from Langton and Das (2007).

TABLE 8-1 Gaps and Incompatibilities Between IOS Models

Type	Definition
Interface	Mismatch between the data types of different models or outputs of one model and inputs of another, e.g., real number vs. Boolean
Ontological	Different relationship structures, naming schemes, etc., in ontologies for different models
Formalism	Different logic and inferencing mechanisms and procedures for different models
Subdomain gaps	Differing domains and dynamics between PMESII model dimensions, e.g., economic vs. social

SOURCE: Langton and Das (2007).

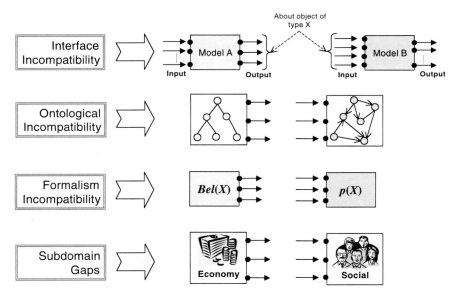

FIGURE 8-1 Illustration of gaps and incompatibilities between IOS models. SOURCE: Langton and Das (2007).

ally based on one approximation method or another. The reverse process is much more difficult, going from a low-resolution output to a high-resolution input, since, in effect, missing input attributes have to be inferred or approximated and filled in. A number of approaches can be used to resolve the interface incompatibility. These are described in the section on interoperability recommendations below.

Ontological Incompatibility

The second problem illustrated in the figure is ontological incompatibility between models, which arises due to differing vocabularies and expressive power in their respective ontologies.[2] Different teams of engineers and subject matter experts with a diverse range of expertise, knowledge, and cognitive capabilities independently creating models will inevitably develop and use different underlying ontologies, which in turn will give rise to incompatibilities across models. Initially, one might suggest the development of a common ontology for the set of all possible models; however, many failed efforts in this direction make it clear that developing a universal ontological standard for model creation is impractical, if not theoretically impossible. Moreover, if models are to be built rapidly, analysts should ideally be free to use a model-building environment of their own choosing without assistance from knowledge engineers. The analysts should not be constrained by a predefined ontology to express their knowledge, which usually inhibits their expressive flow. Hence, rather than proposing to develop a common ontology for the model space, one approach is to focus on facilitating better mapping capabilities between differing ontologies. For example, there are tools that can map ontological terms from one domain to another by solving the problems of synonymy and polysemy;[3] these clearly offer hope for translating differing ontologies used in the models. In some cases of incompatibility between the underlying ontological structures of the models (e.g., semantic networks versus logical expressions), one domain can be mapped to another by providing a more expressive ontological structure for one of the models (e.g., semantic networks can be mapped to first-order logical sentences). Therefore, some parts of the ontological incompatibility problem can be addressed via automated techniques. A number of approaches can be used to resolve the ontological incompatibility, described below.

Formalism Incompatibility

While ontological incompatibility creates problems due to multiple ways of designating an entity, the formalism incompatibility shown in Figure 8-1 is concerned with multiple ways of instantiating the object entity

---

[2]An ontology, for the purposes discussed here, is "a systematic arrangement of all of the important categories of objects or concepts which exist in some field of discourse, showing the relations between them. When complete, an *ontology* is a categorization of all of the concepts in some field of knowledge, including the objects and all of the properties, relations, and functions needed to define the objects and specify their actions" (http://www.answers.com/ [accessed July 2007]).

[3]Synonymy refers to one referent (concept) with several words that can denote it (plain English examples: big, large); polysemy refers to one word denoting multiple referents (plain English examples: break; park).

computationally represented in the model. For example, uncertainty can be expressed not only in terms of probability values, but also via various other formalisms, such as certainty factors, the Dempster-Shafer measure of beliefs (Shafer, 1976), and numerous other qualitative dictionaries. These are fundamentally incompatible with each other, both in terms of their underlying conceptual representation of uncertainty and probabilistic reasoning, and in the sense of having different types of scales. Conversion between two such formalisms often requires deep understanding of the models and their formalisms, thus breaking the simple I-O black box idea of encapsulation. Specialization of formalism is often appropriate to map one approach to another. For example, probability theory is a special case of Dempster-Shafer theory that allows beliefs to be expressed only on singleton sets, facilitating development of a mapping from probability models into Dempster-Shafer models.

Subdomain Gaps

If one wants to feed the output from a model in one domain to another, it will require an analyst or domain expert with knowledge of both domains to bridge the subdomain gaps. This is due not only to the ontological gaps between the domains being considered, but also to differing dynamics between the domains. Addressing this problem requires the skills of experts from the respective domains or ideally ones who are expert in both domains.

A number of approaches can be proposed to bridge such gaps, by highlighting possible correspondences between concepts and variables across domains, described below. Recommendations are also made for more comprehensive approaches that could be part of a long-term development effort.

Figure 8-2 provides an illustration of model interoperability—focusing on political, military, economic, social, information, and infrastructure (PMESII)-related issues—with interactions among three layered models: one focusing on the social structure, one on the community infrastructure, and a third on the underlying information models, respectively from top to bottom.

The *infrastructure model* in the middle models a stabilization and reconstruction operations (SRO) model, developed by the AFRL/IF, (Robbins, Deckro, and Wiley, 2005) using a system dynamics modeling approach (see Chapter 4), and captures a sequence of influences among variables, starting from the power supply at an electrical power substation. The generated power is fed into an industrial water plant, which produces water consumed by oil field work. An oil field produces crude oil to be refined by a refinery. Refined fuel is used to generate power, which in turn is supplied to various power substations, thus forming a loop. It is especially

276

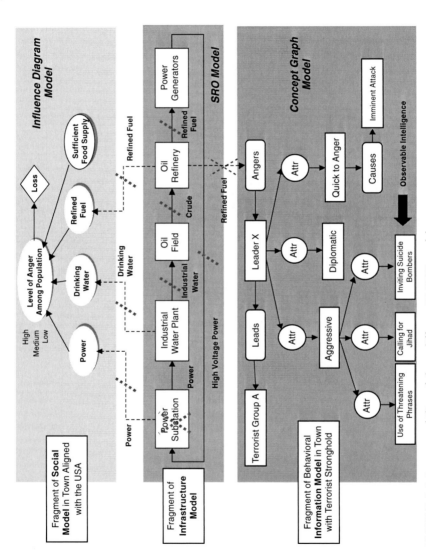

FIGURE 8-2 Interoperability of three different PMESII models.
SOURCE: Langton and Das (2007).

difficult to reason with these types of graphs, containing such loops spanning many variables, as it creates an additional burden for discounting the
variables' self-influence.

The *social model* at the top of the figure captures the impact of these
infrastructure-related variables on the society, using influence modeling
technology (see Chapter 6). The model specifically captures the influence
of the four variables of power, drinking water, refined fuel, and sufficient
food supply on a variable representing the level of anger of the population
in a town aligned with coalition forces. The dynamics of the social model
are that short supply in any one of these three consumable products will
increase the level of anger among the local population. In fact, if a terrorist
organization became aware of the mid-layer SRO model sequence in the
infrastructure, then the power substation would assume heightened importance in the eyes of the terrorist strategists: an attack on a substation would
not only cripple other services in the loop, but would also drive the senti-
ment of the local population against the coalition. Note that the diamond
box represents the expected mission utility in line with the level of anger.
The utility (although difficult to quantify here) should go up when the anger
level is down and vice versa.

The *behavioral information model* at the bottom of the figure illustrates how a model of a terrorist leader can be built using a concept graph
approach (Sowa, 1984) in which concepts are represented by rectangles
(e.g., [Person: Leader X] and [Behavior: Aggressive]), and conceptual relations are represented by circles (e.g., has Attributes) and soft-cornered
rectangles (e.g., Leads, Causes). An analyst can query such a model to
determine who the terrorist leader is and the nature of the leader based
on various observable intelligence. Such a leader X, who leads the terrorist group A, can possess different types of behavior attributes, including
aggressive, diplomatic, quick to anger, etc. If the leader is quick to anger
and there are some stimuli to make the leader angry, then an attack on
friendly targets may be imminent. One such stimulus would be coalition
forces stopping the supply of oil to the region, as indicated by the link to
the SRO model above.

The key issue here is the interoperability among the models. Note
that although an I-O connection has been made between the two variables
Oil Refinery and Refined Fuel of the top two models, they are ontologi-
cally incompatible as defined earlier. However, they can be made compatible by recognizing that the term "Oil" is synonymous with "Fuel," and
"Refined" and "Refinery" have a common base word. Another difficult
compatibility problem is illustrated by the fact that there is no input for the
variable Sufficient Food Supply in the social model, illustrating the interface incompatibility described earlier. One can envision, however, that this
"sufficiency" concept could be automatically computed from the supply of

food previously recorded in available databases to bridge this last gap. A number of recommendations for resolving specific model incompatibilities and functionality gaps are provided below. More general approaches to resolving more than one of these gaps simultaneously are a current area of study (Langton and Das, 2007).

Recommendations for Resolving Gaps in Model Interoperability

A number of approaches can be taken to maintain, adapt, and integrate diverse models in the context of the interoperability gaps just defined.

Dealing with Interface Incompatibility

Interface incompatibility generally refers to two or more models having different types of data for their inputs and outputs and thus not being able to interoperate without some form of data conversion. There are at least three types of interface incompatibilities:

1. *I-O format incompatibilities:* string versus binary, real versus integer, fixed versus floating point, numeric versus Boolean, incompatible scale, incompatible zero point, date-time format, color format.
2. *Logical incompatibilities:* number of I-O points (e.g., three outputs versus four inputs—RGB to CMYK is a trivial example), I-O timing (e.g., fast output versus slow input).
3. *Model persistence format incompatibilities:* XML versus YAML, OWL versus RDF, etc.

One way to deal with these issues is via a development interface that provides a basic set of translation functions that can learn from user interaction over time. A graphic user interface (GUI) would allow users to explicitly modify, add, and remove interface translation functions, as illustrated in Figure 8-3. Users could also specify these translation functions within an ontology or the XML schema of a model, based on specifications derived, for example, from an evolved, global ontology. A full-scope GUI would then allow users to explicitly modify, add, and remove interface translation functions. A number of potential translation functions are described below in the context of the type of incompatibilities each addresses.

Dealing with I-O Format Incompatibilities

Many interface incompatibilities fall within this category, and most solutions can be resolved by some combination of the following:

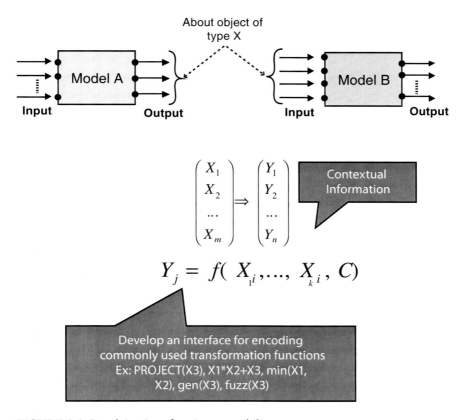

FIGURE 8-3 Resolving interface incompatibility.

- *Normalization:* mapping any value to lie between 0 and 1 relative to its minimum and maximum possible values.
- *Weighting:* scaling a value, typically in relation to other values.
- *Fuzzification:* randomly generating a number to lie within some constraining interval (e.g., some random number between 0.3 and 0.6).
- *Discretization:* "binning" values according to their range and a range they must fall within—somewhat like rounding—sometimes taking their distribution into account (e.g., 0.5 within a range between 0 and 1 can be discretized to 1 for a range of only 0 or 1).

XML schemas often exist to support model file persistence. These schemas define the elements of a model along with the possible values they can take on. XSLT can then be used along with a number of standard translation functions for integrating inputs and outputs of two models on relevant nodes or links. These functions can also be adapted according to user interaction over time.

Dealing with Logical Incompatibilities

In some cases, one model may have more outputs than another's inputs or vice versa. When integrating models, we therefore need methods for addressing these situations. For an overabundance of values, we can simply use some form of aggregation. Again, a model schema or ontology can specify how this aggregation should be performed, or the user could specify this through the above-mentioned GUI. In the case in which we have only one value but must map to more than one, we can simply duplicate the value or partition it according to any context provided in the model schema or ontology.

In some cases, the sample rate of inputs and outputs may differ. One way of dealing with this is through smoothing and resampling.

Dealing with Model Persistence Format Incompatibilities

In essence, this issue really mirrors the greater task of integrating models. The existence of a standard schema or ontology for different models would immediately resolve this issue. However, we cannot now depend on such a standard or on adherence to it. A partial solution may be to evolve or derive a standard schema or ontology. In either case, most effective solutions will entail the use of XML and XSLT for the translation of one model format to another.

Dealing with Ontological Incompatibility

Ontological incompatibility refers to two models having different structures, including the entities they specify and the relationships between them. For instance, a rules system model may have several pairs of nodes connected by one link (precedent and consequent), whereas a Bayesian net typically has more of a tree structure. Nodes can have different names, graphs can be directed or undirected, and two models representing the same system can be at different resolutions and thus include a different number of nodes and links. The principal issue of this incompatibility is determining which entities, nodes, or links in different models should map to one another for interoperation.

Syntactic heuristics: The labels and descriptions of nodes and links in differing models can be compared on the basis of their raw string content. If these string components match, then the nodes or links may be a match as well. For instance, "runway16" may map to "runway." A threshold for how many characters must match to infer a string match must be specified. This type of matching can also include matching nodes/links based on the range, cardinality,[4] and other attributes of their possible values.

Semantic heuristics: Nodes and links from different models can be compared on the basis of the semantics of their labels, descriptions, and any other textual metadata specified in an XML file, XML schema, or ontology. Elements from different models that have a semantic similarity can then be mapped to one another for model integration. For instance, a node with the name "airport" in one model may be mapped to a node with the name "runway" in another model on the basis of the semantic similarity of their labels. Semantic similarity is determined by the relations between two words as derived from statistical usage, ontologies, thesauruses, dictionaries, etc. There are both service-oriented architectures and application program interface specifications for this purpose, including WordNet (Al-Halimi and Kazman, 1998) and Lexical Freenet (Beeferman, 1998).

Relation mapping: Relation mapping can be used to address ontological incompatibility by mapping nodes from one model to nodes of another based on their relations (how they are connected) within their individual models. With this information, we can then suggest potential mappings between nodes of different models based on the similarity of their relations within their respective models. Consider the nodes α of model A and β of model B. Although these nodes may have very different names, they may have very similar relations. For example, both could influence five other nodes and be influenced by four other nodes. Based on their similarity, we may be able to deduce that these nodes can be mapped together for model integration. It is important to note that relations encompassing a node are not merely all of its incoming and outgoing links; they also include features identifying how the node affects any other nodes in the model. While this approach should rarely be used to draw links automatically, it could be used to make effective recommendations.

Model node aggregation: Model aggregation can be used to address ontological incompatibility by identifying how sets of nodes in different models with differing cardinalities may be mapped to one another. It may be the case that a node α in model A maps to a subset of nodes N in model B, resulting in incompatible ontologies. For example, consider α to be the node *airport* and N to be the subset of nodes *runway, plane, radar,* and *air traffic control.* The question is, which nodes should *airport* be mapped to for model integration? We can use the semantic similarity of the labels on the nodes of N (e.g., interfacing with WordNet for ontological inference)

---

[4]In mathematics, the cardinality of a set is a measure of the "number of elements of the set" (Wikipedia, see http://en.wikipedia.org/wiki/Cardinality [accessed Feb. 2008]).

to aggregate N into one meta-node: *airport'*. Using semantic similarity between *airport* and the constituent nodes of *airport'*, we can then infer that these two entities should be mapped for model integration. More specifically, the inputs to *airport* could potentially be mapped as inputs to all nodes of *airport'*.

We can also infer the pairing of *airport* and *airport'* using relation mapping. To continue the example, consider relations between *airport* and the other nodes of model A, and *airport'* and any remaining nodes of model B. If their relations are similar, then *airport* could be a candidate for integrating with all of the nodes of *airport'*. For instance, both *airport* and airport' could be connected to the nodes *passenger*, *ticket*, and *pilot*. With ontological inference, we can see that the relations of *airport* and *airport'* are similar within their respective models (even though the relations of *airport* are not similar to the relations of any of the constituent nodes of *airport'*). We can then deduce that these nodes should be mapped for model integration.

Dealing with Formalism Incompatibility

Established formalism mappings. There has been a great deal of research on mapping between different algorithm formalisms, and there are a number of established standards. Table 8-2 shows a matrix in which the following illustrative formalisms appear in the outer cells of both the X and Y axis: Bayesian probability, Dempster-Shafer, fuzzy logic, possibilistic theory, certainty factor, and symbolic dictionary. Each internal cell denotes the mechanism used for mapping between associated formalisms on the outer cells of the X (to) and Y (from) axes. Note that the mechanism for mapping from X to Y may not necessarily be the same mechanism used for mapping from Y to X. Shaded cells represent established mechanisms for mapping between different formalisms, while nonshaded cells represent potential mapping approaches. In other words, there are known and established algorithms for mapping between the formalisms that are joined by a shaded cell.

Using XML schemas and ontologies for formalism mapping. Ontologies can explicitly identify mappings between formalisms in the attributes of links within a model. For such formalisms as Bayesian networks and argumentation networks, relations correlate with the links between nodes. We can add an attribute to each link in a Bayesian network XML schema, declaring a "causes" relationship between a parent and child node it connects. Furthermore, we can specify that types of links declaring a "causes" relationship should be mapped to the "entails" relationship of a rule-based model. When mapping from a Bayesian network model to a rule-based model, we can then infer from the ontology that a "cloudy" node in a Bayesian network that

TABLE 8-2 Mappings Between Modeling Formalisms

	Bayesian Probability	Dempster-Shafer	Fuzzy Logic	Possibilistic Theory	Certainty Factor	Symbolic Dictionary
Bayesian probability	X	Generalization	Membership degree interpretation of probabilities	Transformations based on consistency principles	Mapping from probabilities to certainty factors	Probability-to-symbolic mapping
Dempster-Shafer	Bayesian approximation (transferable belief model)	X	Via Bayesian approximation	Via Bayesian approximation	Via Bayesian approximation	Belief value-to-symbolic mapping
Fuzzy logic	Normalization of membership degrees	Via normalization	X	Possibility measure interpretation of membership degrees	Mapping from membership degree to certainty factors	Membership degree-to-symbolic mapping
Possibilistic theory	Transformations based on consistency principles	Belief interpretation of possibility measures	Membership degree interpretation of possibility measures	X	Mapping from possibility measures to certainty factors	Possibility measure-to-symbolic mapping
Certainty factor	Normalization	Via normalization	Via normalization	Via normalization	X	Certainty factor-to-symbolic mapping
Symbolic dictionary	Symbolic-to-probability mapping	Symbolic-to-belief value mapping	Symbolic-to-membership degree mapping	Symbolic-to-possibility measure mapping	Symbolic-to-certainty factor mapping	X

SOURCE: Langton and Das (2007).

"causes" a "rain" node should be mapped to a rule that has the variable "cloudy" as a precedent and "rain" as a consequent.

Subdomain Gaps

Subdomain gaps can be addressed by learning ontologies and ontological evolution in which the relationships between models are implicitly specified as models are built. Mixed initiative approaches can also be used to address all of the interoperability gaps, including differences in subdomains. For example, the system might offer suggestions about what nodes or links should be related between two models. The user could then accept, edit, or ignore these suggestions. One simple mixed initiative approach would be reinforcement learning (Kaelbling, Littman, and Moore, 1996) guided by these user selections. If a user accepts a suggestion, the system could increase the number of related suggestions. If a user rejects a suggestion, then the system should learn not to make similar suggestions in the future. Even devoid of other heuristics, this approach would allow the storage of historical information as to what input and output types the user typically maps together and offer these mappings as suggestions for subsequent model integration efforts.

In summary, there are no currently agreed-upon and widely used standards for model integration and interoperability. The field of IOS modeling is fragmented, with models being developed from different perspectives, at different levels of detail, and using different theoretical frameworks and architectures. To address these issues, we suggest improvements in "translation" interfaces, schemas, or ontologies that could guide integration, as well as mixed initiative efforts in which model developers and users from different perspectives work together to create models. Architectures and standards for that would support the development of integrated interoperable federated models identified as a key area for future research in Chapter 11.

FRAMEWORKS AND TOOLKITS

General Issues and Requirements[5]

Earlier chapters have described many IOS modeling and analysis techniques, but it is generally accepted that no single approach or model-

[5]Much of the work described in this section was performed by Karen A. Harper, Jonathan D. Pfautz, Chen Ling, Sofya Tenenbaum, David Koelle, and Marc Sageman, with support from the Air Force Research Laboratory, Human Effectiveness Directorate (AFRL/HE) under contract FA8650-06-C-6731, and by Karen A. Harper and John Bachman with support from the AFRL/IF under contract FA8750-06-C-0078, to Charles River Analytics.

ing formalism can or should be applied to capture all of the complex dynamics of modern military missions and activities. The previous section has described some of the fundamental modeling issues that arise when we consider linking up or "federating" different models, and it is clear that considerable progress will have to be made at the conceptual level before such activities become commonplace in the IOS modeling and simulation (M&S) community. In the meantime, there has been and continues to be progress in the development of more specialized (and therefore less globally encompassing) frameworks and toolkits that attempt to address some of the more practical issues of model development, verification and validation (V&V; see following section for more complete discussion), and integration across modeling concepts and instantiated simulations. In general, these efforts attempt to provide an integrated development environment (IDE) that enables

- the development of simpler, more focused submodels to represent specific features of the behavior of interest to the analyst, using the most appropriate tools for modeling those features;
- the straightforward integration of those submodels into a cohesive and sophisticated representation of the overall operational environment; and
- the effective accounting of the complex interdependencies between modeled variables within the integrated system.

Most of the work in developing frameworks and tools has occurred in the individual "stovepipes"—some refer to these as "cylinders of excellence"—that characterize each modeling community. Perhaps the best funded over the longest development history is the OneSAF System (One Semi-Automated Forces; see Chapter 2) of the Department of Defense (DoD). OneSAF is an M&S environment with a strong Army legacy that models combined arms tactical operations up to the battalion and brigade level, at variable levels of resolution ("entity") from the individual soldier on up. A key driver in its development was to ensure "composability," which is another way to say that the associated development environment provides for user-specifiable systems, entities, units, and associated behaviors (with variable "dial-in" levels of fidelity). This is accomplished via a Product Line Architectural Framework, illustrated in Figure 8-4, a layered architectural approach that allows for "plug-and-play" modules at many different levels and via a model-developer suite of GUIs that provide the following functionalities to the M&S developer:

- *System composer:* High-level control and testing of the overall simulation.

286

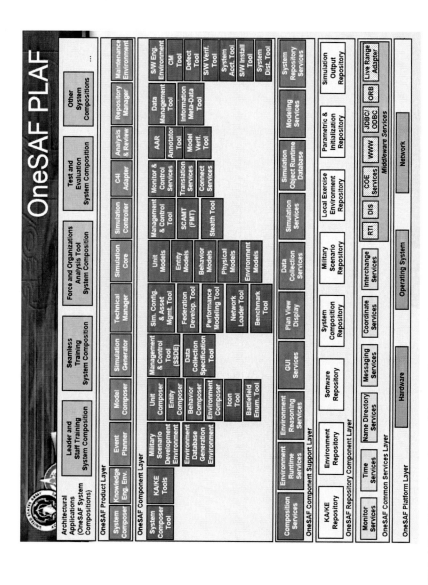

FIGURE 8-4 OneSAF product line architecture framework.
SOURCE: See http://www.peostri.army.mil/CTO/FILES/SmithR_GeneralFrameworkInterop.pdf, p. 12. [accessed April 2008].

- *Entity composer:* Specification of hardware components (weapons, sensors, etc.).
- *Unit composer:* Specification of the organizational structure.
- *Behavior composer:* Specification of the entity behaviors in terms of a task-network branching structure comprised of conditional branch points and behavior primitives (this is the heart of the OneSAF behavior model).
- *Management and control tool:* High-level control of the mission objectives, order of battle, route plans, etc.

OneSAF's capabilities are impressive, buts its focus on the ground war, its limited repertoire due to prescripted behavior primitives, its inability to model deep cognitive or social interactions, and its narrow focus on military missions (in contrast to more encompassing PMESII considerations) all point to the need for further model development in this area. Its focus on bringing the modeling out of the hands of the programmers and into the hands of the analysts and users, via a focused effort on IDE development, is commendable and should serve as a model for parallel efforts now ongoing in other M&S communities.

Another modeling community working on IDEs is the group of researchers and model developers focusing on the behavior of the individual human, often based on the framework of the particular cognitive architecture underlying the model (see Chapter 5 for additional discussion). Table 8-3 provides a sampling of some of the individual behavior models discussed earlier, along with their associated IDEs. As can be seen, the IDEs are very specific to each cognitive modeling paradigm; can range from highly generic programming language development environments (e.g., CLOS) to very specific model development environments (e.g., iGEN); assume varying levels of expertise on the part of the model developer, from general programming expertise to "drag-and-drop" graphic construction skills; and provide a range of developer support, from little beyond the basic programming IDE to extensive debugging, logging, and visualization. Again, this is not meant to be a survey of such IDEs, but rather an illustration of the variety of IDEs in use by the development community.

Yet another modeling community engaged in developing frameworks and toolkits is the widespread and diverse group of researchers, model developers, and applications specialists focusing on group and organizational models. One of the best clearinghouses for gaining an overview of available models and tools is maintained by the Computational Analysis of Social and Organizational Systems Center at Carnegie Mellon University. Although there exists some conflation of models and the associated IDEs for their development, the site provides useful pointers to a number of model

TABLE 8-3 Selected Cognitive Architectures and Their Development Environments

Model	Development Environment	Comment
ACT-R 6.0	No formalized development environment, since model developers work directly with the different ACT-R frameworks: LISP ACT-R, Python ACT-R, and jACT-R	• LISP ACT-R is the "baseline" version and requires a knowledge of LISP programming • Python ACT-R makes ACT-R available to a wider audience (i.e., non-LISP programmers) • jACT-R is a Java version • All have associated programming IDEs but require programming skills—and theoretical knowledge of the underlying cognitive architecture constructs—significantly beyond drag-and-drop model-building activity
COGNET	iGEN	• Workbench-based development environment with a collection of high-level agent-building tools • GUI for defining program logic and knowledge, without programming • Application program interface for integration of iGEN cognitive agents within existing applications using standard languages/protocols
D-OMAR	OmarL, OmarJ	• OmarL is a LISP-based environment for knowledge representation and the definition of agents and their behaviors. The languages are extensions of the Common LISP Object System (CLOS) • OmarJ is a Java-based agent development environment that provides tools for creating and managing systems of agents operating in a distributed computing environment. OmarJ provides most of the features of OmarL with an improved external communication layer that uses Jini for internode communication and the ability to break out of simulation mode and run agents in a non-time-controlled environment
EPIC		• IDEs associated with original LISP version of EPIC and with current C++ version
SAMPLE	AgentWorks™	AgentWorks™ consists of: • Perceptual, cognitive, and communications modules including neural networks, fuzzy logic, Bayesian belief networks, expert systems, and argumentation engines • Advanced processing capabilities supporting planning, learning, and distributed applications • Enhanced usability components for construction, validation, and visualization of agent processes
Soar	SDB	• Soar Debugger (SDB) is an XDB-like debugger for the Soar programming language, including functionality, such as deep structure inspection, watches, and breakpoints and a graphical interface to common Soar commands

development tools and frameworks at Carnegie Mellon and elsewhere, as illustrated by:

- Construct (http://www.casos.cs.cmu.edu/projects/construct/info. html), a multiagent model development environment
- OrgAhead (http://www.casos.cs.cmu.edu/projects/OrgAhead), an organizational structure analysis tool
- DyNet (http://www.casos.cs.cmu.edu/projects/DyNet)
- BRAHMS Composer (http://www.agentisolutions.com/products/composer.htm), the IDE for BRAHMS, an agent-based organizational modeling framework
- SimVision (http://www.epm.cc/solutions/simvision.htm), a bundled software environment and methodology for organizational design
- CONNECT (http://www.cra.com), a social network analysis tool for organizational modeling and simulation
- DDD (Distributed Dynamic Decision-making; http://www.aptima. com/a-sim.php), a simulation building and execution environment for predicting and assessing team performance

A quick perusal of these tools (and others) makes it clear that, like the cognitive modeling frameworks described earlier, the associated IDEs run the gamut in sophistication, from those demanding high levels of user expertise in the underlying theory and the associated modeling language, to those stressing ease of use, but imposing limited applicability for selected domains. Considerable work is still needed to bring these highly specialized models out to the general user community, via IDEs that provide wide applicability as well as usability.

In the general area of developing representations for "soft" problems in IOS behavior,[6] such as modeling the evolution of a terrorist organization or understanding the multiple possible paths in nation-state rebuilding—and the interplay of critical diplomatic, information, military, and economic (DIME) and PMESII variables—little has been accomplished in the way of developing associated IDEs to support the DIME/PMESII M&S community. This is primarily due to the fact that such nascent modeling efforts are still grappling with the conceptual underpinnings of representation; considerations of model development infrastructure and user- (developer-) friendliness are still considered a secondary objective. However, the lack of such environments may actually be hampering conceptual development,

---

[6]Soft in the sense of heavily driven by human and social rules of behavior, as opposed to more readily modeled problems that are well constrained by generally accepted physical, economic, or doctrinal factors.

because unwieldy development environments slow the "test-and-evaluate" spiral cycles that must inevitably occur in this field.

As noted earlier, modeling of military activities in which IOS behaviors dominate outcomes (e.g., asymmetric threats embedded in urban environments) demands a clear understanding of the complex sociopolitical context. This translates to the analysis of the potential effects that a given set of DIME actions will have across the full range of the PMESII context. Within the context of the Integrated Battle Command program of the Defense Advanced Research Projects Agency (http://www.darpa.mil/sto/solicitations/ IBC/), these analyses are viewed in two ways, as shown in Figure 8-5. From left to right, the figure shows a causal analysis in which, given a set of possible DIME actions to be taken, a system of complex and integrated behavior models is used to predict the potential effects those DIME actions may have across the PMESII dimensions. From right to left, the figure shows a diagnostic analysis in which, given a set of desired PMESII effects in the operational domain, the same system of integrated behavior models is used to identify the candidate sets of DIME actions that might be applied to achieve those desired effects. By conducting both types of analyses—ones that move well beyond the limits of conventional military "metal-on-metal" modeling embodied by OneSAF, for example—commanders will be able to develop significantly deeper insight into the dynamics of the big picture operational context (see additional discussion later in this section).

The key to successfully executing such encompassing analyses lies in the development of the embedded behavior models representing the full range of PMESII variables and how they can be individually and collectively affected by specific DIME actions. For example, as described earlier in this chapter, the SRO model (Robbins et al., 2005) analyzes the organizational hierarchy, dependencies, interdependencies, exogenous drivers, strengths, and weaknesses of a country's PMESII systems using a complex set of interdependent system dynamics representations. While approaches like this have demonstrated some success in modeling subcomponents of the PMESII environment, it is generally accepted that no single approach or modeling formalism can or should be applied to capture all of the complex dynamics of modern asymmetric warfare; in other words, it is not necessary to stick with a single modeling formalism (e.g., system dynamics modeling) to model something as complex as a nation-state undergoing political upheaval, foreign intervention, or civil war.[7]

A better approach is to provide for an IDE that enables the interconnection of disparate modeling methods representing DIME/PMESII features using the most appropriate method to model those features. A key issue

---

[7]A brief overview of potentially useful modeling paradigms for DIME/PMESII modeling and analysis issues is given in Appendix C.

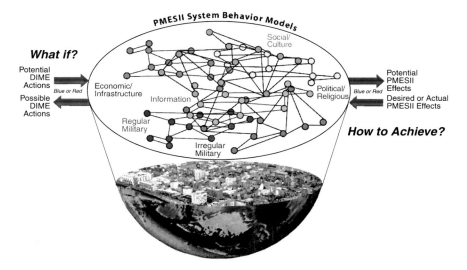

FIGURE 8-5 Predicting and analyzing PMESII effects of DIME actions.
SOURCE: Adapted from Allen (2004).

is providing compatibility across models and their underlying modeling formalisms, such as the models described in Chapters 4 through 6 and the generic formalisms presented in Table 8-2. As noted above, this is a conceptually difficult problem to solve theoretically, but some progress can be made with the development of sufficiently flexible IDEs.

IDE Development Goals and Examples

An ideal IOS IDE, especially one targeted for the complex task of developing DIME/PMESII models, would include

- an intuitive graphical model development environment supporting the specification of heterogeneous submodels using a variety of modeling formalisms (Bayesian reasoning, fuzzy logic, system dynamics models, rule-based expert systems, etc.).
- a suite of model integration tools enabling user-driven sharing of data and information among constituent DIME/PMESII models.
- a suite of model V&V tools enabling user-driven verification of individual and integrated DIME/PMESII model behavior as well

as the large-scale data collection required to support validation of model behavior against empirical data.

- a model analysis infrastructure that enables user-driven causal and diagnostic reasoning within the integrated modeling framework using sampling techniques and sensitivity analysis, respectively.
- a suite of multiresolution modeling tools and supporting infrastructure to support the user-driven specification of DIME/PMESII submodels at multiple levels of modeling fidelity.
- a model management infrastructure that enables the capture, distribution, and maintenance of large libraries of DIME/PMESII submodels.

We describe here an exemplar effort in developing such an IDE—the Human and System Modeling and Analysis Toolkit (HASMAT) developed for AFRL/HE (Bachman and Harper, 2007; Harper et al., 2007)—and describe two specific modeling efforts conducted with this framework, to illustrate how nonconventional modeling problems—specifically counterterrorism and military recruiting—can be addressed within such frameworks. HASMAT is intended to be representative of efforts under way to develop frameworks in this area. We close with a description of generic DIME/PMESII analysis capabilities that also need to be part of such frameworks.

Human and System Modeling and Analysis Toolkit

HASMAT is designed to support predictive analysis of behavioral and organizational dynamics by integrating existing and mature technologies. The HASMAT functional system architecture is shown in Figure 8-6. HASMAT is used by a modeler to create a model representing human behavior at multiple levels, from societal behavior down to individual cognitive decision-making behavior. To create these models, the modeler can use a variety of modeling methods (e.g., social network modeling, Bayesian belief networks, rule-based systems, fuzzy logic, case-based reasoning). The modeler can also use a number of different methods for model integration, defining ontologies, data schemas, and mappings between individual modeling components or between a model and an external environment (e.g., a decision aid, a simulation, a real-world data source). A modeler also has access to tools for model management, including version control methods for existing or newly created models, and libraries of models and model templates that can be adapted to a particular domain or situation. All of these capabilities are accessed via the modeler interfaces, which provide GUIs to specific toolkit features. All of these components are integrated into a software system architecture designed to support reconfigurability, integration, and incorporation of new capabilities.

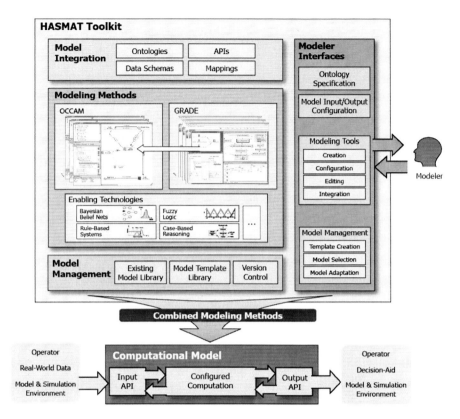

FIGURE 8-6 HASMAT system architecture.
SOURCE: Harper et al. (2007).

Figure 8-7 provides an overview design vision for the PMESII model analysis tools designed in the HASMAT environment. In the upper left of the graphic is shown a simple selection tool for the analyst to select from among the range of available models defining the PMESII environment for execution and analysis. The selection of a specific model results in the input fields for that model being captured from the selected model (via its XML schema-based I-O specification) and populating the tabular input sets shown on the left. On the right side of the figure are displayed the outputs of the model's execution, in this case, showing the national and regional SRO model outputs. This could be generalized as other potential representative structures, including a map-based overview of the model-generated results. As the user selects specific outputs in the callout datasets, the overview map shows the comparison of that value set across the modeled

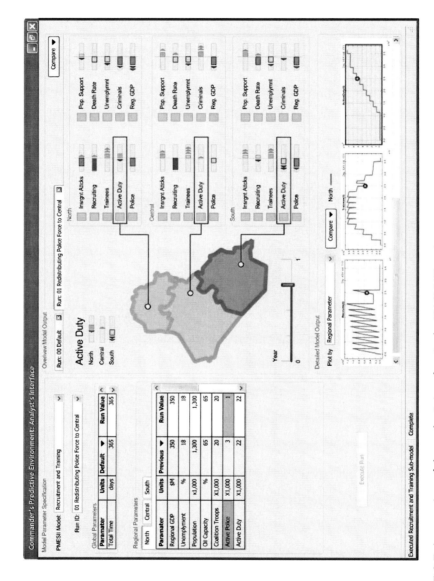

FIGURE 8-7 An overview of the analyst's interface in the HASMAT environment.
SOURCE: Harper et al. (2007).

regions through fill color. As the user drags the timeline back and forth, the output datasets will display the values as described for that point in time. Finally, in the lower right, the analyst is also provided with graphical displays of selected model outputs in time-series data plots, a feature that allows for easier access to trend data throughout a model run for more detailed analysis. As the user manipulates the timeline, these data plots shift a marker to identify the value at the selected time.

Modeling Terrorist Network Evolution

Figure 8-8 illustrates the software integration strategy that was used to generate a HASMAT-based modeling framework of terrorist organization activity (Harper et al., 2007). The model constructed within the integrated HASMAT framework consists of a social network representation of an organization or loosely connected set of groups or individuals of interest to the counterterrorism analyst. Each node within the social network can represent an individual (e.g., a key leader in the community of interest that has been the target of specific intelligence-gathering activities), a group (e.g., a set of individuals representing a cohesive entity in the community of interest), or an event. The links within the social network represent relationships between nodes in the modeled community, in which these relationships are defined at the outset by known intelligence (e.g., individual X is a known leader of group Y). This social network representation enables the analyst to build up the network over time based on intelligence products.

In typical social network analysis applications, this static representation would be used and analyzed to infer structural elements or features of the organization that, for example, might be exploited by counterterrorism specialists to capture further intelligence or to infiltrate a known group of interest. In HASMAT, however, this social network topology provides only the first step of the modeling capability. In HASMAT, each node of the social network is then populated by a "behavioral agent" representing the dynamic behavior of the modeled individual or group. These agents can be configured by the analyst based on gathered intelligence information. Thus, these agents are not static representations of individual or group "profiles," although they do contain representations of such information. Instead, they provide dynamic simulations of behavioral responses of the modeled individuals or groups within the social network to events and actions that are "injected" into these models based on evolving simulations of the social network dynamics. For example, an agent representing a given individual can "react" to incoming information (e.g., the invasion of Iraq by U.S. forces, the introduction of a new leader into a group of interest) and generate new events that propagate out to the social network (e.g., the establishment of new or strengthening/weakening of existing relationships

296

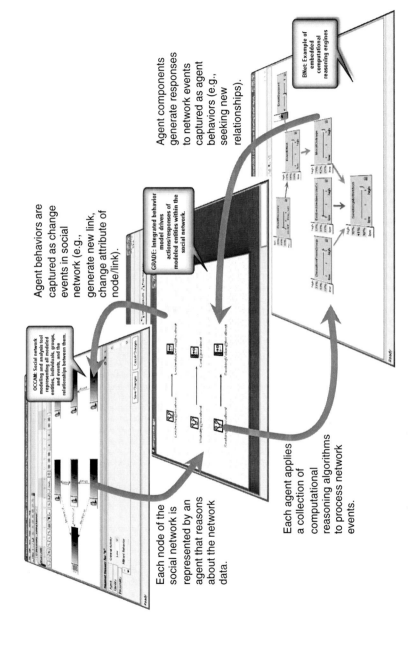

FIGURE 8-8 Overview of HASMAT software integration strategy.
SOURCE: Harper et al. (2007).

within the network). The result is an emergent, evolving representation of the organizational dynamics of the modeled community, driven by modeled reactive behaviors of individuals and groups.

Finally, at the bottom of Figure 8-8, we show the supporting modeling technologies that are assembled by the model developer (the analyst, a third-party social scientist, etc.) to provide the detailed representations of modeled individual and group behaviors. These detailed components capture and generate the simulated responses of a modeled individual or group based on injected or simulated stimuli. These simulated responses are then pushed back up to the social network representation as "change events" within the social network itself. Such change events include the generation of new links (i.e., relationships), the deletion of existing links, the adjustment of profile characteristics of the modeled node (e.g., an increase or decrease in an individual's radicalism), and the adjustment of a link attribute (e.g., strength or nature of a relationship).

This modeling framework was then used to model the well-documented terrorist activity leading up to the 2004 Madrid train bombings (Sageman, 2004, 2006; Harper et al., 2007), including organizational relationships among individuals associated with the attacks and their evolution over time (Telvick, 2007). Many interesting dynamics were seen in the data and modeled in the HASMAT environment. One goal of the effort was to model the outcomes that a group can take—it can talk or boast about operations, or it can actually take action. There are many factors that contribute to the final shift to action, including how much they have boasted of action so far, the easy access to weapons, the required skills, an external missive or deadline, and past criminal history—a predisposition to act. This and several other model outcomes were compared with the available data to assess model fidelity to the real world, to support rapid spirals of hypothesizing, developing, and validating, in an effort to understand the underlying dynamics of the terrorist network's behavior in terms of fundamental behavioral "primitives" (Harper et al., 2007). Without the rapid development environment afforded by HASMAT (and similar IDEs now beginning to be used in the community), these rapid spirals and exploration of possibilities would not be possible.

Modeling Iraqi Recruiting Activity

A similar software integration strategy was applied to generate a HASMAT-based modeling framework of Iraqi recruiting and training activity (Bachman and Harper, 2007). The SRO model (Robbins et al., 2005) was constructed within the integrated HASMAT framework, consisting of a system dynamics model representation of key PMESII components of Iraq, including demographics, coalition and insurgent activities, critical infrastructure, etc. This allowed for full communication between two heterogeneous modeling environments and the development of specialized models that were best served by the system dynamics paradigm (e.g., SRO model), or by a suite of computational intelligence components (e.g., agents that incorporate fuzzy logic, belief networks, expert systems).

A major objective of the development effort, besides developing a framework for integrating heterogeneous modeling paradigms, was to provide the analyst with an aggregated, accessible view of model results that could be used to support decision making by a commander or other decision maker. Such a tool would allow the commander's staff to easily generate model inputs (representing DIME actions that could be taken) and monitor model responses over time in a presentation framework that would be more intuitive than the PMESII IDE itself. For example, in the context of the recruiting and training SRO model, the analyst might specify a specific allocation of troops to support recruiting and training of specific capabilities in a given region of Iraq. The models would then be executed against this input set, and the analyst could monitor the overall effects on high-level PMESII variables (unemployment, economic stability, crime rates, etc.) in an intuitive graphical interface. This would insulate the analyst from the detailed outputs and implementation details that would be of interest to the model developer, while providing the analyst with intuitive and targeted real-time decision support leveraging the models constructed using the HASMAT framework.

Advanced Analysis Capabilities

Making predictions using DIME/PMESII models requires two types of what-if analysis, depicted in Figure 8-5 earlier in this chapter. The first type, causal reasoning, enables analysis from causes to effects. This allows the user to consider the effects of potential DIME actions on the PMESII models under consideration. The second type, diagnostic reasoning, enables reasoning from effects to causes. This allows the user to specify the desired (or actual) PMESII effects and determine the DIME actions that are most likely to achieve this result while minimizing undesirable second- or third-order effects. Supporting these two types of reasoning using PMESII models requires specific statistical sampling and analysis techniques.

Causal reasoning. In this type of analysis, the user specifies a set of DIME actions and the analysis indicates how these actions would influence the given PMESII models. Due to the nonlinearity of the systems being modeled and the incompleteness of information about system state, it is unreasonable to expect that PMESII models will provide high-fidelity predictive capabilities. Instead, the predictive value of the models lies in their ability to generate the distribution of plausible outcomes across multiple courses of action. One approach to this is to use Monte Carlo sampling, which refers to a family of algorithms that approximate a function f by calculating f(x), for a randomly chosen x, over many iterations. Sampling is a useful approximation technique in cases in which the function to be computed is difficult or impossible to calculate exactly. For complex nonlinear models such as PMESII models, randomized sampling provides an effective approach to approximating model outputs because it is independent of the underlying formalisms being used by the model.

Sampling can be used to analyze any model that incorporates both (1) a representation of the cause-effect relationships between model elements and (2) a specification of the relative likelihoods of inputs or initial states of model elements for which such conditions are not explicitly specified by the user. The first condition requires only that the model being sampled have some predictive capability. For example, hidden Markov models, belief networks, neural networks, and rule bases all meet this condition; a purely analytical tool such as a topic tree or a concept map does not. The second condition requires that the model specify a distribution of initial conditions for model elements, including the likelihood that various actions (either blue or red) will be observed. This allows the sampling algorithm to select random inputs according to a plausible distribution.

Given that the DIME/PMESII models in the system meet these two criteria, a user would perform a causal analysis using the analysis sampling tool in the following manner:

1. *Specify conditions.* The user first specifies the set of assumptions to be evaluated by the analysis; this includes not only the DIME actions of interest but also assumptions about the state of hidden variables in the models. The user also specifies the number of iterations to be performed by the sampling algorithm.

2. *Select data collection parameters.* The user then selects the elements in the models for which state data will be collected.

3. *Begin the simulation.* The IDE samples the PMESII models repeatedly. At each iteration, the states of variables not explicitly set in step 1 are randomized to permissible states given information about the relative likelihood of the initial states of the variable. The effects of the model inputs are propagated through the model, and the framework collects system state data for the variables selected by the user in step 2.

4. *View collected data.* The user then views the data collected in step 3, viewing the relative frequencies of various outcomes.

The envisioned IDE would allow the model analyst to specify initial conditions for input variables and view the resulting simulation data in a graphi-

cal format for analysis. By performing this type of simulation-based analysis
for multiple DIME actions, the user would be able to determine which actions
result in a greater likelihood of achieving the desired effects as well as which
actions result in a greater likelihood of causing undesired effects.

Means-ends and sensitivity analysis. The second type of reasoning of interest to a DIME/PMESII modeler is means-ends analysis: for a given effect or
system state, what are the actions that can be taken to achieve the desired
state? This type of analysis is very difficult to do using heterogeneous
models and is an area in need of further work. Outlined here are some
potential approaches to supporting this type of analysis.

One approach would be to perform a forward-chaining analysis for
each set of actions under consideration; the set of actions most likely to
achieve the desired result could be selected empirically based on the results
of each analysis. Such an approach is clumsy and inefficient, however, since
the forward-chaining reasoning process is itself computationally expensive.
Also, performing a brute-force means-ends analysis in this fashion, with the
large number of action sets that are likely to be possible, would quickly
become prohibitively complex and computationally expensive.

One solution is to reduce the search space of possible actions or input
states using a technique known as sensitivity analysis. Sensitivity analysis
computes, typically using black box sampling techniques, how variability
in the output of a model depends on variation in its inputs. Because it uses
sampling, sensitivity analysis can also be applied to any type of model formalism: only the inputs and outputs are observed. In the case of reasoning using
DIME/PMESII models, we can use sensitivity analysis to determine which
actions or input variables are most relevant in determining the outcome or
effect in which we are interested. Once we have identified a subset of relevant
actions, we can then perform brute-force means-ends analysis in the manner
described above to determine the optimal combination of those actions.

To illustrate this process further, consider the following example.
Suppose a group of modelers have developed a network of DIME/PMESII
models specifying the interrelationships between the economic and political
elements of a particular country. A user of the envisioned framework wishes
to use the aggregated model to gain insight into the types of actions that can
be taken to boost public confidence in the existing government. Because of
the complexity of the model and the number of possible inputs and actions,
the user performs a sensitivity analysis and determines that the factors
most critical in determining public confidence are the supply of electricity,
the visibility of police in the community, and the price of gasoline. Having
identified this subset of factors, the user performs a brute-force means-
ends analysis and determines that public confidence can be maximized by
increasing electricity supply by 20 percent, maintaining the current high
level of police forces, and reducing taxes on gasoline by 3 percent.

Because sensitivity analysis determines the variability of model output according to its inputs, it can provide results of interest other than just the relevance of an input. For example, the *rate of change* of model output as a result of input may be of even greater significance for a model user in selecting an optimal course of action. For example, if the model indicates a strong nonlinearity or "tipping point" in the output variable under consideration, this would indicate the importance of gathering additional information to determine how close to this tipping point the system being modeled actually is. Or the model may indicate that the results of an action on an output variable may be highly variable, with a large standard deviation; this would indicate a higher risk associated with the action, especially in cases in which the impact of the actions being taken is difficult to control.

In summary, a variety of frameworks and toolkits are in development, although the choices for IOS models are much more limited than for cognitive models of individuals, for which there are a number of well-known, tested, alternative architectures in widespread use. It is a recommendation of the report (see Chapter 11) that diverse frameworks for IOS models be supported and further developed—it is too early to tell which approaches will be most useful for different purposes.

VERIFICATION, VALIDATION, AND ACCREDITATION

In this section we describe some of the significant issues involved in the VV&A of IOS models: the ways in which they differ from physics-based models, the special challenges of forecasting human behavior, given the huge number of variables that can combine to determine it, and other thorny issues. We introduce the term "action model" and argue that military requirements for IOS models often include models for action as well as for understanding and exploration, and that the validation of such models cannot be done without a clear specification of the purpose for which the model is being developed. We also discuss the ways in which the military approaches VV&A and provide some examples of VV&A issues specific to various model types discussed in previous chapters. Finally, we make recommendations for dealing with IOS VV&A challenges.

General Issues: Validation for Use

All models are wrong, but some are useful.
G.E.P. Box (1979)

V&V are challenging issues for social science M&S. As generally understood, verification is the "process of determining that a model implementation accurately represents the developer's conceptual description and specifications." Validation is the "process of determining the degree to which a model is an accurate representation of the real world from the perspective of the intended uses of the model" (ITT Research Institute, 2001, p. 10). Stated more intuitively (ITT Research Institute, 2001, p. 10), verification asks "Did I build it right?" Validation asks "Did I build the right thing?"

In building it right, there are two elements: the degree of real-world representation and the intended uses of the model. They are related but are not the same thing. A realistic representation may not meet the intended use. It is a frequent error to put primary emphasis on a realistic representation, assuming it will meet the purpose. The result is an unending quest for realism without considering the intended use or purpose of the model. When one begins with the intended use or purpose, then the degree of real-world representation follows. Depending on the type of understanding that is needed or the action that might be taken, we can determine the degree of realism required. Here we develop an action approach to validation that begins with the model purpose—to take action.

V&V are necessary to support the goal of building and applying purposeful models and model simulations for understanding and exploration as well as for real-world actions. Research program managers frequently see V&V as a drain on resources. In contrast, practitioners or model users typically view the V&V process as a worthy investment of time and effort, since it can prevent the costly consequences of using incorrect models and simulations. If the intended use is not fully considered, then the model is not as useful as it might be. When the intended purpose is to take action—to do something—not just to understand or describe the world, the degree of realism needed is determined by the actions that can be taken in the situation.

This section stresses validation issues. Validation can be approached in two different ways within the larger V&V process. The first way is to begin with verification, proceed to validation, and then to the intended purpose. This ordering of concerns may result in a model that is verified and validated yet fails to be useful for its intended purpose. The second and recommended way is to begin with the intended purpose, proceed with verification, and then to validation in relation to intended purpose (Burton and Obel, 1995; U.S. Department of Defense, 1995).

First and foremost, without a prior specification of intended purpose, there are no clear-cut a priori criteria for deciding which features of a phenomenon to stress in its modeled representation. Indeed, multiple models that represent different aspects of a given phenomenon might be desirable and even necessary to achieve different purposes. For example, given a potentially unstable situation, a model constructed to describe the situation will in general differ substantially from a model constructed to guide the selection of an intervention action to stabilize the situation. That is, a model for understanding may not be a good model for action.

Moreover, each model purpose entails its own unique model validation requirements. In particular, the model purpose determines the appropriate trade-off between predictive accuracy, the appropriate formulation of dynamic processes, and the appropriate treatment of idiosyncratic and stochastic elements of real-world processes.

For example, models are frequently criticized for lack of realism—that is, not describing the world as it is observed or leaving out some aspect—but realism to what purpose? The continued addition of realistic features makes the implications of a model more difficult to understand, requiring increasingly sophisticated statistical and analytical techniques. Eventually, the continued addition of realism will result in a model's exhibiting such complexity that it has all of the interpretation problems of the real world itself, problems that presumably motivated the modeling effort in the first place. Extreme realism might also require an impractical amount of data to build the model or to specify parameter values and run the model. Consequently, if a simple model serves the intended purpose, then it should be preferred. Action models require some degree of realism for action, but realism is not a good test for action models. For action models, the purpose is to support decisions to take action—particularly when there is considerable uncertainty about the world.

As stressed by Marks (2006), the assertion that a model is validated when it is determined to be useful for its intended purpose is vacuous until "purpose" is defined. The purpose of a model could be to explain an observed phenomenon, to forecast a range of future phenomena that might occur without an intervention, or to guide the taking of actions in some specific problem context. For example, purposes might include behavioral description, behavioral explanation, behavioral prediction, exploration, normative advice and implications, training, and decision making (Burton and Obel, 1995). A different model would typically be required to meet each of these different purposes.

The first type of model purpose—explanation—requires what might be termed an understanding approach to validation. The latter type of model purpose—guidance for action—requires what might be termed an action approach to validation. Both purposes typically involve forecasting. The next section briefly reviews the understanding and exploration approach, commonly adopted in academic research. Following that, the next section elaborates the action approach, in which the purpose is to take action or intervene.

Validation for Understanding and Exploration

Consider first the case in which validation is undertaken for the purpose of explanation or of understanding the system that is modeled in order to gain new insights. Intervention is not the purpose here. Ideally, an explanation of a phenomenon would entail a complete understanding of both the necessary and sufficient conditions for its occurrence. In practice, compromise is essential. Any model will fall short of a complete understanding. There is no limit to further refinement for a more complete understanding.

As stressed by Haefner (2005), one possibility is that a model of a given phenomenon is incomplete in the sense that it is not capable of explaining all aspects of the phenomenon deemed to be important for an intended purpose. At the other end of the spectrum, multiple distinct models could offer different competing explanations for a given phenomenon, none of which could reliably be eliminated on the basis of currently available empirical evidence (observational equivalence).

An intermediate possibility stressed by Epstein (2006) is that a model has been constructed that is capable of reliably generating a particular phenomenon of interest (generative sufficiency). Such a model offers one candidate explanation for the phenomenon. Intensive experimentation could then be used to judge the robustness of the generative explanation to perturbations in the model specifications (Judd, 2006). If this process could somehow identify the entire class of models capable of generating the phenomenon, then the ideal but elusive goal of necessary and sufficient explanation would be achieved.

Consider next the case in which the purpose of validation is forecasting to identify a range of possible outcomes and estimate the likelihood of each. As Marks (2006) notes, this is a simpler purpose than explanation, in that only sufficient conditions for the occurrence of a phenomenon are sought. That is, one wants a model capable of generating reliable forecasts of outcomes (or outcome distributions) under various possible circumstances in some specified problem domain of interest. Whether this model is capable of elucidating all possible circumstances under which these outcomes would occur is not an issue of concern.

However, what is of concern for forecasting is whether a model is inaccurate. Does the model predict outcomes with misleading likelihoods? In particular, does the model predict outcomes that could never actually be observed? Prediction is a very important element of the action approach, as we explain below.

An important use of models is for exploration and the generation of nonobvious insights into complex phenomena that could not have been obtained without the model. A classic example is Schelling's (1971) tipping point model, which showed that neighborhood segregation could occur even if most people are racially tolerant. In these cases, the focus is not so much on the "accuracy" of the model, but on "unexpected" results. However, these models are also driven by their purpose—to provide new and important insights, where "new" and "important" are in the eye of the beholder—and their validity cannot be assessed without a deep understanding of that purpose.

Validation for Action

There are many aspects of an action model. An action model needs to relate actions of interest to outcomes of interest. The model does not necessarily need to reveal deep understanding. However, an action model must be timely and accurate relative to its purpose. For example, a model that predicts a hurricane's landfall is useful only if it provides predictions that are timely enough to allow for evacuation in advance of landfall and accurate enough to be taken seriously by those who need to evacuate. An action model is context specific in terms of available resources that help define what is feasible at this time and this place. In the illustration to follow, these issues are fundamental.

Validation for action begins with the purpose of the model. Prediction without and with intervention is an important element of an action model. Consider, now, an action model whose intended use is to provide guidance for the taking of actions in an uncertain environment. The validation process for an action model is necessarily different, but it does incorporate aspects from the validation processes described in the previous section for explanatory and exploratory models. In particular, prediction is important to action.

Specifically, the validation process for an action model must include a careful consideration of the modeled action choices, including no intervention. For example, have these action choices been specified in a suitably realistic or feasible way? And have enough action choices been included in the action domains of decision makers to permit them to display a realistic degree of flexibility in the face of changing and possibly unanticipated conditions? Appropriate modeling of action choices will not eliminate the uncertainty inherent in a situation, but it should help to clarify the possible action alternatives and hence provide useful guidance regarding the best action to take.

We start by considering the validation of a simple forecasting model with no action domain. This model is then generalized to an action model, and the implications for validation are considered.

A simple forecasting model with no action domain. The simplest situation is one of pure prediction in which there is no action to be taken. As an illustration, consider a corn farmer who lives in an area where it might rain or not and who wishes to predict the weather. This weather prediction problem for this corn farmer prediction model can be parsed into a number of distinct modeling issues.

First, what exact form could a weather prediction take? For example, the farmer could focus solely on rain, or he could also take sunlight into account. If the focus is solely on rain, the farmer could consider a simple probability distribution consisting of probability assessments for rain or no rain, or he could consider a more sophisticated probability distribution consisting of probabilities spanning the range of possibilities from no rain to a great deal of rain. The farmer might also choose to collapse this probability distribution into a simple prediction (forecast) concerning whether it will rain or not.

Alternatively, it could be that the farmer ultimately cares only about his corn yield, and he cares about the weather only to the extent that he believes the weather affects his corn yield. The farmer might express this belief by postulating an if-then relationship between weather and corn yield of the form "if A, then B." The contingency condition A might be either "rain" or "no rain" and the result B might then be a specific conditional probability distribution $Prob(b|A)$ for the corn yield b conditional on the realization of A. For example, it could be that the contingency condition "rain" is postulated to result in a two-thirds chance of a high yield and a one-third chance of a low yield, whereas the contingency condition "no rain" is postulated to result in a 50-50 chance of a high or low yield.

If-then relationships permit the formation of compound predictions. For example, continuing with the above illustration, constructing a corn yield prediction requires the farmer to assess and compound the uncertainty arising from two distinct types of events: the weather A, and the corn yield b conditional on the weather A. By the Bayes rule, the joint probability $Prob(A \cap b)$ that a specific weather event A and a corn yield b both occur is given by $Prob(A)Prob(b|A)$. Each probability assessment—$Prob(A)$ and $Prob(b|A)$—requires its own form of validation.

Second, what exact form should a weather prediction take, given the purpose that drives the corn prediction model? If the farmer wants only a rain forecast, then a simple assessment of the probability of rain versus no rain might suffice. If the farmer wants a more sophisticated understanding of the weather, he might assess a finer range of probabilities spanning a range of rainfall amounts. If he is interested in constructing a compound prediction of corn yield, then the fineness of his weather probability assessments will presumably depend on the postulated impact of different weather events A on corn yield b; there is no need to separately assess the probability of no rain and very light rain if both events are postulated to have the same effect on yield. Moreover, in addition to forming probability assessments Prob(A) for weather events A, he will need to assess the conditional probability Prob(b|A) of each possible corn yield b conditional on each possible weather event A.

Third, if an outcome is inherently uncertain, then point predictions regarding this outcome cannot be made with certainty. In the simple prediction model at hand, there is no way to eliminate the inherent uncertainty about the weather, nor should there be. A model cannot eliminate the inherent unknown of whether it will rain or not. A model describes what a modeler thinks he knows, but it also highlights what he does not know. A farmer might be able to say with confidence that the probability of rain is 0.3. Or, more generally, he might be able to provide a complete description of all possible rain events A in terms of probability assessments P(A) or by means of a nested sequence of confidence intervals. But he cannot say for sure whether it will rain or not.

To be valid for situations with inherent uncertainties, a model should reflect honestly what is knowable and capture well what is known. It is misleading at best, and quite possibly damaging, to use point estimates as if it is known with certainty what will happen. It is inappropriate to build more into a model than is knowable for the situation.

The corn farmer prediction model at hand is a pure prediction model; the farmer is not faced with the need to choose an action. The following section considers the implications for validation when this simple model is generalized to include action choices for the farmer.

A simple illustrative action model. Consider, now, a corn farmer action model that represents a simple extension of the previous corn farmer prediction model. The corn farmer now has the option of adding fertilizer to his field or not. Consequently, the farmer's action domain consists of two possible action choices: add fertilizer to the field or do not add fertilizer to the field. (This action validation model is based on a decision theory model, similar to those discussed in Chapter 5.)

Some structural aspects of the farmer's problem are assumed to remain the same: the probability that it will rain or not (which assumes the weather is independent of the farmer's action choice) and the possible corn yields realized under rain or no rain in the absence of fertilization. However, the impacts of fertilization on corn yield—for example, bushels per acre—when it rains and when it does not rain must now also be considered. Specifically, as depicted in Table 8-4, the farmer needs to specify what the corn yield would be under rain and no rain should he choose to add fertilizer to his field or not. This results in four distinct "compound" contingency conditions (combined weather and action states) that could impact corn yield.

As Table 8-4 shows, the corn farmer action model has the same general framing as the corn farmer prediction model, except that the if-then rela-

TABLE 8-4 Contingency Table for the Corn Farmer Action Model

	Outcome	
Action	Rain	No rain
Add fertilizer	What will be the yield with rain and fertilizer?	What will be the yield with no rain and with fertilizer?
Do not add fertilizer	What will be the yield with rain and no fertilizer?	What will be the yield with no rain and no fertilizer?

tionships must now be generalized to indicate what will happen for various actions or interventions inserted into the natural order of things.

An important point to stress is that the validation of an action model does not necessarily require the model to generate highly accurate predictions. For example, for the problem at hand, the corn farmer might be able to deduce that the addition of fertilizer to his field will profitably increase his corn yield whether or not it rains, because of a government support program that reimburses farmers for all of their fertilizer costs—that is, fertilizer is free. In this case the farmer's best (most profitable) action choice is clear; he should choose to fertilize his field. He does not need to predict with high accuracy the probability of rain, the probable effects of rain on his resulting corn yield, or the price of corn in order to take the best action.

In short, the validation process for action models is sharply distinct from the validation processes for explanatory, purely predictive, and exploratory models. The primary focus is on taking the best action rather than on the realism of the model or the ability of the model to generate accurate predictions. On one hand, a purely predictive model might not say much about which action to take—a rain forecast by itself does not tell us whether to fertilize or not. On the other hand, good understanding and predictive power could be essential requirements for achieving a useful action model. A good understanding of weather and how weather affects crop yield under different fertilization conditions could be essential for deciding whether to fertilize or not.

We now illustrate a more involved situation, in which the validity of an action model depends critically on the model's descriptive and predictive accuracy.

A more complicated action model. Consider a more complicated action model involving the following hypothetical decision: Should a military force enter a potentially hostile village in order to establish a relationship with the local militia leaders, and if so, which entry mode should be chosen?

COMMON CHALLENGES IN IOS MODELING

The outcome resulting from each possible choice of a village entry mode depends on the degree of hostility of the mayor and the presence (or not) of a local resistance group. The purpose or goal is simply to occupy important terrain and minimize casualties—those of both the military force and the villagers.

This village deployment action model could be entirely framed in terms of if-then relationships connecting compound contingency conditions to ultimate outcomes, in which each compound contingency condition involves a village entry choice together with a mayor hostility level and the presence or absence of a local resistance group. To focus attention on action choice, however, it is useful to separate out the entry choice from the latter two nonaction contingency condition aspects. In particular, it is useful to think of these nonaction contingency condition aspects as constituting a scenario conditioning the choice of an action. The contingency table for this model is depicted in Table 8-5. Note that only three of the four possible scenarios are depicted for ease of exposition.

Validation of this village deployment action model involves three critical considerations. First, how appropriate is the action domain listed down the left-hand side of the table for the problem at hand? The action domain is the set of possible actions that can reasonably be taken in this situation. There are two possible kinds of errors here. The action domain might be poorly specified. For example, are the village entry choices in the table truly relevant and feasible given the available resources, the situation constraints, and the time available to take action? Alternatively, the action domain might be incompletely specified. For example, the village entry choices in the table are classified only by degree of armament, ignoring possibly critical timing issues (e.g., enter at dawn versus enter at night). In addition, other ways to enter the village (e.g., with an initial leaflet drop or bombardment) might be feasible.

Second, are the scenarios listed along the top in Table 8-5 appropriately specified for the situation at hand? The scenarios are the set of possible conditions that might arise that we do not control or determine. In particular, are the contingency condition aspects that form the basis for these scenarios both reasonably accurate and reasonably complete? For example, it might be the case that the initial attitude of the mayor and the existence (or not) of a resistance cell are not the only important aspects to consider for the characterization of the initial conditions. Another attribute of equal or greater importance might be whether the civilians (i.e., the inhabitants of the village) are religious or not.

Even assuming the initial attitude of the mayor and the existence (or not) of the resistance cell are correctly identified as the two most important aspects to consider in conjunction with village entry mode, only three of the four possible combinations of these two aspects are analyzed for the village

TABLE 8-5 Contingency Table for Village Deployment Action Model

Action	Outcome When the Mayor Is Hostile and There Is a Resistance Cell	Outcome When the Mayor Is Hostile and There Is No Resistance Cell	Outcome When the Mayor Is Not Hostile and There Is No Resistance Cell
Do not enter	[None]	[None]	[None]
Enter with firepower evident; return defensive fire only after receiving sporadic gunfire from snipers	The mayor does nothing to stop active resistance; fire is returned; it is likely that a few civilians are killed; one or two soldiers are wounded	The mayor will organize a demonstration; a few civilians are roughed up; no casualties	The mayor attempts to negotiate while the citizens resist passively; a few civilians are detained; no casualties
Enter with firepower evident	The above with a lower probability of killing villagers	Do not know	Same as above
Enter with small group to negotiate	The mayor negotiates and there is a high probability that the small group will be held captive; no casualties; no terrain occupied	The mayor negotiates in the town square and finally lets the small group go; no casualties; no terrain occupied	The mayor negotiates for food and medical supplies; no casualties; no terrain occupied
Enter with food and medical supplies	The mayor forbids the distribution of the food and medicine and the cell initiates an exchange of fire; no casualties; terrain occupied	The mayor forbids the distribution of the food and medicine and demands that the troops leave the area; no casualties	The mayor is welcoming and negotiates for more food and medicine; no casualties; terrain occupied

in the table. A fourth possible combination—a nonhostile mayor together with the existence of a resistance cell—is not considered. By excluding this fourth combination, the modeler is effectively concluding either that it is impossible or that it is so improbable that it is not worthwhile to consider.

Third, are the if-then relationships mapping the contingency conditions (scenario-action pairs) into possible outcomes appropriately specified and explained? The cells in Tables 8-4 and 8-5 contain the outcomes for the scenario-action pairs. In the corn farmer action model, it is assumed that fertilizer (or not) and rain (or not) are the only two unknown conditions of importance and that corn yield is the only important outcome variable. For the village deployment action model, however, the if-then relationships mapping the contingency conditions into possible outcomes are inherently much more complicated.

Specifically, as seen in Table 8-5, the village deployment action model is assumed to have three important aspects making up each contingency condition: the village entry mode, the village mayor's initial hostility level, and the existence (or not) of a local resistance cell. Given any particular combination of the latter two aspects, the military chooses a village entry mode. Given any particular combination of these three aspects, there is then a response by the mayor and a response by the resistance cell (if present). The overall combination of all of these events then determines an ultimate effect on civilians and the mission of entering the village. Also, the if-then relationships connecting contingency conditions to possible outcomes in Table 8-5 might not be appropriately specified. For example, it is assumed that a definite outcome results under each possible contingency condition when, in fact, a great deal of residual outcome uncertainty might remain (e.g., the level of civilian losses might still be uncertain). Moreover, these outcomes might be incompletely specified (e.g., the casualty rate among soldiers is currently ignored in some cells).

A template for validating action models. As developed above, the validation of an action model should begin with the purpose of action—not the model itself. One cannot assess the validity of an action model without first knowing its purpose. Next, the validation of an action model will typically be demanding, involving specification of an action domain, scenarios, and if-then relationships. The best approach to follow for carrying out this validation will depend on the purpose of the model.

In brief, validation should establish the purpose of the model, list the possible actions or interventions for the purpose, specify the scenarios that depict the uncertainties or unknowns that are inherent and cannot be eliminated, and develop the if-then relations between the possible actions and the possible uncertainties—that is, predict the outcomes, which might be multidimensional and uncertain as well.

The validation of an action model is a challenge that does not lend itself to a well-specified set of detailed procedures; the action model gives a template for which the details must be filled in. Nevertheless, the examples presented in the previous sections suggest a reasonable order of concerns, which is summarized below.

Before an action model can be validated, the purpose must be specified. One cannot proceed with the validation without this specification. In the corn farmer example, the purpose of the farmer is to make a high profit. In the military force example, the purpose of the military force is to occupy the village with a minimum of casualties.

First, is the action domain appropriately specified for the situation with regard to an operational specification of each action and the completeness of the possible actions? For the farmer example, is to fertilize or not an appropriately complete specification of the farmer's possible actions? For the military force example, are the four ways to enter the village an appropriately complete specification of the force's possible actions?

Second, are the considered scenarios appropriately specified? Are rain and no rain the appropriate scenarios for the farmer? Are the mayor's and resistance cell's actions appropriate? Does the range of considered scenarios cover the range of situations in which actions might actually have to be taken?

Third, is each if-then relationship connecting a contingency condition to a possible range of outcomes specified with an appropriate level of realism and prediction? For the farmer, the corn yield is important, as is the price. For the military force, the mayor's reaction and then the resulting effect on the occupation and casualties are important.

The specifications of these three key model features (action domain, scenarios, and if-then relationships) are interdependent, and all three aspects need to be carefully considered for the overall validation of the model. Each presents a different problem for the modeler. The specification of an appropriate action domain requires a deep understanding of what actions are feasible and reasonable for a situation. The specification of appropriate scenarios requires a deep understanding of what is likely to be known and not known (and what is knowable) about a situation at hand. The specification of appropriate if-then relationships requires a deep understanding of the causal structure connecting contingent conditions (scenario-action pairs) to potential outcomes.

The specification and validation of if-then relationships is particularly difficult for systems involving multiple interacting human beings with capabilities for learning and social communication. The largest source of uncertainty in social systems is behavioral uncertainty, that is, uncertainty regarding what other people will do. It is for this reason that the "then" parts of the if-then relationships postulated for social systems will generally have to be in the form of multidimensional subjective probability assessments giving likelihoods for a range of possible outcomes. These probability assessments will inherently be subjective judgments based on an understanding of individual and group behavior gleaned from observations, surveys, human subject experiments, and biological and physical considerations.

There are a number of common errors that are to be avoided. It is a frequent error to develop a simple predictive model that assumes no action or intervention and that does not explicitly specify the scenarios. The possible mistakes are, first, that the model might be used for action or intervention without consideration of the critical outcomes; second, the model does not specify the scenarios, and it is easy to assume inappropriately that the model applies to all scenarios. Without a specification of the actions and the scenarios, a model is quite limited.

The uncertainty inherent in most action models involving individuals and organizations makes validation difficult. Inability to conduct repeated experiments is a key issue. We can observe rainfall and corn yield over many years. Consequently, statistical averages might be both available and useful for a corn farmer contemplating these types of events. However, the general lack of repeated experience for those contemplating the entry of a village might make it impossible to use simple descriptive statistics based on averaging. At best, there will be a range of possible outcomes with subjective estimates about what will occur. Because the largest source of uncertainty in social systems is behavioral uncertainty, the if-then relationships postulated for social systems will typically have to be stochastic, giving a range of possible outcomes for each contingency condition together with a probability assessment for each outcome.[8]

Military Approaches to Verification, Validation, and Accreditation

The Defense Modeling and Simulation Office (DMSO) has devoted considerable effort to the development of definitions, processes, and tools for V&V of models and simulations; formal definitions of terms and concepts are given in DoD Directive 5000.59-M, Glossary of Modeling and Simulation Terms. Additional information and a larger glossary can be found at the website devoted to VV&A, the DoD VV&A Recommended Practices Guide (http://vva.dmso.mil/).

A simplified sketch of how VV&A is interrelated to the overall process of M&S development is given in Figure 8-9.[9] Ideally, the M&S process begins with the development of a conceptual model, proceeds to the design and implementation of a simulation of that model, and ends with testing and evaluation, allowing for "spirals" of iterative design-development-testing over time. In parallel, the VV&A process begins with validation of the conceptual design, verification that the design and its implementation properly instantiate the conceptual model, and validation of the test results.

---

[8]For a more detailed discussion of model validation for social systems, see Carley and Svoboda (1996), Fagiolo, Windrum, and Moneta (2006), and the extensive resources available at http://www.econ.iastate.edu/tesfatsi/empvalid.htm.

[9]The figure shows the process for new M&S development; the process for VV&A of existing M&S systems is considerably more complicated (http://vva.dmso.mil/Role/VVAgentLegacy/default.htm).

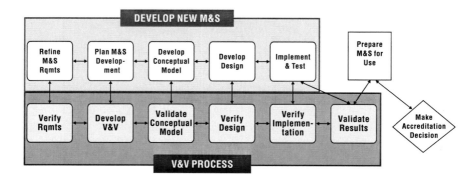

FIGURE 8-9 Interaction between V&V and new development activities.
SOURCE: Adapted from http://vva.dmso.mil [accessed Feb. 2008].

In effect, the validation tasks focus on ensuring that the model adequately represents that portion of the real world being modeled, and the verification tasks focus on ensuring that the simulation adequately implements the model. Accreditation is shown as the last step in the process, the point at which the accrediting agency (the owner of the simulation) places its stamp of approval on the validation results.

The traditional focus for DMSO M&S development has been the physical battlespace environment (terrain, ocean, atmosphere, etc.), with an emphasis on conventional warfare. Simulation and VV&A of the entities that populate the environment has typically been left to the separate services or "accrediting" agencies: the Air Force has been responsible for development and VV&A of F-16E aircraft, the Army for Bradley Fighting Vehicles, etc.

It is also fair to note that the great preponderance of M&S entity development has been devoted to "platform" entities, not the human decision makers who "drive" those entities, either at a one-on-one level (e.g., the pilot of an F-16E) or at a higher command and control level (e.g., the Joint Force Air Component Commander). As a result, corresponding VV&A efforts have been equally unbalanced, with the majority of the effort devoted to the VV&A of systems for which there is a strong conceptual model (e.g., a behavioral law for platform kinematics, such as "distance equals speed multiplied by time") and a well-understood protocol for validating that model against the real-world behavior of the entity being modeled (e.g., measuring time, speed, and distance of the moving platform in the field). VV&A of IOS models is particularly problematic, because of both the lack of clear and generally accepted conceptual models and the difficulty of conducting "clean" human-in-the-loop experiments, unconfounded by large individual differences across individuals, inadequate experimental controls over variables that subtly influence human behaviors, learning and adaptation over repeated experimental trials by individual experimental subjects, etc.

These shortcomings were identified in an earlier National Research Council (NRC) study, which noted as its primary conclusion that the M&S community should adopt a general framework for developing *and* accrediting models over three time horizons (National Research Council, 1998):

- Short term
  - Collect and disseminate human performance data.
  - Support incremental improvement for selected models.
  - Create accreditation procedures for human behavior modeling.
- Mid-term
  - Extend task analysis efforts—for example, STRICOM's Common Model of the Mission Space.
  - Support sustained human behavior modeling development in focused domains (e.g., AFRL's Agent-Based Modeling and Behavior Representation (AMBR) Air Traffic Control Testbed).
- Long term
  - Support theory development and basic research.

At the time that the earlier report was written, the focus of most DoD M&S was on conventional platform-dominated nonurban warfare. The report was similarly focused. It is now clear that considerably more emphasis has to be given to the development of models that span the space from the individual decision maker, to small groups, to urban populations, and even to entire national and transnational populations. As noted in Chapter 5, we need to account not only for "nominal" human behaviors, but also for those colored by individual differences (e.g., personality traits) and ethnic/religious/cultural influences. And, since no individual operates in a vacuum, we also need to account for the influences of the organizational structures and social networks mediating human intercourse.

This is a tall order for the M&S community, but efforts have started. Since the 1998 NRC study, the Air Force has initiated a number of programs:

- AFRL/HE workshops on Adversarial Modeling (2002), Cognitive Engineering (2002), Cognitive Modeling, Science, and Engineering (2003), and Representing Personality and Culture (2003).
- The Aeronautical Systems Center Engineering Directorate's SAMPLE program to develop agent-based pilot models to populate the SIMAF engagement simulation.

- The AFRL/HE AMBR Program, directed at developing intelligent agents to mimic human behaviors.
- A nascent effort in modeling human behavior by the Behavioral Research Branch of the National Air and Space Information Center.

The other services have likewise started comparable efforts in this area, most notably the Office of Naval Research's Affordable Human Behavior Modeling Program, having the following major goals (see http://www.onr.navy.mil/sci_tech/34/342/training_afford.asp):

- Reducing the time-consuming knowledge engineering needed to define the fundamental behaviors to be modeled for a particular operator in a given military role.
- Reducing the M&S construction effort (model concept design and simulation development) required to develop simulations of the desired human behaviors.
- Identifying processes to ensure reusability of models and model components.
- Developing improved V&V techniques for the developed models. This is one of the few programs that directly addresses V&V issues.

Another major finding of the NRC report on human behavior modeling (National Research Council, 1998) was that substantial effort needs to be invested in the development and VV&A of larger scale models that go beyond the representation of individual humans, to begin to address collections of individuals, from small teams, to groups, crowds, urban populations, and even nation-states. This was echoed by the DMSO-sponsored conference on organizational simulation held in 2003 (Rouse and Boff, 2005), which attempted to address qualitative and quantitative changes in the fundamental M&S issues associated with modeling larger groups of individuals. Although a number of novel and disparate approaches were proposed and described, only a very small fraction of the proceedings directly addressed critical VV&A issues for this class of behavioral models, such as:

- What constitutes behavior prediction/forecasting for multiple interacting simulated human entities? Should we relax our need for prediction accuracy and instead be satisfied with robustness in anticipating the range of future possibilities?
- How does one go about validating a conceptual model when the model is still being formulated? Are there different levels of validation that apply?

- How does one verify the resulting model simulation, when so many of the interesting qualities of behavior reflect idiosyncratic and stochastic activities that change over time due to learning by the individual and socialization by the group?
- Who should be the accrediting agency?

There are, however, a number of ongoing efforts aimed at advancing the application of large-scale organizational modeling to DoD questions of interest, most notably the Joint Forces Command's Urban Resolve Program, which focuses on urban operations and how M&S can be used to explore and define urban operations war-fighting capabilities for the future joint force commander. The first phase focuses on intelligence, surveillance, and reconnaissance operations in the urban environment, using a high-entity count simulation of Jakarta built on top of JWARS, which, in turn, is built on top of the existing OneSAF simulation framework (see Chapter 2). Although the development plan for JWARS originally included a detailed VV&A plan for conventional warfare scenarios,[10] it is unclear whether V&V efforts for Urban Resolve went any further than the "looks ok" test. This is not atypical of large-scale simulations in general.

Finally, we note that AFRL/HE has begun the process of attempting to formalize the VV&A process for individual human performance models, cognitive models, and group representations developed or "owned" by the directorate, via the publication of the AFRL/HE Instruction 16-03 (Brinkley, 2003). We believe this is a good start in this particularly difficult area.

Validation Issues Specific to Individual Modeling Approaches

In this section we review the validation challenges and approaches that are specific to various modeling approaches used for IOS models.

Validation of Conceptual Models

Verbal conceptual models are sometimes specific enough that they can be tested and plausibly falsified, using empirical field studies or controlled experiments. For example, Fiske and colleagues have used social cognition experiments to demonstrate that people organize acquaintances in memory according to the dominant model that organizes the relationship, in studies of subjects from Bengali, Chinese, Korean, Vai (Liberia and Sierra Leone), and U.S. cultures (Fiske, 1992), and that for many subjects this classifica-

---

[10]This information is based on the program overview at http://www.msiac.dmso.mil/spug_documents/JWARS_Overview_Brief.ppt.

tion accounts for more variance in recall and substitution errors than personal attributes, such as gender, race, and age.

In contrast to such well-developed conceptual frameworks, broad metaphors (brains as information-processing devices, organizations as cultures) are not really subject to verification or falsification. Whether or not they are used in a particular domain is likely to depend largely on face validity and established precedent. In evaluating the usefulness of a broad conceptual model, the yardstick is often not how well supported the model is, but how much interesting research it inspires. Even when a verbal model seems, in principle, to be subject to falsification, the underspecification of relations and processes often means that a rather broad array of different outcomes can be presented as consistent with the theory. As Harris (1976) noted in his paper entitled "The Uncertain Connection Between Verbal Theories and Research Hypotheses in Social Psychology," theoretical terms often are not defined, boundary conditions are unspecified and, under various plausible interpretations of assumptions or conditions, several well-known theories include internal contradictions and inconsistencies (cited in Davis, 2000).

Validation of Cultural Models

Cultural inventory models rely on ethnographic observation and are therefore both time-consuming to develop and highly subjective. Having multiple independent observers helps ameliorate the subjectivity problem, but it is expensive.

Dominant trait models, such as the Hofstede dimensional models, can involve two sets of data. The first set is used to derive the dimensions. These can be validated by a number of different statistical methods, such as factor analysis. Once these are fixed, another set of data is obtained to score each new culture on the dimensions. These data have to be obtained from willing natives of the culture, and the data have to be updated over time because cultures change.

Validation of Cognitive Models

While there is increasing emphasis on validation of cognitive architectures, validation remains one of the most challenging aspects of cognitive architecture research and development. "[Human behavioral representation] validation is a difficult and costly process [and] most in the community would probably agree that validation is rarely, if ever done" (Campbell and Bolton, 2005, p. 365). Campbell goes on to point out that there is no general agreement on exactly what constitutes an appropriate validation of a cognitive architecture. Since cognitive architectures are developed for a wide variety of reasons, there is a correspondingly wide set of validation (and evaluation) objectives and metrics and associated methods. Lack of established benchmark problems and criteria exacerbates this problem.

Validation of Cognitive-Affective Architectures

In spite of the challenges associated with validation of emotion models and cognitive-affective architectures, progress is being made in the area. A promising trend in emotion modeling is the increasing emphasis on including evaluation and validation studies in publications. As is the case with cognitive architectures, no existing emotion models or cognitive-affective architectures have been validated across multiple contexts and a broad range of metrics. However, some important evaluation and validation approaches and studies exist and are discussed in detail in Chapter 5. Cognitive-affective architecture validation has not yet reached the stage of systematic comparisons that is beginning to be used for their cognitive counterparts. However, given the recent emphasis on validation in the computational emotion research community, such studies are likely to be taking place in the near future.

Validation of Agent-Based Models

Agent-based models (ABMs) are computational frameworks that permit the theoretical exploration of complex processes through controlled replicable experiments (see Chapter 6). In principle, these experiments could be run entirely with artificially generated initial conditions, parameter values, and functional forms. Nevertheless, their ultimate usefulness depends on the extent to which they prove capable of shedding light on real-world systems, that is, their ability to enhance understanding and guide decisions and actions.

When validation of ABM frameworks is attempted, the validation is generally restricted to small areas of performance. A typical approach to validation is to run an experiment using an ABM framework, collect data from this experiment, statistically analyze the results to generate the response surface, and then contrast the response surface with real data. It is easy, even with only a few variables, to generate such a quantity of data from an ABM framework that there are no existing data with which to compare them, no existing statistical package can handle them, and most desktops cannot store them. Therefore, typically only small portions of the overall response surface can be estimated at once. The size of the analyzed response surface is thus often dictated by the user's interests and the critical policy or decision-making questions at issue (i.e., the action domain and the scenarios relevant to that domain, as discussed above).

ABM researchers have recently begun to explore promising new approaches to validation. For example, a number of them are now advocating iterative participatory modeling (IPM) as an effective way to incrementally achieve validation of the structural, institutional, and behavioral aspects of the complex systems they study. For an introductory exposition of IPM, see Barreteau (2003). The essential idea is to have multidisciplinary researchers join with stakeholders in a repeated looping through a four-stage modeling process: (1) field study and data analysis, (2) scenario discussion and role-playing games, (3) ABM development and implementation, and (4) intensive computational experiments.

The new aspect of IPM relative to more traditional participatory modeling approaches is the emphasis on modeling as an open-ended collaborative learning process. The modeling objective is to help stakeholders manage complex problems over time through a continuous learning process rather than to attempt the delivery of a definitive problem solution.

In addition, ABM researchers are also beginning to explore the potential benefits of conducting parallel experiments with real and computational agents for achieving improved validation of their behavioral assumptions.[11] A critical concern is how to attain sufficiently parallel experimental designs so that information drawn from one design can usefully inform the other.

Recommendations for Developing and Validating IOS Models

We have argued that IOS models should be validated beginning with the purpose and then considering the action set, scenarios, and if-then relations in the specific situation. The committee makes a number of suggestions for modeling and simulations that will facilitate the validation of a specific model.

Check with Multiple Experts

Four different experts should examine an IOS model: the users of the model, the scenario experts, the if-then or domain experts, and the modelers themselves. Modelers cannot examine a model by themselves; they tend to focus on the verification with less emphasis on the purpose of the model. For an action model, the user is very important to check the relevance and feasibility of the action set. The scenario expert should examine the uncertainties and unknowns. Domain experts are particularly knowledgeable about the if-then relationships. However, their knowledge is not necessarily

---

[11]See http://www.econ.iastate.edu/tesfatsi/aexper.htm for annotated pointers to ABM research on parallel experiments with real and computational agents; see also the survey by Duffy (2006).

framed in this manner, so some adjustment may be required. For example, domain experts know about "what is" and "what has been" but may be less certain about "what might be" outcomes. However, they are likely to point out errors in the models for what might be and limits of what is known. Each expert can contribute to the validation of an action model. It is unlikely that any single expert can ensure a valid action model alone. The structure and content of the model provide a template for a procedure by which multiple experts can validate different aspects of an integrated action model.

Keep the Model as Simple as Possible for Its Purpose

An IOS model does not have to be complex. Parsimonious models are preferred. The corn farmer action model is simple and does not capture the complexity of weather forecasting or the chemistry of fertilizers. But it is understandable and permits the farmer to make a decision and take action. Action models that are intuitively understandable to decision makers (transparent) are preferred. An action model that is disconnected from a decision maker's intuition and from concepts he or she is familiar with does not permit interplay between the decision maker and the model. In short, complicated, nonintuitive action models require decision makers to accept the implications of the models on blind faith. Action models should aid decision makers, not replace them.

Examine "What Might Be" as Well as "What Is"

"What is" should mimic the real world within limits. "What is" models are a basis for "what might be." A model that has little or no correspondence with the real world is not likely to be relevant for what might happen. What might be is very important for action models—particularly in new situations (Burton, 2003). Many of the relevant action-scenario combinations have not been observed in the past. So the model must be relevant for action beyond what is or what has been to new situations. For example, it would be desirable if the illustrative village deployment action model could be used reliably in other similar situations, say for the withdrawal from a village as well as entry. But it is not likely that the model could be used to help plan an action to disarm a resistance cell. Presumably this would require a more detailed model of the functioning of the cell. Whether it would be desirable to develop one model to handle both entry and cell disarmament or two separate models would presumably depend on economies of scope—Is there anything to be gained by considering both issues jointly?—and on computational implementation costs. IOS models should be developed and examined beyond what is to what might be. At the same time, it is important to examine the limits of the model and not use it in situations in which it might be inappropriate. As suggested above, simplicity is desirable, but it must be balanced so that the action model is useful for its purpose.

IOS models are likely to forecast a range of possible outcomes, some more likely than others, and to incorporate many factors that are highly uncertain and, indeed, unknowable at the time the model is developed. How then can such models be validated? Popper, Lempert, and Bankes (2005) argue that models used to explore policy alternatives for an uncertain future should not be expected to yield predictions that can be tested but rather should be used to explore and compare possible outcomes under a variety of possibilities in order to select strategies that are robust—yielding the best overall results across a variety of possible futures.

Postevent outcomes can also be used to evaluate models, although models are not necessarily incorrect if the actual outcome that occurred was not the one forecast to be most likely. Unlikely events do occur, and many IOS applications do not permit the replication that would generate a distribution of actual outcomes. A very useful approach would be to develop multiple models that take different perspectives and use different theories and data, merge their predictions to create zones of likelihood, and compare their forecasts with the actual outcomes (see Docking below). As with other validation approaches, the value of the model's results depends on its intended use, so the degree to which forecasts need to correspond to reality will depend on the model's purpose.

Use Model Touching for Validation

Model touching is comparison or juxtaposition of models. There are many ways to bring models together. Here is a list:

- Bring experts (as described above) together to develop and examine the model.
- Compare the action model with qualitative studies for the situation or domain.
- Check with other studies that might be empirically based on data from the field or from experiments.
- Compare with computational models that are based on field data.

Docking. Docking is the bringing together of two models—a metaphor borrowed from space exploration. More precisely, docking is an evaluation of the extent to which two or more different models of the same action situation can be cross-calibrated so that they yield the same outcome (or outcome probability distribution) given the same contingency condition

(Axtell, Axelrod, Epstein, and Cohen, 1996). Docking goes beyond model touching to compare in more detail. It can provide a better understanding of the true connections relating the three key elements of an action model: the actions, the scenarios, and the possible outcomes resulting under each contingency condition (scenario-action pair). Docking gives confirmation that we have a reasonable understanding of an action situation, and that our conclusions are being driven by the intrinsic nature of the action situation and not by idiosyncratic aspects of the model implementation. One possible approach is to compare how different models perform under the same benchmark action-scenario combination, which can provide insight into how different models define actions and how they structure if-then relationships. That is, for an action model, take the same action possibilities and the same unknown scenarios, then develop two separate if-then relationship models. Develop and compare the outcome tables for the two models. Are the outcomes the same? If not, why? One must go behind the model outcomes and examine the details of the models to understand their differences. Individuals who are expert in the subject are critical in judging the models and their value. Docking should involve experts throughout the process, as discussed above. Docking of multiple modeling approaches against common benchmark problems using a panel of expert judges has recently been used to provide considerable insight into individual cognitive performance models (Gluck and Pew, 2005).

At this time, there is a need to develop benchmark scenario-action situations that can be used to dock two or more models. This effort will involve action, scenario, and if-then experts. With these benchmarks, docking studies can add greatly to the development of action models.

Given the current state of the art, the participation of experts in the docking process is essential. The next best step in validation is to support docking studies among experts who develop computation-based models. Automated machine docking of two or more models is a very high-risk endeavor at present. At a later stage of understanding, we may be able to develop a computationally based approach to the docking of models. But for now, experts and their judgment are mandatory.

Triangulation. Triangulation goes beyond docking and involves examining the same action domain using an action model, an expert group using a qualitative approach, and reference to quantitative studies in the domain. An action model validated using multiple approaches is more likely to help the decision maker take actions that meet the purpose. However, a large number of triangulations are often possible. We do not know a priori what the best triangulation is for a given situation, but it is quite likely that a good triangulation will be situation dependent.

Exploratory testing of robustness. Miller (1998) proposes active nonlinear tests for complex models to validate the model's structure and robustness. In this approach, automatic nonlinear search algorithms probe for extreme outcomes that could occur within the set of reasonable model perturbations. This multivariate sensitivity technique can find places where a complex model "breaks," that is, produces results that are outside a range of reasonable predictions.

In summary, universal rules about what is the appropriate procedure for validating IOS models are not possible. However, we recommend the validation of models through a three-part triangulation process, based on the purpose of the model. Validation should involve (1) participation by multiple experts who can provide different perspectives on the action domain, the scenarios, and the if-then rules incorporated in the model; (2) docking of similar computational models against one another; (3) comparison to qualitative and theoretical studies and previous quantitative results and exploratory testing for a range of outcomes. A good heuristic would be to begin with the experts as discussed above and move as quickly as possible to docking studies and exploratory testing.

DATA ISSUES AND CHALLENGES

Data can be used in two different ways in modeling. When models are developed inductively from data, the quality of the data is extremely important. In that case the data are broader in scope and limited only in a very general manner. For example, an anthropologist sees different things than an engineer in the same situation. For existing models, the data are prescribed by the model, and the quality of data is extremely important. Here again, the data yield values for the model parameters and make the model specific to a given situation and problem. The data requirements are driven by different modeling needs. For each situation, quality data are needed and are important to the usefulness of the model.

This means that even the most promising, sophisticated, and elegant models may be severely limited or hampered by specific data needs and requirements. Thus, data issues are an essential component for assessing the ultimate success for model development, validation, and applications. A number of potential data factors need to be considered in the course of conceptualizing and developing models. These include but are not limited to the following.

- *Primary/secondary:* Data may already exist (secondary) or may need to be collected (primary). Obviously, models using secondary source data have some advantages because they require little or no data collection. However, models using such forms of data may be limited by the nature and quality of the data that exist. This might mean the model will be constrained by the type of data available, and such constraints may limit the model's ability to address important issues and problems. Models using primary sources of data have more flexibility, given that they can determine exactly what type of data needs to be collected. However, primary data collection involves its own set of limitations that are reflected in the factors described below.

- *Observable/nonobservable:* Some data are directly observable, and this may facilitate ease of collection. Phenomena that are not directly observable may require more extensive efforts to uncover the necessary information (e.g., face-to-face interviews).

- *Distant/close:* Some forms of data can be collected at a distance. This may involve the use of technology, such as cell phones or video links. However, other types of data require actually being there on the ground, as for face-to-face contact or interviews with subjects, respondents, or informants.

- *Representative/nonrepresentative:* Often model assumptions require data to be collected or compiled in some specific manner. The best example of this is the explicit assumptions underlying classical parametric statistical models that require random samples from a population. There are other models that simply require units of analysis to be representative of a given theoretically important category of some type, and it may be the case that any unit of analysis fitting the categorical criteria will suffice. An important consideration is the extent to which units of analysis used in the model need to be derived by either probabilistic or nonprobabilistic methods (see Johnson, 1990).

- *Passive/active:* This is related to some of the factors above in that some data can be collected casually or on the fly. Such data may still require being there but may require only documenting or recording naturally occurring events, conversations, or interactions. In contrast, more direct and active methods of data collection may be necessary and will involve, for example, actually interviewing individuals at events or interviewing them about given conversations or interactions.

- *Tacit/explicit:* Some forms of data require little interpretation or reading between the lines. Other types of data are implicit, and there is a need to make them more explicit. This is particularly true for some forms of human knowledge that are often tacit and may require specific types of elicitation interviewing techniques to extract the requisite information to be used in the model (Johnson and Weller, 2002).

There are certainly other important factors to be considered in terms of relating models to various data requirements. However, the factors described above potentially reflect impediments to the utility and validity of any proposed model. If, for example, models require data involving forms that are tacit, active, representative, close, nonobservable, and, of course, primary, then the data may be costly to obtain and may limit the models' potential effectiveness given the data constraints. But this does not address in any way issues of data quality concerning reliability and validity. We can consider the factors above to reflect elements of how hard data might be to collect or obtain. Although some of these factors are related to issues of reliability and validity, they are not necessarily one and the same. Often the data that are the most difficult to collect (i.e., on-the-ground face-to-face interviews) are the data that have the most reliability and validity, whereas data that are the easiest to obtain (i.e., secondary source data) may be the most problematic. The extent to which one trusts the data will ultimately determine the extent to which one trusts model outcomes or predictions.

In summary, even though quality data are extremely important, the operationalization of quality is different for the different demands of the model. One implication is that we need better quality data. Another implication is that we need a better understanding of how we can model, describe, predict, and explain with less than quality data. This further suggests that a better notion is needed of what is meant by quality data for the various models and needs.

REFERENCES

AFRL/HE. (2002). *Proceedings of the Eleventh Conference on Computer Generated Forces and Behavioral Representation*, May, Orlando, FL.

Al-Halimi, R., and Kazman, R. (1998). Temporal indexing through lexical chainin. In C. Fellbaum (Ed.), *WordNet: An electronic lexical database* (pp. 333–351). Cambridge, MA: MIT Press.

Allen, J.G. (2004). *Commander's automated decision support tools.* Briefing to Proposers' Symposium for DARPA's Integrated Battle Command Program, Dec. 15, Washington, DC.

Axtell, R., Axelrod, R., Epstein, J.M., and Cohen, M.D. (1996). Aligning simulation models: A case study and results. *Computational and Mathematical Organization Theory, 1*(2), 123–141.

Bachman, J.A., and Harper, K.A. (2007). *Toolkit for building hybrid, multi-resolution PMESII models.* (Final report #RI-RS-TR-2007-238.) Cambridge, MA: Charles River Analytics. Available: http://stinet.dtic.mil/cgi-bin/GetTRDoc?AD=ADA475418&Location=U2&doc=GetTRDoc.pdf [accessed Feb. 2008].

Barreteau, O. (2003). Our companion modelling approach. *Journal of Artificial Societies and Social Simulation, 6*(2). Available: http://jasss.soc.surrey.ac.uk/6/2/1.html [accessed April 2008].

Beeferman, D. (1998). Lexical discovery with an enriched semantic network. In *Proceedings of the Workshop on Applications of WordNet in Natural Language Processing Systems*, Association for Computational Linguistics/Conference on Computational Linguistics (ACL/COLING 1998), Montreal, Canada. Available: http://www.lexfn.com/doc/lexfn.pdf [accessed Feb. 2008].

Box, G.E.P. (1979). Robustness in the strategy of scientific model building. In R.L. Launer and G.N. Wilkinson (Eds.), *Robustness in statistics: Proceedings of a workshop*. New York: Academic Press.

Brinkley, J. (2003, May 14). *AFRL/HE instruction #16-03*. Wright Patterson Air Force Base, OH: Air Force Research Laboratory/Human Effectiveness Directorate.

Burton, R.M. (2003). Computational laboratories for organization science: Questions, validity and docking. *Computational and Mathematical Organization Theory*, 9(2), 91–108.

Burton, R.M., and Obel, B. (1995). The validity of computational models in organization science: From model realism to purpose of the model. *Computational and Mathematical Organization Theory*, 1(1), 57–71.

Campbell, G.E., and Bolton, A.E. (2005). HBR validation: Integrating lessons learned from multiple academic disciplines, applied communities, and the AMBR project. In K.A. Gluck and R.W. Pew (Eds.), *Modeling human behavior with integrated cognitive architectures: Comparison, evaluation, and validation* (pp. 365–395). Mahwah, NJ: Lawrence Erlbaum Associates.

Carley, K.M., and Svoboda, D. (1996). Modeling organizational adaptation as a simulated annealing process. *Sociological Methods and Research*, 25(1), 138–168.

Davis, J.H. (2000). Simulations on the cheap: The Latané approach. In D.R. Ilgen and C.L. Hulin (Eds.), *Computational modeling of behavior in organizations: The third scientific discipline* (pp. 217–220). Washington, DC: American Psychological Association.

Duffy, J. (2006). Agent-based models and human subject experiments. In L. Tesfatsion and K.L. Judd (Eds.), *Handbook of computational economics, volume 2: Agent-based computational economics*. Amsterdam, The Netherlands: Holland/Elsevier.

Epstein, J.M. (2006). Remarks on the foundations of agent-based generative social science. In L. Tesfatsion and K.L. Judd (Eds.), *Handbook of computational economics, volume 2: Agent-based computational economics*. Amsterdam, The Netherlands: Holland/Elsevier.

Fagiolo, G., Windrum, P., and Moneta, A. (2006). Empirical validation of agent-based models: A critical survey. *Journal of Artificial Societies and Social Simulation*, 10(2). Available: http://jasss.soc.surrey.ac.uk/10/2/8.html [accessed April 2008].

Fiske, A.P. (1992). The four elementary forms of sociality: Framework for a unified theory of social relations. *Psychological Review*, 99(4), 689–723.

Gluck, K.A., and Pew, R.W. (Eds.). (2005). *Modeling human behavior with integrated cognitive architectures: Comparison, evaluation, and validation*. Mahwah, NJ: Lawrence Erlbaum Associates.

Haefner, J.W. (2005). *Modeling biological systems: Principles and applications*. New York: Springer.

Harper, K.A., Pfautz, J.D., Ling, C., Tenenbaum, S., and Koelle, D. (2007). *Human and system modeling and analysis toolkit (HASMAT)*. (Final report #R06651.) Cambridge, MA: Charles River Analytics.

Harris, R.J. (1976). The uncertain connection between verbal theories and research hypotheses in social psychology. *Journal of Experimental Social Psychology*, 12(2), 210–219.

ITT Research Institute. (2001). *Modeling and Simulation Information Analysis Center (MSIAC). Verification, validation, and accreditation (VV&A) automated support tools: A state of the art report, Part 1-Overview*. Chicago, IL: Author.

Johnson, J.C. (1990). *Selecting ethnographic informants*. Thousand Oaks, CA: Sage.

Johnson, J.C., and Weller, S. (2002). Elicitation techniques in interviewing. In J. Gubrium and J. Holstein (Eds.), *Handbook of interview research* (pp. 491–514). Thousand Oaks, CA: Sage.

Judd, K.L. (2006). Computationally intensive analyses in economics. In L. Tesfatsion and K.L. Judd (Eds.), *Handbook of computational economics, volume 2: Agent-based computational economics*. Amsterdam, The Netherlands: Holland/Elsevier.

Kaelbling, L.P., Littman, M.L., and Moore, A.W. (1996). Reinforcement learning: A survey. *Journal of Artificial Intelligence Research, 4*, 237–285.

Langton, J.T., and Das, S.K. (2007). *Final technical report for contract #FA8750-06-C-0076*. Air Force Research Laboratory, May 16, Rome Research Site, Rome, NY.

Marks, R.E. (2006). *Validation and complexity*. Working paper, Austrailian Graduate School of Management, University of New South Wales, Sydney.

Miller, J.H. (1998). Active nonlinear tests (ANTs) of complex simulation models. *Management Science, 44*(6), 820–830.

National Research Council. (1998). *Modeling human and organizational behavior: Application to military simulations*. R.W. Pew and A.S. Mavor (Eds.). Panel on Modeling Human Behavior and Command Decision Making: Representations for Military Simulations. Commission on Behavioral and Social Sciences and Education. Washington, DC: National Academy Press.

Popper, S.W., Lempert, R.J., and Bankes, S.C. (2005). Shaping the future. *Scientific American Digital,* April. Available: http://www.sciamdigital.com/index.cfm?fa=Products.ViewIssue&ISSUEID_CHAR=8E3EADFC-2B35-221B-6A8455DA45AE8B50 [accessed Feb. 2008].

Robbins, M., Deckro, R.F., and Wiley, V.D. (2005). *Stabilization and reconstruction operations model (SROM)*. Presented at the Center for Multisource Information Fusion Fourth Workshop on Critical Issues in Information Fusion: The Role of Higher Level Information Fusion Systems Across the Services, University of Buffalo. Available: http://www.infofusion.buffalo.edu/ [accessed Feb. 2008].

Rouse, W.B., and Boff, K.R. (2005). *Organizational simulation*. Hoboken, NJ: John Wiley & Sons.

Sageman, M. (2004). *Understanding terror networks* [web page, Foreign Policy Research Institute]. Available: http://www.fpri.org/enotes/20041101.middleeast.sageman.understandingterrornetworks.html [accessed August 2007].

Sageman, M. (2006). The psychology of Al Qaeda terrorists: The evolution of the Global Salafi Jihad. In C.H. Kennedy and E.A. Zillmer (Eds.), *Military psychology: Clinical and operational applications* (Chapter 13, pp. 281–294). New York: Guilford Press.

Schelling, T.C. (1971). Dynamic models of segregation. *Journal of Mathematical Sociology, 1*, 143–186.

Shafer, G. (1976). *A mathematical theory of evidence*. Princeton, NJ: Princeton University Press.

Sowa, J.F. (1984). *Conceptual structures: Information processing in mind and machine*. Reading, MA: Addison-Wesley.

Telvick, M. (Producer). (2005, January). *Al Qaeda today: The new face of the global Jihad*. Arlington, VA: PBS Frontline.

Telvick, M. (2007). *Al Qaeda today: The new face of the global jihad* [web page, PBS Frontline]. Available: http://www.pbs.org/wgbh/pages/frontline/shows/front/etc/today.html [accessed August 2007].

U.S. Department of Defense. (1995). *Modeling and simulation (M&S) master plan*. (DoD #5000.59-P.) Washington, DC: Office of the Undersecretary of Defense for Acquisition and Technology.

9

State of the Art with Respect to Military Needs

In this section we review the state of the art in individual, organizational, and societal modeling against the military modeling needs outlined in Chapter 2 and discuss the major shortfalls in meeting those needs. The five representative problems described in Chapter 2 are used as a structure to organize the review.

DISRUPT TERRORIST NETWORKS

One potential use of cultural and organizational models is to fuse partial and uncertain information from multiple sources to develop a model of the network structure of a terrorist organization and to use that model to evaluate alternative strategies for disrupting that organization, for example, through disconnecting leaders or interrupting recruiting. This goal can be supported by many of the modeling approaches described earlier, and each has its own limitations for attacking the problem. Table 9-1 summarizes the capabilities that would provide advantages for each modeling approach and the major limitations of the approach in addressing this problem.

Network models provide a promising approach for this problem, but a general limitation across the network modeling approaches is the lack of data for model development. Lack of data is the primary challenge for using models to understand and disrupt terrorist networks. The data availability problem is compounded by the classification levels for existing data, the control of those classified data by multiple organizations, and the inconsistencies in data structure and content. For example, the same individual may be identified by different names in multiple databases. The

TABLE 9-1 Modeling Approaches and Limitations for Disrupting Terrorist Networks

Problem 1: Disrupt a terrorist network.
How can we fuse uncertain and partial information from multiple sources to identify the dynamic network structure of a terrorist organization? How can we then best disrupt this network?

Advantages of Approach	Major Limitations
Conceptual models indicate how removal of a leader will affect the network.	Need ethnographic data on terrorists to see if models apply in the specific culture.
Cognitive/affective models could predict reactions of members of network if leader is removed.	Extensive time and effort are required to develop specific models from open sources.
Organizational models of existing network could predict impact of changes in network on performance.	Data are not available. Existing data are in multiple databases controlled by different organizations, with inconsistent structures and contents.
Link analysis could identify network. Social network analysis models can predict how changes in leadership will affect network structure and change power and centrality.	Link data are difficult to acquire.
Dynamic network analysis models could predict who the emergent leader will be if current leader is removed.	Requires resource and activity data, which are very difficult to acquire.
System dynamics models could predict whether change in leadership might lead to an increase in violence in the community.	High-level model would not predict for individuals.
Model could be tested in massively multiplayer online games (MMOGs).	Behaviors in the MMOG might not resemble those in the real world.
All approaches could make recommendations for action.	No facility for rapid development of models or provision of easy-to-understand guidance to war-fighters.

more specific the predictions that could be made by the model, the more data are required. A way to mitigate this problem (see Chapter 11) would be to make unclassified representative databases more widely available for model development and evaluation.

There is an additional challenge in how to test and validate these models and how to clearly communicate the uncertainty surrounding model forecasts. If models could be developed more quickly and easily and in closer collaboration with both subject matter experts and the ultimate users, they could be a more useful thinking tool for decision makers.

FORECAST ADVERSARY RESPONSE TO COURSES OF ACTION

Models could be used to forecast the responses of adversaries to friendly force actions over a range of responses, with estimates of the likelihood of each. This is especially needed in urban operations and operations other than war in an asymmetric warfare environment in which conventional methods for predicting adversary behavior are often not relevant. For example, what are the likely reactions of both noncombatants and local insurgents to friendly force movements, and can those reactions be affected by the diffusion of information or disinformation? Table 9-2 summarizes the contributions that could be made by models and their major limitations.

In the example of forecasting enemy response to a disinformation campaign about troop movements, network models are clearly applicable conceptually—the limitations are in the details that would make the model useful for a specific cultural context. First, we lack theory regarding cultural differences in the believability of a message as a function of its source. Multi-agent models could predict information diffusion patterns, but they would require data on transmission links (media channels, literacy rates, etc.) that are specific to the location and culture being modeled as well as ways to predict the believability of the message when received. Finally, there are no well-defined outcome variables for assessing the validity of the model's

TABLE 9-2 Modeling Approaches and Limitations for Forecasting Enemy Response to Disinformation

Problem 2: Forecast adversary response to blue actions.
Predict the likely response of noncombatants and local insurgents to friendly force movements, basing, logistics, and courses of action. Can disinformation be used to partially protect our intentions? What is the most effective point of insertion of the disinformation?

Advantages of Approach	Major Limitations
Persuasion theory suggests the need to disconnect message from the source. Research exists on direction of change in message with diffusion.	No models for how individuals are likely to transform/distort messages as a result of cultural and cognitive factors. There are no individual models to predict interpretation or believability of messages, especially taking cultural factors into account.
Use multiagent model of networks to decide where to drop information into rumor mill (e.g., viral marketing).	Multiagent models would require models of culture and technology to estimate speed of transition. Need to link multiple types of models together.
Use model to maximize predicted diffusion of information.	Straightforward conceptually but there is a lack of empirical country-specific data.

predictions. Clearly the uses to which the model is to be put, including the details of the cultures and location of interest, must drive both its structure and content. As in the previous example, the major recommended mitigation strategies (see Chapter 11) include the development and dissemination of detailed datasets and the development of a close collaborative relationship between model developers, subject matter experts, and model users.

SOCIETAL FORECASTING

There is an urgent need for models that can forecast attitudes and behaviors at a societal level as a function of alternative courses of action particularly for diplomatic, information, military, and economic (DIME) campaigns. For example, models are needed that can forecast the stability of civilian governments and the incidence of violence as a function of these DIME factors. Table 9-3 summarizes the possible contributions and limitations of models for this goal.

Agent-based models (ABMs) on a large scale appear very promising for modeling societal behaviors and forecasting responses to diverse DIME courses of action (COAs). However, there are a number of limitations that currently constrain what can be done with these models. First, such models are time-consuming to build and often require data collection on a massive level. Second, computational power limits the cognitive complexity that can be built into individual agents if tens of thousands of agents are to be included in the model. Third, predicting the response to integrated DIME COAs requires multidisciplinary expertise in military planning, economics, and political science. The need to integrate models across disciplines is the primary challenge faced in this problem area. Theories and models in these diverse disciplines are not currently integrated, and experts in these areas often have little opportunity or incentive to collaborate. Hybrid federated models that combine different models at different levels of detail for different factors offer considerable promise for tackling large-scale societal prediction, but these models are more an idea than a reality at present and will require the multidisciplinary development of architectures and standards for federation. To mitigate these limitations, Chapter 11 recommends an extensive multidisciplinary research program focused around common challenge problems and datasets. Finally, as with other models, the predictions of these models will have a high degree of uncertainty. Model development must include collaboration between end users and diverse subject matter experts to ensure that the predictions provided are relevant and that their limitations are well understood.

TABLE 9-3 Modeling Approaches and Limitations for Societal Forecasting

Problem 3: Societal prediction.
Forecast the effects of alternative diplomatic, information, military, and economic (DIME) courses of action on the attitudes and behaviors of residents in areas of interest.

Advantages of Approach	Major Limitations
System dynamics models can predict the effects of changes in DIME factors on outcome variables of interest, such as level of violence.	Sufficient theory is lacking to identify the key variables to be included in the model and to specify the connecting links between actions and outcomes.
ABMs that capture resource use and economic interactions as well as social links can predict the effects of economic factors within a social and cultural context.	ABMs need both theory and data to make useful predictions. Large-scale societal ABMs are time-consuming and costly to develop.
ABMs can include tens of thousands of agents to capture complex interactions at the societal level.	Agents must be cognitively simplistic to make large-scale ABMs computationally feasible.
Historical data can be used to develop ABMs that predict societal effects.	Prediction of the future, not the past, will involve inherent uncertainty.
Federated models could integrate multiple types of models at different levels of detail to capture diverse DIME factors.	There is a lack of infrastructure, architectures, and standards for federated models. DIME factors are studied by different disciplines and few integrated models exist that attempt to combine them. There is little theory or data on how DIME factors interact.
MMOGs can provide an environment for data collection on a large scale to support model development and testing.	The environment created by the MMOG may not reproduce the key elements of a real-world society.

CROWD CONTROL TRAINING

Virtual training environments offer an opportunity for troops involved in peacekeeping operations to learn best practices for crowd control. Such training will require models of noncombatants that respond to trainee actions in a way that is realistic and appropriate for a specific location and cultural environment. Table 9-4 summarizes the state of the art in this area.

The development of models for crowd control training is perhaps the most advanced of the five representative problems considered. This can be done now. The only major issue is whether the models can produce behavior that is close enough to that of a real crowd in a specific environment to provide useful training. The extent to which cultural factors create differences in crowd behavior in different locations deserves further study, but

TABLE 9-4 Modeling Approaches and Limitations for Crowd Control Training

Problem 4: Crowd control training.
Use models of crowd behavior to create a virtual training environment in which soldiers can learn to take the appropriate action.

Advantages of Approach	Major Limitations
Cultural models can provide theory on how crowd members in a specific culture are likely to react to different actions.	Theory linking attitudes and behaviors may not be specific enough for the environment for which training is needed.
Cognitive and affective models can represent individual reactions to soldier behavior.	Individuals may react differently as part of a crowd than they would alone. The effects of cultural context on behavior are not fully understood.
ABMs can capture the interactions among crowd members that cause them to act collectively in ways in which they might not have acted individually.	Cultural variability in crowd dynamics is not completely understood.
MMOGs can provide an interactive environment for quickly testing model behavior as well as an environment for the training.	Model behavior should be reviewed by subject matter experts for believability.

crowd behavior models can be implemented in virtual environments, such as MMOGs, and their behavior can be reviewed by subject matter experts in a specific culture to ensure that they are not behaving in unrealistic ways that would result in negative training.

ORGANIZATIONAL DESIGN: FORCE COMPOSITION AND COMMAND AND CONTROL ARCHITECTURE

Because of the rapid changes in mission requirements, the military services are moving toward modular expeditionary forces that are readily reconfigurable for different types of missions. Making the best use of these modular forces requires not only a recommended force composition (systems, equipment, units, and personnel) but also a command and control (C2) architecture that is most effective for the force as constituted. The best force composition and C2 architecture for a conventional military operation may be quite different from that for a peacekeeping or disaster relief mission. Table 9-5 summarizes how organizational models could help in this process.

TABLE 9-5 Modeling Approaches and Limitations for Organizational Design

Problem 5: Organizational design: Force composition and command and control architecture. Use organizational models to develop optimal force composition packages and C2 architectures for different mission types.	
Advantages of Approach	Major Limitations
Use organizational models to develop the force composition, structure, and processes that are predicted to best meet mission requirements. Simulate organizational performance for different structures in different mission scenarios.	Requires detailed data on the tasks to be performed in the mission and the resources available.
Use simulation and ABMs to identify the points in the mission and the organization at which the most intensive cooperation will be required, the points of maximum workload, and the potential information bottlenecks.	Requires detailed data on task information and workload requirements.
Use MMOGs as a testbed for organizational structures.	May be difficult to replicate realistic mission tasks and conditions.

Models have proven to be a useful tool for understanding, designing, and testing organizations. Although people rarely think about "designing" organizations in a systematic way, the military faces the need to do just that as it develops new flexible, adaptive structures for rapidly changing missions. One example of an attempt at systematic organizational design based on modeling is the recent effort at the National Aeronautics and Space Administration (Carroll, Gormley, Bilardo, Burton, and Woodman, 2006), using an ABM and a heuristic rule-based model. Modeling and simulation can be used to develop and adapt force composition for changing missions and to suggest the best C2 structure for accomplishing the mission. The best C2 architecture is especially challenging for coalition operations, in which different types of forces may be involved, and for peacekeeping and disaster relief operations, which require close coordination with nongovernment organizations. The major limitation for this work is the need for detailed information on the tasks to be accomplished in the mission and the resources required to accomplish those tasks. The recommended mitigation strategy (see Chapter 11) is the development of common challenge problems and datasets and the use of collaborative workshops to ensure that that operational users and modelers have a shared understanding of what can be done through modeling.

REFERENCE

Carroll, T.N., Gormley, T.J., Bilardo, V.J., Burton, R.M., and Woodman, K.L. (2006). Designing a new organization at NASA: An organization design process using simulation. *Organization Science, 17*, 202–214.

Part III

ADDRESSING UNMET MODELING NEEDS

10

Pitfalls, Lessons Learned, and Future Needs

Chapters 3 though 9 summarized the state of the extensive work under way to develop a variety of individual, organizational, and societal (IOS) human behavioral models and how that work both contributes and falls short in solving representative military problems. In this chapter we take a step back from this detail and summarize the findings of the committee in the form of lessons learned and future needs if IOS models are to live up to their potential for delivering useful results.

Given that most IOS models are in early phases of development, a clear set of best practices has not yet emerged. We can, however, identify some lessons to be learned from the initial approaches that have been taken and some of the pitfalls that have occurred on the road toward developing effective IOS models. Awareness of these pitfalls should help those developing models to avoid wasting valuable time and effort relearning the same lessons. Avoiding these pitfalls will therefore improve the probability of success for new initiatives.

When particular programs or efforts are mentioned as examples, this is not meant to suggest that those involved made choices that were known to be wrong at the time; in fact, many of the authors of this report have fallen into one or more of these pitfalls in our own modeling efforts. Hindsight often brings clarity, revealing that choices that seemed reasonable at the time have had undesirable results. Our goal is to support the field in gaining maximum benefit from the "tuition paid" thus far in intellectual effort, hard work, and taxpayer money. For each pitfall we summarize the lessons learned, and on the basis of those lessons we identify the key needs to be met in order to move forward. Chapter 11 then presents our recommended plan to meet those needs.

PITFALLS IN MATCHING THE MODEL TO THE REAL WORLD

The following problems are created either by inattention to the real world being modeled or by unrealistic expectations about how much of the world can be modeled and how close a match between model and world is feasible.

Model-Problem Mismatch

Modelers should choose variables based on what theory and experience suggest will be most useful in characterizing the problem of interest. Poor characterization of the specific domain of problems to be addressed, failure to attend appropriately to the dictates of the problem domain, or failure to consult theory to assess which IOS variables are most likely to matter can lead to serious model-problem mismatches. Practical considerations of availability, for example, can lead modelers to select "off the rack" components just because they are available, even if they are inappropriate to the problem at hand.

For example, using the Hofstede dimensions (see Cultural Models in Chapter 3) as generic representations of culture is inappropriate unless there are good theoretical reasons to believe that the specific dimensions chosen are relevant for behaviors that are important to the application. When modeling adversarial reasoning, specific cultural variations in inference and dialectical reasoning may well be informative, while Hofstede dimensions such as masculinity-femininity are generally irrelevant.

Another source of poor choices is the pull of familiarity. Those versed in game theory are often attracted to representations of culture as a distribution of strategies for playing stylized games, such as the prisoners' dilemma (see Game Theory in Chapter 5). For some applications, such as predicting the adversarial responses of enemy organizations to courses of action, game theory approaches to culture may be a good match. For other applications, such as predicting broader societal reactions, game theory approaches to cultures can lead to such problems as characterizing cultures as necessarily moving toward or existing in an equilibrium state or assuming that conflict involves two parties. In reality, cultures are rarely in equilibrium; changes in technology, resources, and migration all impact culture; and the implications of conflict for a particular society rarely involve just two parties. This particular model-problem mismatch has resulted in misleading policy advice and a tendency to overlook major shifts in culture, resulting in policy makers and commanders being surprised by emergent factors in the situations they face (see also Illusions of Permanence in this chapter).

Another example of this pitfall comes from network research. Over the years, network researchers have used the same network properties (e.g., betweenness, centrality) as independent variables in a wide variety of different contexts (see Network Models in Chapter 6). Yet there is little reason to believe that these properties are relevant for all applications for which network analysis may be informative. Even when correlations are found, it is difficult to design effective interventions and changes to a system because the variables are not necessarily explanatory. The key network properties relevant to prediction and control of any particular problem domain should generally be derived from the problem itself (Borgatti and Everett, 2006).

Lessons Learned and Future Needs: The modeler should tie model choices to the application, which assumes that the application domain and the class of problems to be addressed are clearly specified. Subject matter experts in the application domain should lead or be represented on modeling teams, or at the very least they should be extensively consulted regarding the appropriate choice of variables and assumptions for a particular problem domain.

In general, a tighter connection is needed between model developers and the operational personnel who will use the models being developed. Shared understanding between developers and users should result in a clear specification of model purpose. Better communication—including sharing of both theory and data—is also needed across the many disciplines that may contribute to model specification. Based on the purpose of the model and the application domain, more collaborative cross-disciplinary efforts in an integrated community of interest are needed to ensure that model developers do not simply rely on the set of variables with which they are most familiar.

All-Purpose Models That Ultimately Serve No Purpose

Universal scope or "Swiss army knife" models attempt to solve, via large-scale software development, an entire set of wide-ranging concerns. In most cases, attempting to build universal scope models for the Department of Defense (DoD) has led to failure, to the loss of years of modeling and simulation effort, and to the expenditure of large amounts of money. The classic universal scope model, JSIMS (2 to 5 million lines of code and 6 years to develop), had the following scope (Bennington, 1995, p. 805):

> The mission of JSIMS is to develop a Joint Simulation System that will provide readily available, operationally valid synthetic environments for use by the CINCs, their components, other joint organizations and the

Services. JSIMS has five major objectives: integrate the range of missions of the Armed Forces within a common modeling and simulation (M&S) framework that includes live, virtual, and constructive M&S capabilities: provide a training environment which will also accommodate space, transportation and intelligence requirements: establish a common simulation support structure which enables harmonious sharing of simulation resources, processes, and results among users; enable simulation users to readily create or access a simulation environment which supports their requirements: and enable joint simulation users to interact freely with elements of their command structure, supporting/supported organizations and other simulation centers or users. While the initial focus of JSIMS is joint planning and training activities, as the system matures, JSIMS will be available to the DoD community at large for the analysis of doctrine, organization, system and material alternatives.

With such a large and broad scope, the likelihood for successful implementation was, in retrospect, near zero. JSIMS ran from December 1995 to December 2002, at a development cost of $1.8 billion. In the end, DoD decided to fall back to smaller scale models and attempt to make those models interoperable (Office of the Secretary of Defense, 2004).

Interoperability concerns in DoD also fell prey to the universal scope model syndrome with the high-level architecture (U.S. Department of Defense, 1996). In 1996, DoD decided that a big bang, Swiss army knife solution to the interoperability of models and simulations was the way forward for defense models and simulations. Instead of building an architecture that was dynamically extensible and semantically interoperable,[1] DoD built a monolithic, black box piece of software that required everything to be defined ahead of time statically.

The consequence is that, to make modeling and simulation (M&S) systems interoperable, the source codes of the systems must be modified and their definition files updated. For any subsequent M&S system to be integrated, the source code for *all* systems must be modified along with their definitional files.

In retrospect, it is temptingly easy to build static interoperability solutions if most of the information transferred is physics-based. As one moves into the realm of modeling human and organizational behavior and begins to include cultural, network, emotional, cognitive, and psychological models, one needs to build models as encapsulated smaller model components that can be dynamically linked together rather than trying to create one large source code component (Pratt and Henninger, 2002).

---

[1]Dynamically extensible means that the structure of the model allows new components to be added without rewriting the source code. Semantically interoperable means that the language of the two models permits them to be put together in a meaningful and theoretically consistent way.

Lessons Learned and Future Needs: Monolithic, static approaches are inappropriate to IOS modeling. Flexible, adaptable components and semantically interoperable models will potentially do much to avoid this pitfall. In order for this to happen, advances are needed in federated model standards and architectures in order to allow different types of IOS models, at different levels of detail, to interoperate in meaningful ways.

VERIFICATION, VALIDATION, AND ACCREDITATION

The DoD M&S community has always lived with the specter of verification, validation, and accreditation (VV&A). We say specter because sometimes VV&A of a model is used to shut down further discussion and consideration of it, particularly if it has not yet gone through a VV&A process. VV&A is an important issue in IOS modeling, as in other types of modeling (see Burton, 2003, and Chapter 8 for an extended discussion of VV&A issues). With respect to the modeling of human and organizational behavior, however, rigorous VV&A (as it has been defined for validation of models of physical systems) is difficult if not impossible to fully achieve.

VV&A for a model means the developers have verified that the model implements processes as intended, they have validated the model against empirical data, and they have accredited the model for use for particular circumstances, usually for a particular service requirement. Early M&S efforts usually modeled physical properties exclusively, so verification consisted of being able to look at the source code and say, "yes, the source appears to implement the mathematics of the physical model." For IOS models, there is no easy path to verification. One can look at the source code but cannot say "yes, the source appears to implement the mathematics of the human or organizational model" because the techniques typically used for such models are code-based and not closed-form mathematics.

Historically, models of physical phenomena have been validated by comparing the results of running the models with observations from the real world. If one builds a model of a tank being hit by a particular weapon, one can go out into the field and shoot that particular weapon at a tank and say "yes, the model is close to the results of the real world" and stamp the model as validated. For IOS models, it is typically not possible to validate the model against the real world in this way. For example, suppose one builds a network model of insurgency formation. One cannot then take real-world inputs into such a model so as to predict precisely what will happen next, as in the film "Minority Report." At best, it is possible to run the model against historical data and see how well the model accounts for the observed events. Perhaps, at the end of such a validation, it will be possible to state that the model provides a possibility space or set of potential outcomes that are useful to consider in the analysis of the next course of action. Valid IOS

344 BEHAVIORAL MODELING AND SIMULATION

models do not predict exactly what will happen in the future (see also Illusions of Permanence in this chapter) but rather provide a set of potential outcomes to consider. So when writing up the model usage document for such a model, it is feasible to state something like "this model is useful for analyzing situations that have the following characteristics and will provide outputs that allow you to consider the set of things that may happen within the following limits." Of course, this assumes valid inputs: data that are reasonably accurate, acceptably complete, and that match the requirements of the model (see Chapter 8 for a more detailed discussion of data issues).

Accreditation is the final step in the VV&A process. Basically accreditation means that a sponsoring organization is willing to bless the model for a particular use, which generally occurs after someone has verified the model to an acceptable degree and some validation has been performed. Model accreditation is usually specific to a service office or the Office of the Secretary of Defense, and accreditation means that the office has determined that the model is sufficiently robust for some operational deployment. Many physics-based models have been accredited but, as far as the committee knows, no models of human and organizational behavior have been accredited. Such models are perhaps too new to have yet made it through the accreditation paperwork process. However, we think that accreditation decisions for IOS models should be based on a better understanding and explication of the limitations and usages for such models, as well as a set of VV&A requirements that are appropriately tailored to the special nature of such models.

Lessons Learned and Future Needs: Failure to appreciate the extent to which IOS models differ from physical models has led to inappropriate expectations regarding VV&A for IOS models. Rather than trying to apply an inappropriate VV&A process, an IOS model needs to be deployed with a strong set of guidelines that describe the limitations of the model and that remind users that the purpose of the model is not definite point predictions but rather indications regarding what possible outcomes will likely result from any particular course of action.

Better standards are needed for IOS models, including appropriate VV&A guidelines. This report recommends an action validation approach (see Chapter 8) that requires a clear specification of the purpose of the model and validates the usefulness of the answers provided by the model against that purpose. We also recommend triangulation, in which models are reviewed by multiple types of experts, compared with qualitative and theoretical studies as well as quantitative results, and similar models are compared with each other (docking). Appropriate IOS model validation approaches need to be further developed and promulgated among model developers, users, and funding agencies through a widespread multidisciplinary community of interest.

PROBLEMS IN DESIGNING THE INTERNAL STRUCTURE
OF A MODEL

The following tactical design pitfalls are sometimes generated by unwarranted assumptions about the nature of the social, organizational, cultural, and individual behavior domain and sometimes by a failure to deliberately and thoughtfully match the scope of the model to the scope of the phenomena to be modeled. These pitfalls reveal the challenge of making wise choices of simplifying assumptions about the highly complex domain of IOS structure and behavior.

Pitfall of Unvalidated Universal Laws

Modelers who are accustomed to dealing with physical objects that behave according to well-known physical laws are especially prone to this pitfall. Comparable universal laws of human behavior and social structures have yet to be discovered, codified, and supported by empirical data. Even should they be discovered, it is unlikely that they could be represented as closed-form equations. Furthermore, human behavior involves freedom of choice, and the results of the model themselves, if widely publicized, might affect those very behaviors that they were intended to forecast. Modelers fall into this pit when they model particular structures or processes in fixed form because they mistakenly believe that these structures are universal. As an example, some modelers have subscribed to the notion that all evolved networks are scale-free (i.e., they have a degree distribution that is well described by a power law). However, because the behavioral capabilities of nodes in a network make a demonstrable difference (people networks are different from gene networks), the data do not support the assumption of abstract commonalities across *all* networks. While the assumption of a scale-free network may well be warranted for particular types of networks and nodes, building this assumption in as a fixed feature of the model will limit its application in ways that may not be recognized by end users. Instead, the network structure should be treated as a model parameter.

Lessons Learned and Future Needs: Beware of assumptions that any particular structure or process is universal in any IOS domain. Consult with subject matter experts to be sure empirical data provide very strong support for any such claims before relying on them in designing a model. Set up the model so that users are explicitly reminded that they are making an assumption when they select a particular structure or process to represent a domain. A better integrated multidisciplinary community of interest in IOS modeling, with greater availability of empirical data and more extensive docking of alternative models around common applications, could protect against the unthinking and mistaken assumption that universal laws from other domains apply to IOS models.

One-Dimensional Models

Modelers should beware of inappropriately limiting themselves to a single independent variable and using it to account for an array of different processes and outcomes. For example, there is a tendency in network research to focus exclusively on structure as represented by a few network variables, while completely ignoring other information that is available about the nodes, the processes that are going on, and other contextual factors. Such models ignore possible influences, for example, the possibility that the behavior of the nodes not only is influenced by the network structure, but also can alter that structure. Modelers may encounter this pitfall when operating under the sway of a strong structuralist position that views (network) structure as far more important than other variables, like culture or psychology. As a result, we often see standalone network models that do not incorporate cognitive, cultural, or other processes. Another example of a heavily structuralist approach that ignores process would be a model that uses Hofstede's Big Five personality structure (see Cultural Models in Chapter 3) as the sole predictor for a broad array of cognitive and behavioral outputs.

We think the relative importance of structure should instead be treated as an empirical question, which can of course be investigated only in models that include more than one input variable.

In any modeling enterprise, simplifying assumptions are necessary, and parsimony is an important scientific principle. However, the emphasis should be on parsimony for a purpose—for example, to conduct a focused investigation of whether a particular input variable is plausibly related to an array of different outputs. The decision to exclude other candidate input variables should be based on careful deliberation rather than on unexamined assumptions.

Lessons Learned and Future Needs: Focus is good; myopia is unwise. Better methods are needed to decide which variables are relevant for inclusion in a model. The specification of the variables to be included in a model should be based on a clear specification of the purpose of the model and, depending on that purpose, should take into consideration the judgment of multiple subject matter experts, theories drawn from multiple disciplines, empirical data if they exist, and prior work on similar problems. Comparative studies are needed that address the same problem from multiple perspectives to determine which set of variables offers the most useful results.

Kitchen Sink Models

IOS modelers who appreciate the complex nature of human and organizational behavior and who wish to avoid the pitfall just described may back themselves into a different pit by adding variables to a model in a hodgepodge fashion. Modelers may be especially vulnerable to this pitfall if they are operating outside their area of expertise (for example, people with no training in anthropology or cultural psychology attempting to model culture), or not relying on strong theory for guidance. Modelers who are not well versed in a field will have little basis for choosing appropriate variables and will be especially vulnerable to suggestions to add this or that variable to increase realism. Sometimes the addition of variables is motivated by a desire to improve prediction by adding features and variables so that model output more closely matches a particular set of cases for which the modeler has data. This is actually postdiction (see Dibble, 2006, and Gauch, 2003, on postdictive versus predictive accuracy). The kitchen sink tactic is based on a misconception about the relation between model features and variables and about the model's ultimate usefulness for providing information about behavior in cases beyond those used for testing.

Agent-based models of human and organizational dynamics are often suggested as an effective way to approach the IOS domain. However, the costs of developing, verifying, calibrating, and running complicated agent-based models can be extraordinarily high in relation to our ability to trust what we learn from them. Such a model may have so many degrees of freedom that it often overfits to sample outcomes at the expense of providing an accurate characterization of the full population of potential outcomes that are important for effective insights and decisions. Predictive models are useful to the extent that they provide trustworthy insights and guidance about a particular population of potential outcomes.

A related pitfall is pouring energy into model development, with endless tuning and adjustments, and never using the model in a rigorous fashion to generate insights, answers, or predictions about the probability of different plausible outcomes. What matters most is what new information can be learned from the model and to what degree and under what conditions what is learned can be trusted. In computational laboratories, research, developing, testing, refining, and calibrating a useful and trustworthy model should represent a modest fraction of the time, effort, and expense of putting the model through its paces to answer the important questions that motivated its development. Answers and insights are the ultimate goal, not the model itself.

Lessons Learned and Future Needs: Models can become unwieldy when weighed down by a proliferation of features and variables. Strong theory and a clear specification of purpose should guide subject matter experts in the choice of features and variables to be included and excluded, based on the specific questions and problems to be addressed. Model development should not become an end in itself. As with the extreme parsimony pitfall of one-dimensional models, better methods, including comparative studies of alternative models for a common problem, are needed to determine which variables should be included in a model to generate the most useful results.

PITFALLS IN DEALING WITH UNCERTAINTY AND ADAPTATION

The problems in this section are based on unrealistic expectations of how much uncertainty reduction is plausible in modeling human and organizational behavior, as well as on poor choices in handling the changing nature of human structures and processes.

Unrealistic Expectations

A validated model of weapons delivery could be reasonably expected to predict the exact location where a bomb will fall when dropped from a specific height and location from an aircraft traveling at a specified speed and heading, with specified wind conditions, along with a trustworthy estimate of error in prediction based on likely measurement errors. Plugging in the numbers for the specified variables will supply the user with the desired prediction. It is, however, unrealistic to expect comparable model outputs when the outcome to be predicted is behavior by a human, an organization, a nation-state, or other social entity. We illustrate this problem using the behavior of an individual person, but the caveat also applies to predicting the actions of particular governments or other specific organizations.

Models that purport to specify the exact actions of any given individual human being after plugging in a list of values (for example, nationality, group membership, gender—for a nation, this might be values on a look-up table of cultural traits, estimated military strength, and known alliances) are misleading and seriously incomplete.

Unrealistic expectations are often based on a misconception about what sort of prediction a human behavior model can actually produce. In most situations of interest, there is a range of plausible behaviors, and within the same situation, different people will behave differently, and the same person may also behave differently at different times. Rather than generating a single definitive prediction of behavior, a good human behavior model will instead identify the space of possible outcomes, give probability assessments for these behaviors, and specify some of the factors that could alter these probability assessments.

This pitfall does not necessarily apply to targeted profiling of a particular identified individual, when highly specific idiographic information about that individual and a specified context for behavior are available. The data demands of such models are typically very high, however, and it remains plausible that even a very carefully profiled individual will do something completely unexpected. Hence, even for such profiles, predictions that are couched in terms of probabilities are more complete. For example, "John Doe is likely to do X, with probability estimate of 60 percent, but may do Y or Z instead (model estimate of 10 percent each) or take some other action not covered by the model (20 percent)" is a more informative and less misleading guide to planning and action than a point prediction: "John Doe will do X."

Unrealistic expectations can lead developers to reject a model as useless if postdictive accuracy is not very high. Yet any model that aids in decision making and understanding and that measurably reduces uncertainty can have practical value. The primary contributions of some models are to suggest the space of possible outcomes, reduce the likelihood of surprise, and support systematic analysis. Bronowski (1953) discusses criteria for determining the usefulness of what one might learn. Users do need to know that they can trust what they are learning from the model, but it may be possible to support and test such trust without necessarily expecting the model to replicate observed outcomes in the real world, especially when modeling phenomena that are rare, infrequent, or otherwise nearly impossible to observe and compare with the model. For example, Cronbach and Glaser (1965) produced results that were useful for personnel selection and placement because they represented an improvement over the systems that were then in place. It was not the validity coefficient of the results that mattered, but the meaningful gain in prediction that they represented.

Lessons Learned and Future Needs: When actions must be taken in social situations, IOS models can potentially be used to highlight the range of possible outcomes associated with each considered course of action, together with probability assessments clarifying the likelihood of these possible outcomes. Point predictions are generally misleading and incomplete. The value of IOS models should be measured in terms of the reduction of uncertainty they achieve. Better methods are needed to define the inherent uncertainty in model results and communicate that uncertainty more clearly to users.

Illusions of Permanence

All models include variables, adjustable parameters, and constants. Even when a goal of modeling is to guide the choice of intervention intended to change structure and behavior (for example, to change organizational structure or culture), which implies that the target of intervention is mutable, the feature to be modified is sometimes modeled as fixed. This strategy appears in models of culture (with culture treated as a static set of attributes) and in many social network models (with the network treated as fixed). This misleading approach encourages users of the model to overlook how the modeled structure may be changing in ways that dramatically alter the impact of an intervention and also forecloses the modeling of how a system or structure adapts and adjusts after an intervention.

For example, using a model of terrorist networks to guide decision making about which members of a network to target for removal can mislead users if the network is modeled as a fixed structure from which nodes will be deleted, rather than as a dynamic network with a trajectory of change that will be altered by the deletion of a node, in ways that could either weaken or strengthen the effectiveness of the network. Models used to characterize adversary choices (e.g., game theory models) should explicitly allow the strategy of the adversary to change in response to (or in anticipation of) one's own strategy choices.

Lessons Learned and Future Needs: When feasible, treat IOS structures as variables or as parameters that can be adjusted, rather than as hard-coded fixed attributes that can be altered only by rewriting the source code. Parameters and assumptions will change as a situation evolves, including adversary knowledge of the assumptions. Better methods are needed to build variability over time into models and to communicate the model results (with their accompanying uncertainty) to users.

PROBLEMS IN COMBINING COMPONENTS AND FEDERATING MODELS

The last three pitfalls we discuss arise from the way in which linkages within and across levels of analysis change the nature of system operation. They arise when creating multilevel models and when linking together more specialized models of behavior into a federation of models.

Moving from Individual to Collective Action

Social entities such as groups, organizations, and societies are made up of social beings. Yet many individual-based models do not include social capacities. Merely assembling such agents together into a group model will not enable the understanding of teams, the prediction of collective actions, or coordinated group decisions. To model the most rudimentary forms of social behavior, agents need the means to track the behavior of other agents and rules for adjusting their own attitudes or behavior accordingly. Depending on the application, the rules can be quite simple. Traffic models, for example, can model the interactive behavior of a collection of agents effectively by assuming that each agent acts to pursue an individual goal (getting to a destination in a reasonable time without colliding with others) and chooses among possible actions based on the presence, position, and density of other agents, who are also trying to get to their preferred destinations.

Collective action, however, such as group decision making, requires further rule structures that specify how agents communicate and coordinate their preferences (see Voting and Social Decision Models in Chapter 6). Models that represent changes in attitude or behavior based on social influence need to incorporate rules for how social influence operates.

In deciding what social capacities need to be explicitly modeled, relevant theory should be consulted. In modeling crowds, for example, social science theories (see Conceptual Models in Chapter 3) suggest that changes in behavior are driven either by a weakening of normative regulation or by emergent norms that become salient to crowd members. While flocking models of crowds that treat human beings as analogous to birds and fish may well be useful in capturing some aspects of crowd behavior (particularly a crowd in flight), such models are unlikely to be adequate to inform interventions designed to strengthen social structures that help prevent a crowd from becoming a mob. Moreover, a flock of birds is not the same as one big bird; crowds of people do not necessarily behave like one big person. Models of large organizations should draw on the extensive existing theory and research on how organizational levels are defined and how they relate to each other (Klein and Kozlowski, 2000).

Lessons Learned and Future Needs: For social dynamics to operate in multiagent systems, social capacities of agents, such as communication, must be explicitly modeled. For collective action, collective structures such as rules, norms, and social decision schemes are needed. More work is needed to determine the level of detail at which individuals and groups need to be modeled in order to provide useful results. Comparative studies are needed that examine the contribution of models at different levels of granularity to a common challenge problem, and better methods are needed to link models of individuals to models of larger groups and organizations.

Using Collective Attributes to Predict Individual Action

Just as modeling problems arise in moving from the individual to the collective, inferences made in the opposite direction also pose special problems. Incorporating cultural information in models of human behavior is a positive step toward explicitly modeling the heterogeneity of people. Modelers need to keep in mind that modeling people from the same culture as homogeneous is also a simplifying assumption. First, the boundaries of nation-states are not necessarily the appropriate cultural boundary. In relatively homogeneous nations, such as Japan, the nation-state boundary may well be a good choice. For multiethnic countries, such as Iraq and Afghanistan, tribal boundaries based on ethnic identities, such as Kurd or Pashtun, may be more appropriate. Second, people in the same group also exhibit considerable variability. The extent to which shared culture results in more or less homogeneous behavior depends strongly on the situation and the type of behavior involved. Third, people have multiple social identities and belong to multiple groups of different sizes, all of which have cultural norms and practices that shape behavior. Membership in a group such as the military, for example, may influence individual behavior more powerfully than membership in a national and ethnic group, so that soldiers in two countries may behave more similarly in a large variety of domains than soldiers in either country compared with civilians belonging to the same nation and ethnic group. Finally, one must also be aware that characteristics of a higher level of unit of analysis (macro indices) may not be characteristic of behavior at a lower level of analysis (individuals at the micro level). For example, members of a rioting crowd may smash windows, set vehicles alight, and violently attack innocent bystanders even if hardly any of the individuals involved would behave that way on their own.

Lessons Learned and Future Needs: Be aware of the limits and boundary conditions that apply in predicting individual behavior from information about groups, organizations, and cultures. Behaviors vary in how strongly they are regulated by cultural norms; people belong to multiple groups, all of which have cultural features; and the unit boundary used for modeling may not be the most appropriate one (see Cultural Modeling in Chapter 3). Better methods are needed to represent variable and shifting cultural identities, and comparative studies are needed to assess the benefits of modeling cultural affinities dynamically in providing useful results for a representative challenge problem.

Assemblage of Parts

Recognizing the problems inherent in universal scope models (see above), many in the modeling community have embraced the goal of linking together component models (which may focus, for example, on a specific aspect of human affect or on culture) to create a more comprehensive IOS model. The logic is for subject matter experts to build the parts separately and eventually snap the component models together to yield the complex behavior of the whole.

The need for creating such federated models is a fundamental challenge. It is not possible to build a large universal model without a federation of models. So the challenge is to develop systematic ways to federate models so that the federated result is valid for its own purpose.

Good federated models require (1) an understanding of the purpose of the federated model, which might require a deep understanding of the problem domain; (2) a good understanding of the individual federated components; and (3) assessment of the validity and limitations of the relationships between the individual federated components and the resulting federated model.

In creating such federated models, modelers need to be aware that straightforward snap-together assembly will yield sensible results only when assumptions of additivity and functional independence are tenable, and they often are not. Complex system analysis has shown repeatedly that the connections among the federated components are themselves components to be modeled. The nature of these connections is part of the structure that yields the behavior of the whole. Moreover, the internal structures of the federated components might themselves need to change dramatically when an additional component is connected into the federated model.

An example from cognitive modeling will help illustrate the problem. In the early days of cognitive science, many believed that it would be possible to simply piece together separate models of reasoning, auditory processing, visual processing, memory, etc., to build a reasonable model of the human. However, it became clear that due to complex interactions, a more holistic approach was needed. Separate models could be connected together as components of a federated model only if the connections themselves were included as part of the federated model and if the internal structures of the components were adapted to the presence of these connections.

The same is true for complex social modeling. Along with models of individuals, the nature of links among the actors and the connections of individuals with larger level units, such as groups and organizations, need to be modeled to yield adequate models of both individual behavior in social context and the behavior of social entities, such as groups and nation-states.

As the complexity of models and federations of models grows, it may create the need for "wrappers" that help human beings understand the implications and dynamics of the models. New analytic components, perspectives, and tools will be needed to support understanding and use. The complex interactions that are typical of social science models, as discussed above, will make this a challenging area for research.

Federation also has implications for the VV&A process (see Chapter 8). A federated model formed by combining two models that have previously been individually validated should not be automatically viewed as validated; the federated model must be validated on its own (Burton, 2003).

Lessons Learned and Future Needs: When linking component models, appropriate theory needs to guide the modeling of the linkages as a new component in the resulting federation of models. Systems of systems theory (see Systems Analysis in Chapter 4) can help guide the process of federation, and standards are needed for validating the federation itself. Standards, guidelines, methods, and architectures are needed to improve the state of the art in model federation, addressing semantic interoperability issues that go beyond simple syntactic interoperability. Issues to be addressed include the compatibility of definitions and levels of abstraction, time scale resolution, and treatment of uncertainty in the models to be federated.

SUMMARY OF FUTURE NEEDS

Social, cultural, and organizational modeling is a complex, emerging science with roots in many different disciplines: psychology, sociology, economics, anthropology, systems theory, and computer science, among others. The advancement of a scientific field typically requires that researchers maintain awareness of each other's work and build on each other's results. The multidisciplinary nature of IOS modeling, however, has created a fragmented field with researchers in different disciplines often unaware of each other's relevant work and failing to make use of relevant existing theory and data. In order for the field to advance, researchers need better frameworks and forums in which to compare, discuss, and evaluate their results. The field currently features a multitude of complex models created using different data and different theories to address different problems, making comparative analysis nearly impossible. Common datasets and challenge problems are needed in order to learn which modeling approaches and sets of variables are most useful for specific types of problems.

It seems clear that no single model or approach will meet everyone's needs. There is no single right model and probably will never be. The committee thinks that a federated modeling approach, in which different models at different levels are linked together and component submodels can be swapped in and out, are promising for attacking complex IOS modeling problems. Considerable research needs to be done to make this federated vision a reality, however. Standards, architectures, methods, and tools are needed to lower the barriers for developing, linking, and validating federated models.

Different modeling purposes require different types of models. In the committee's judgment, the purpose of the model should drive the appropriate variables to be included in the model. To do this successfully requires a clear specification of model purpose and criteria for usefulness for that purpose, which in turn requires that model developers work closely with the eventual users of the model.

The committee also recommends validation for action, in which the purpose of the model drives its validation criteria. IOS models cannot be validated "in general"—they must be validated for a specific use. Research is needed with a cross-disciplinary community of interest to establish and promulgate accepted standards for validation of IOS models. Triangulation methods that combine expert judgment, qualitative results and theoretical work, and quantitative results should be further refined and more widely used. Common challenge problems and datasets are needed to facilitate docking of models for comparative purposes.

Finally, models of human beings and their individual and collective behaviors must necessarily include a large amount of inherent uncertainty. This uncertainty is not a flaw of the model and cannot be designed out of the model. Human behavior is dynamic and adaptive over time, and it is impossible at the moment (and into the foreseeable future) to make exact predictions about that behavior. What is needed are ways to estimate the probability of plausible outcomes and express those estimates in ways that are clear and meaningful to model users, who can then judge whether the results meet their needs. It is important also to avoid raising expectations about the capabilities of IOS models beyond what they can realistically deliver.

REFERENCES

Bennington, R.W. (1995, May). Joint simulation system (JSIMS): An overview. In *Proceedings of the IEEE 1995 National Aerospace and Electronics Conference (NAECON 1995)* (pp. 804–809). Available: http://ieeexplore.ieee.org/iel3/3912/11342/00522029.pdf [accessed Feb. 2008].

Borgatti, S.P., and Everett, M.G. (2006). A graph-theoretic framework for classifying centrality measures. *Social Networks, 28*(4), 466–484.

Bronowski, J. (1953). *The common sense of science.* Cambridge, MA: Harvard University Press.

Burton, R.M. (2003). Computational laboratories for organization science: Questions, validity and docking. *Computational and Mathematical Organization Theory, 9*(2), 91–108.

Cronbach, L.J. and Glaser, G.C. (1965) *Psychological tests and personnel decisions, second edition.* Urbana: University of Illinois Press.

Dibble, C. (2006). Computational laboratories for spatial agent-based models. In L. Tesfatsion and K.L. Judd (Eds.), *Handbook of computational economics, volume 2: Agent-based computational economics.* Amsterdam, The Netherlands: Holland/Elsevier.

Gauch, H.G., Jr. (2003). *Scientific method in practice.* Cambridge, England: Cambridge University Press.

Klein, K.J., and Kozlowski, S.W.J. (Eds.). (2000). *Multi-level theory, research and methods in organizations: Foundations, extensions, and new directions.* San Francisco: Jossey-Bass.

Office of the Secretary of Defense. (2004). *Training capabilities: Analysis of alternatives, final report, volume 1.* Washington, DC: Author.

Pratt, D., and Henninger, A. (2002). A case for micro-trainers. In *Proceedings of Interservice/ Industry Training, Simulation and Education Conference (I/ITSEC)*, Orlando, FL.

U.S. Department of Defense. (1996). *DoD high level architecture baseline approved.* (Report #537-96.) Washington, DC: Author.

11

Recommendations for Military-Sponsored Modeling Research

This report has reviewed the state of the art in individual, organizational, and societal (IOS) modeling and the ability of current modeling approaches to meet military needs; assessed the common pitfalls and problems associated with this type of modeling; and pointed out areas in which additional work is needed. This chapter summarizes the committee's recommendations for advancing behavioral modeling capabilities to meet the military's current and anticipated needs.

There are many challenges in advancing the science of human behavioral modeling. The theory on which to base the models is often fragmented and incomplete, failing to specify key links that are needed to answer the questions of interest. Data for testing theories and models (or for deriving empirically based models) are also sparse and often lacking in detail for exactly those factors that are critical for the model. Because of the scale of many behavioral models, it is rarely possible to generate useful data from controlled laboratory experimentation (as, for example, is often possible for models of individual cognition and behavior). Furthermore, there are often no well-defined criteria for success in these modeling efforts and no widely accepted definitions and methods for validation of IOS models. Finally, the research and development efforts are being conducted in many different disciplines. Modelers currently use different types of data, at different levels of detail, to model different types of behavior in order to answer different kinds of questions. Little effort is devoted at present to comparison or integration of models from different perspectives.

How, then, can this fragmented field best advance? Our recommendations focus on cross-disciplinary information exchange and the comparison and integration of models, structured around well-defined challenge problems and common datasets, with independent research thrusts recommended for those issues that are most critical.

Recommendations fall into three broad categories: (1) large-scale, integrated cross-disciplinary research programs, focused around representative challenge problems and common datasets; (2) research in six independent areas that will advance the capabilities to address these integrated problems; and (3) multidisciplinary conferences, workshops, and other information exchange forums, with attendees to include not only model developers but also government program managers and military decision makers.

INTEGRATED CROSS-DISCIPLINARY RESEARCH PROGRAMS

We suggest the funding of multiple large-scale, multiyear research programs that focus on comparing and, if appropriate, integrating models from different disciplines, different perspectives, and different levels of detail. This funding would provide incentives for researchers in diverse disciplines to work together on military-relevant problems. The goal would not be to pick the best model but rather to create a level playing field on which the capabilities of different approaches could be compared and the strengths of each assessed (see Gluck and Pew, 2005, for a description of a similar research program conducted for individual cognitive models). The ultimate goal is to move IOS modeling science forward through the process of comparison, docking, and integration.

It is essential for all participants in each program to focus on the same well-defined challenge problem instantiated in a common testbed and to use a common program dataset. At the heart of each program would be a representative problem that is critical for military operations, defined in detail. The five representative problems described in Chapter 2 provide a possible starting point for choosing the problems to be addressed.

The definition of challenge problems is a difficult but essential step for the recommended IOS modeling research program, and it should be the first step in such a program. Initial grants should fund challenge problem development, and continuation of the program should be contingent on success in defining these problems. Operational users must be involved in defining the challenge problems, and the criteria for modeling success should be clearly specified as part of problem definition. What type of model results—discovery, understanding, or forecasting—will be relevant for the problems being defined? What actions may be taken based on the model results? Criteria for model usefulness in the challenge problems should be clearly defined up front.

The research teams for these efforts should be multidisciplinary, and the program team should also include military users with operational experience in the domain for which the models are to be developed. These users will be ultimate judges of whether model results are useful (which we argue is the ultimate criterion for validation; see Chapter 8) and will provide advice on how the model results can best be presented for immediate comprehension and relevance. The use of a common challenge problem and a common testbed will facilitate the docking of the different models for purposes of comparison.

These integrated programs will encourage mutual education between modelers and operational users. Researchers will learn about the military domain and about user expectations. Users will learn about the scientific limitations to understanding of the basics of human behavior and what is feasible to represent in models (and implement in usable simulations) and will develop an understanding of the level of uncertainty associated with model forecasts or predictions. Results should be presented at workshops for program participants and other interested parties and at public conferences, published in the open literature for the research community at large, and presented "up the chain" to the program managers who rely on these models for operational, training, and mission rehearsal uses.

INDEPENDENT RESEARCH THRUSTS

In support of the integrated programs we recommend, we have identified six independent areas in which research is needed. Progress in each of these areas could increase the ability to develop the integrated modeling capabilities that are needed to address military problems. In each area, we suggest the funding of multiple research teams approaching the work from multiple perspectives, with periodic workshops for researchers to exchange results. We also suggest that operational users as well as government program managers participate in these workshops to draw on their areas of expertise and to gain better insight as to model capabilities and limitations. The funding structure of the programs should support and enable the participation of individual researchers or smaller laboratories in both academic institutions and industry and not be limited to large institutions, as is often the case in collaborative projects supported by the Department of Defense (DoD).

Thrust 1: Theory Development

Models should be conceptually correct and grounded in the underlying fundamentals of what is known about individual human and group social behavior. However, current theory in this area does not answer all of the questions needed to structure models that address relevant issues. Basic research is needed for theory development, especially for the low-level social behaviors (e.g., choosing friends) that are the building blocks for larger scale social behavioral patterns (e.g., joining a terrorist group). Since affective states and traits represent a key component of individual motivation and play a critical role in interpersonal behavior and group and organizational decision making, basic research in emotion and emotion-cognition interactions should be emphasized. This theory development work must involve multiple disciplines and perspectives with periodic workshops to exchange results.

Theory development challenge problems should be defined to guide the work, but these can be nonmilitary and need not involve the level of military detail necessary for the integrated problems discussed above. A series of workshops should be conducted with researchers to identify key theory gaps. We recommend working backward from a set of operational problems (as defined for the integrated programs) to identify areas in which lack of theory is impeding modeling progress. These theory gaps can be used to define theory challenge problems.

Academic institutions are key players for theory development, but they need information, incentives, and funding to address these theoretical issues. There is a need to educate academic researchers in military domains, establish conferences and journals in which their results can be presented, provide postdoctoral support, and provide funding that allows researchers to spend time learning about military domains in depth. Funding for graduate students is a key part of this thrust as a cost-effective way to bring about shared understanding and progress.

Thrust 2: Uncertainty, Dynamic Adaptability, and Rational Behavior

Models must deal with the inherent uncertainty (nondeterminism) and the dynamic adaptation (nonstationarity) that characterizes human behavior. Models must also be capable of modeling both rational and nonrational behavior.

Basic research is needed in each of these areas. Issues include

- How should models capture the "uncertainty-in-the-small" associated with individuals and small groups? How can model structures and parameters capture this variability, and how much of this variability must be included for the purposes of the model?
- How should models capture the "uncertainty-in-the-large" associated with populations and variations in population distributions? For example, to what extent should models be based on mean values versus capturing effects from the tails of a distribution? How much variability must be included for the purposes of the model?
- How can models capture adaptation and learning over time and in response to actions by others? For example, models of cultural

groups often assume that cultural identity is static and unitary. In fact, people have multiple overlapping identities and allegiances that vary in their influences over behavior. How can these be captured in a model, and how can one estimate the effects of actions and events on the primacy of these multiple allegiances as they affect decisions and actions?

- What are the factors that contribute to rational, adaptive behavior and what factors induce behavior that appears nonrational? Historically, emotions and affective factors have often been adduced to explain irrationality, but recent research in psychology and neuroscience has demonstrated that emotions also play a critical role in rational, adaptive behavior. Likewise, behaviors viewed as purely cognitive—including habit, bounded rationality, the range of beliefs unfamiliar to the observer, and ignorance, as well as behaviors with strong cognitive and affective components, such as fanaticism—can lead to what appears to be irrational behavior. Models of both rational and irrational behavior must therefore capture all the key factors—cognitive, affective, cultural, and contextual—that motivate and shape behavior of specific individuals in specific situations.

Better techniques are needed for understanding the implications of diversity and variability for model-based sensitivity analysis. Combinatorial explosion of possible combinations of parameters is a challenge, and better automated technology is needed to put the model through its paces to explore the parameter space effectively and produce robust results.

Thrust 3: Data Collection Methods

The difficulty of obtaining data is an ongoing challenge for IOS modeling. Research is needed to develop better data collection processes through field studies, experiments, and potentially by using massively multiplayer online games (MMOGs).

Although a variety of ethnographic data collection techniques are currently in use, they need to be better tailored to the needs of IOS models. For field data collection, it is necessary to bring modelers and data collectors together to develop data ontologies, joint specifications, and data collection methodologies and tools that are specifically tuned to IOS models.

MMOGs are an untapped resource for collecting social and behavioral data on a large scale. We recommend the creation of an MMOG facility that could serve as a testbed for exploratory research and model testing and the funding of basic research to determine if MMOGs can be used to test, verify, and validate IOS models. We recommend that funding be put into developing the science of MMOGs rather than in developing additional artificial worlds. The research agenda for this facility should be developed through workshops that convene both IOS modeling scientists and game experts. Note that funding MMOGs is a risky endeavor, with no guarantee that games that are useful for research purposes will find the widespread interest necessary for extensive data generation, but we think that the potential benefits outweigh the risks.

Given the critical role of emotions and affective personality factors in organizational decision making and behavior, it is also important to enhance the current methods for collecting affective data. Emotions and moods are notoriously difficult to assess accurately, particularly in naturalistic and field settings. Yet recent progress has been made in using multimodal approaches to affect assessment, including physiological monitoring and indirect assessment of these transient states via diagnostic tasks and performance tracking. We recommend that funding be allocated to the continued refinement of these methods and to the development of standardized assessment instruments, particularly in naturalistic settings.

Thrust 4: Federated Models

It is a fundamental conclusion of the committee that no single modeling approach can provide all the capabilities needed by DoD. We recommend a federated approach in which modeling components are created to be interoperable across levels of aggregation and detail. For example, a federated model might include a detailed representation of a few key individuals, linked to group-level models of different cultural groups and terrorist organizations, linked to geographic sector–level models of the level of unrest in a city. This approach is flexible and extensible, allowing the addition or subtraction of models at different levels of detail as needed for the problem to be addressed.

Combining model components to create federated models in the sense being recommended is not simply a matter of specifying and using interface-level syntactic compatibility protocols. It requires deep semantic interoperability (i.e., theoretical consistency). To create semantic interoperability, it is necessary to recognize that the links among components are themselves elements of the model. Components created at different levels of detail and for different purposes do not simply snap together to produce meaningful results.

Assuming that the interface protocol issues will be solved by others (e.g., enterprise database developers), research is needed to answer the following questions:

- What is the best way to ensure that the models being federated embrace compatible assumptions regarding concept abstractions, entity resolution, time scale resolution (tempo), uncertainty, adaptability, docking standards, input-output semantics, etc.?

- How should the components of the federated model be encapsulated, and which elements must be exposed to other components?
- How should specific classes of models be linked (e.g., cognitive models to social network models)?
- How can developers ensure dynamic extensibility?

These issues are not unique to IOS modeling. In addressing them, IOS modelers should maintain awareness of research and development in model federation in the larger modeling and simulation community.

Thrust 5: Validation and Usefulness

Current verification, validation, and accreditation (VV&A) concepts and practices were developed for the physical sciences, and we argue that different approaches are needed for IOS models. Specifically, we recommend the use of a "validation for action" approach that assesses the usefulness of a model for the specific purposes for which it was developed. Although promising work has been done in testing IOS models through triangulation among multiple types of expertise and multiple data sources, and some work has been done in docking different models for comparison, these approaches are not widespread. We recommend organizing a national workshop to agree on appropriate processes for VV&A of IOS models and to outline a roadmap for developing improved VV&A processes and standards. On the basis of the results of this workshop, we recommend that a DoD-wide authority develop and disseminate VV&A processes and standards for IOS models. These standards should be developed de novo, not as an adjunct to conventional VV&A standards.

Basing model validation on the usefulness of the model for specific problems requires that model purposes be clearly stated by model users and clearly understood by model developers. This is an area in which mutual education is needed. We suggest that, as part of developing a VV&A standard for IOS models, clear guidelines be developed for specifying model purpose.

Thrust 6: Tools and Infrastructure for Model Building

It is important to reduce the barrier to entry for developing models, modeling tools, frameworks, and testbeds. Scientists should be able to build and validate models without the large overhead currently associated with many DoD modeling and simulation investments. It should be possible to tailor existing models easily for specific purposes.

Sharing of IOS modeling knowledge across disciplines, as facilitated by the conferences and workshops recommended below, will support this goal. Work is also needed to develop an infrastructure for IOS modelers, including a national network of possible collaborators, common databases for model development and testing, and frameworks and toolkits for rapid model development. There is also a need for web-based repositories of information about existing models and, later, model components.

To facilitate the development and use of shared ontologies and model components, funding must also be allocated to the refinement of existing markup and modeling languages, as well as the development of new languages for particular domains or tasks.

The limited data that exist for IOS models are often not accessible to model developers. We recommend the funding of national web-accessible data repositories that are open to researchers who seek to inform and test models. For militarily relevant domains in which some data are classified, we recommend an investment in automated tools to sanitize potentially sensitive military data.

Often, the IOS models themselves are not readily accessible *or even known* to researchers or practitioners. Researchers are often unaware of efforts under way in DoD that are not reported on in conventional conferences and journals, and military developers are likewise unaware of progress being made in the research community. Or if they are, the typical user can face great difficulty in assessing the applicability of one approach or model over another, given their particular problem at hand. The occasional studies that attempt to survey the community and categorize development efforts and associated models, such as this one (see, e.g., Table 2-1 and Table 8-3) and its predecessor study (National Research Council, 1998), take small steps in this direction, but they are not meant to be exhaustive surveys and are only snapshots in time, become stale rather quickly, and fail to offer easy electronic access directly to the rich and evolving world of IOS models and associated simulations.

We therefore recommend the development and maintenance of an online web-based catalog of general approaches, models, simulations, and tools. The notion is to develop something along the lines of the Defense Modeling and Simulation Office's (DMSO's) Modeling and Simulation Resource Repository (MSRR) at http://www.msrr.dmso.mil/, maintained by the Modeling and Simulation Information Analysis Center, or the clearinghouse at Carnegie Mellon's CASOS site (see http://www.casos.cs.cmu.edu). But to be effective, the envisioned site needs careful consideration in terms of the following:

- *Organization:* The model ontology and site structure need to be carefully thought out, both from the researcher's perspective (e.g., foundational concepts underlying the particular model in the repository) and from the user's perspective (domain of application, limitations, simulation requirements, etc.).

- *Content:* Considerations need to be given to what is maintained on the site, ranging from simple descriptive abstracts to full-fledged downloadable simulations and "read me first" instructions.
- *Currency:* Once set up, the maintainers must devote effort to constantly updating the site, by tracking changes to existing models, adding new models that arise on the scene, and, certainly of equal importance, removing defunct models, or at least moving them to the archival section of the site to support historical surveys and the like. Failure to maintain currency will be the death knell of the repository, as it is with most websites today. One approach that should be considered is a Wikipedia-based model.
- *Usability:* The site design needs to ensure ease of use for all authorized visitors, including contributors, users, and occasional viewers. Procedures need to be in place to vet content modifications or additions, to support ease of navigation and internally searching for what the user is seeking, and to keep the site fresh and attractive to the larger community.

It is clear that this cannot be a one-time effort like DMSO's MSRR, nor an unfunded academic effort like Carnegie Mellon's CASOS site. It needs significant startup funding and continued support over its lifetime.

MULTIDISCIPLINARY CONFERENCES AND WORKSHOPS

A number of the issues and problems identified by the panel were the results of the failure of different disciplines to exchange information, or they resulted from misunderstandings among government funders of model development efforts, military users of models, and model developers. Because of the diversity of this group, there is no natural forum for them to exchange information, as there would be in conferences and journals for members of the same academic discipline or professional group. We therefore recommend the organization of special-purpose workshops around the integrated research programs recommended above as well as workshops for the independent research thrusts described above.

IOS modelers who are interested in working on military-relevant problems need to be educated on:

- The nature of the military decisions for which models are relevant and of the operational situations in which the decisions must be made.

- Desired model functionality.
- The most useful form(s) for presenting model results.
- The value of work performed by others outside their discipline.
- Feasible and appropriate VV&A approaches for IOS models.

Operational users and managers need to be educated on:

- The value of multidisciplinary approaches and the need for review of models from multiple perspectives.
- The inherent uncertainty associated with IOS model predictions.
- The value of models for sensitivity and trade-off analysis (versus the one right answer).
- The design of virtual experiments to assess results over a range of conditions.
- Reasonable definitions of validation for IOS models, feasible approaches for VV&A testing, and why these approaches differ from those used for physics-based models.

The recommended workshops should involve model developers, operational military users of the models, and government personnel who make funding decisions regarding model development. Issues to be discussed include methods for clearly specifying model purpose, criteria for judging the usefulness of models (i.e., what does it mean to validate a model), reasonable expectations for the certainty of model predictions, and methods for most clearly communicating model results.

ROADMAP FOR RECOMMENDED RESEARCH

The committee's recommendations are based on the concept of use-driven research. As defined by Stokes (1997), use-driven research combines elements from both basic and applied research. Like applied research, use-driven research seeks to solve a practical problem—in this case, the development of IOS models that can serve military purposes. But, like basic research, it also asks "why" in a fundamental way—Why do some methods work and others not work? What are the principles that underlie success or failure?

Figure 11-1 illustrates the major elements of a use-driven research program for IOS modeling. The process starts with challenge problem definition, which includes a clear specification of the use to which a model is to be put. This specification should be based on the needs of the model users, expressed in terms that are meaningful to the IOS modeling community. The challenge problem definition step is critical, and the funding of the remainder of the program should be contingent on its success. The

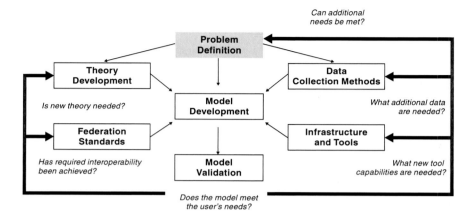

FIGURE 11-1 Elements of use-driven research for IOS modeling.

purpose of the model drives the theory to be applied, the data to be used, and the model development. Model development is made easier by modeling tools and infrastructure and relies on federation standards to ensure the interoperability of model components. Once the model is developed it is validated by asking the question: Is the model useful for its intended purpose?

As shown in Figure 11-1, the problem specification and model development process is cyclical. Based on the answers to the question "Is the model useful?" new models may need to be developed, new theory and new data (and new types of data) may be needed, and new interoperability standards, tools, and infrastructure may be required. Depending on the results, the problem itself may need to be redefined, clarified, or expanded.

Figure 11-1 lays out the areas in which research and development are needed for IOS modeling and shows how they are interdependent. Figure 11-2 organizes the suggested research areas into a roadmap that shows lines of activity and the interrelationships among them, repeating yearly in a cyclical fashion to advance the state of the art in meeting military IOS modeling needs.

As in all use-driven research, the recommended activities start with a clear definition of the purposes to which IOS models are to be put. We recommend that the initial activity for the program (the first six months) be spent on developing a clear definition for selected representative challenge problems (the problems listed at the end of Chapter 2 can provide a starting point) in close collaboration with operational military users. Concurrent with problem definition, the first six months should be spent in developing

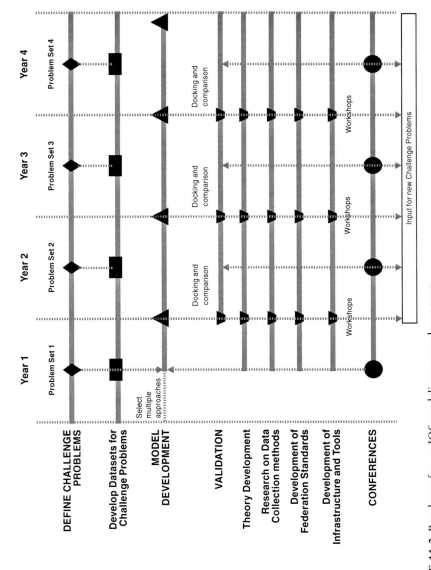

FIGURE 11-2 Roadmap for an IOS modeling research program.
NOTE: Only the first four years are shown.

datasets for these challenge problems. The challenge problems will provide common themes that tie together the diverse research and model development efforts.

There is currently no single approach that is clearly dominant for IOS modeling. Our recommendation is to select and fund a number of modeling teams that take different approaches for each challenge problem. In addition to the modeling teams, we recommend a series of specialized research thrusts focused on theory development, data collection methods, federation standards, and the development of infrastructure and tools. These thrusts will be aware of the challenge problems and will use the problems to focus their research, but their charter is broader and covers the entire field of IOS modeling.

The modeling teams and the research thrusts will come together in a conference at month 6, to learn about the challenge problems and the datasets associated with each problem. Conferences that involve the entire program will be scheduled yearly, with workshops for the individual research thrusts at the intervening 6-month intervals. The yearly conferences will also provide the forum for the presentation of new challenge problems, based on the results obtained in the prior year.

At the end of year 1, the models that have been developed for the challenge problems will be presented and discussed at a validation workshop, and docking and comparison activities will follow during the next 6 months, with results to be reported to the whole program at the yearly conference. These validation workshops should involve representative model users for each challenge problem. These users will assess the extent to which model results are useful for their intended purpose, as defined in the challenge problems. This process will repeat in subsequent years. As shown in Figure 11-1, the intention is that the results of the validation effort will inform all of the research thrusts as well as model development for the next cycle.

Although not shown in the timeline, it is assumed that a concurrent effort will be focused on the development and maintenance of an online web-based catalog of general approaches, models, simulations, and tools, as described earlier. This will serve not only as a repository of current theories and models, but also as a common record of the results of the execution of the roadmap.

The roadmap structure proposed in Figure 11-2 is intended to provide the field of IOS modeling with the common ground and forums for sharing information that will allow it to advance in a systematic way. Development and testing of models against a common set of challenge problems will avoid the current proliferation of specialized models for specialized purposes with no common framework for comparison and validation and therefore no foundation for scientific progress. Figure 11-2 shows the research cycle repeating over a four-year period, but we recommend that the program continue well beyond four years, with each year assessing the progress that has been made and increasing the complexity of the challenge problems based on the increasing capability of the modeling technology. New participants should be added to the funded programs and conferences each year, as new approaches and tools are developed and tested.

REFERENCES

Gluck, K.A., and Pew, R.W. (Eds.). (2005). *Modeling human behavior with integrated cognitive architectures: Comparison, evaluation, and validation.* Mahwah, NJ: Lawrence Erlbaum Associates.

National Research Council. (1998). *Modeling human and organizational behavior: Application to military simulations.* R.W. Pew and A.S. Mavor (Eds.). Panel on Modeling Human Behavior and Command Decision Making: Representations for Military Simulations. Commission on Behavioral and Social Sciences and Education. Washington, DC: National Academy Press.

Stokes, D.E. (1997). *Pasteur's quadrant: Basic science and technological innovation.* Washington, DC: Brookings Institution Press.

APPENDIXES

Appendix A

Acronyms and Abbreviations

MODELS, MODELING TOOLS, FRAMEWORKS

AASPEM	Advanced Air-to-air System Performance Evaluation Model
ACT-R	Adaptive Control of Thought–Rational
ADC	Air Defense Commander
AMBR	Agent-Based Modeling and Behavior Representation
BioWar	Biological Warfare agent-based model (ABM)
BNet	family of Belief Network tools
BRAHMS	Multiagent modeling and simulation environment
C3GRID	Command, Control, and Communications (C3) on Grid model
C3HPM	C3 Human Performance Model
C3TRACE	C3 Techniques for the Reliable Assessment of Concept Execution
CART	Combat Automations Requirements Testbed
CASTFOREM	Combined Arms and Support Task Force Evaluation Model
CBS	Corps Battle Simulation
CCTT	Close Combat Tactical Trainer
CLARION	Connectionist Learning with Adaptive Rule Induction ON-line
CLIPS	C Language Integrated Production System

CLOS	Common LISP Object System
Cmap	Concept Map tools
COGENT	Cognitive Objects within a Graphical EnviroNmentT
COGNET	Cognition as a Network of Tasks
CONNECT	A social network analysis tool for organizational modeling and simulation
Construct	A multiagent dynamic network model
CORES	Complex Organizational Reasoning System
CSSTSS	Combat Service Support Training Simulation System
DDD	Distributed Dynamic Decision-making
DIAS	Dynamic Information Architecture System
DISCUSS	A process simulation model of jury decision making
D-OMAR	Distributed OMAR (Operator Model Architecture)
DyNet	Dynamic Network (M&S tool)
EAAGLES	Enhanced Air-to-Air and Air-to-Ground Linked Environment Simulation
EADSIM	Extended Air Defense Simulation
EAGLE	Air-to-air combat simulation model
EMA	A computational appraisal model
EPIC	Executive Process/Interactive Control
FLAMES	Flexible Analysis Modeling and Exercise System
GRADE	Graphical Agent Development Environment
Graphvix	Graph Visualization Software
GTA-SA	Grand Theft Auto: San Andreas (a game)
HASMAT	Human and System Modeling and Analysis Toolkit
HLA	High-Level Architecture
HOS	Human Operator Simulator
IBC	Integrated Battle Command
ICET	Integrated Concept Evaluation Tool
ICEWS	Integrated Crisis Early Warning System
IDE	Integrated Development Environment
iGEN	COGNET
IMPRINT	Improved Performance Research Integration Tool
INTERMEDIATE	An anthropometry model
IUSS	Integrated Unit Simulation System
IWARS	Infantry Warrior Simulation

Jack	An anthropometric model for system design
JANUS	Physics-based ground combat simulator
JCATS	Joint Conflict and Tactical Simulation
JCM	Joint Conflict Model
JSAF	Joint Semi-Automated Forces
JSIMS	Joint Simulation Systems
JWARS	Joint Warfighting Simulation
MAMID	Methodology for Analysis and Modeling of Individual Differences
MASON	Multi-Agent Simulator of Networks
MATREX	Modeling Architecture for Technology Research and Experimentation
MicroPsi	Agent architecture founded on Psi theory
MicroSAINT	Microprocessor-based Systems Analysis of Integrated Networks
MIDAS	Man-machine Integration Design and Analysis System
MINDS	Modeling Individual Differences and Stressors
ModSAF	Modular Semi-Automated Forces
MTWS	Marine Tactical Warfare Simulation
NetLogo	A cross-platform multiagent programmable modeling environment
NetWatch	A multiagent model/framework
NM	SROM National Model
OCC	Ortony, Clore, and Collins emotion appraisal model
OCCAM	Organizational and Cultural Criteria for Adversary Modeling
OMAR/D-OMAR	Operator Model Architecture/Distributed OMAR
OneSAF	One Semi-Automated Forces
OOS	OneSAF Objective System
ORA	Organizational Risk Analyzer
OrgAhead	Computational Model of Organizational Learning and Decision Making
OrgSim	A multiagent model/framework
OTB	OneSAF Testbed
PCAS	Pre-Conflict Anticipation and Shaping
PMFServ	Performance Moderator Function Server
PRISM	Platform for the Representation of and Inference over Situation-theoretic Models
Ptolemy	System Dynamics Modeling Framework

RAID	Real-time Adversarial Intelligence and Decision-making
RDEBBSM	Crowd model based on diffusion kinetics
REPAST	Java-based framework for agent-based socioeconomic modeling
SAMPLE	Situation Awareness Model for Person-in-the-Loop Evaluation
SEAS	Synthetic Environments for Analysis and Simulation
SIAM	Situational Influence Assessment Model
Soar	Simulation of Adaptive Response
SPECTRUM	A sociocultural training system
SROM	Stabilization and Reconstruction Operations Model
STELLA	A simulation-based training environment for Information Operations, also a graphically oriented front end for the development of System Dynamics models
SWARM	Framework for agent- and individual-based modelers
TACBRAWLER	Tactical Air Combat Simulator
TACSIM	Tactical Simulation System
VISEO	Visible/Electro-Optical Detection Analysis System
VISTA	Visualization of Threats and Attacks
WARSIM2000	Warfare Simulation 2000
XCON	The eXpert CONfigurer program
Xerion	neural network simulator (also known as the University of Toronto simulator [UTS])

OTHER ABBREVIATIONS AND ACRONYMS

ABM	Agent-based models
AFAMS	Air Force Agency for Modeling and Simulation
AFOSR	Air Force Office of Scientific Research
AFRL	Air Force Research Laboratory
AI	Artificial intelligence
API	Application Programming Interface
ARI	Army Research Institute for the Behavioral Sciences
ARL	Army Research Laboratory

BBN	Bayesian Belief Network; also Bolt, Beranek, and Newman, Inc.
BCT	Brigade combat team
BDA	Battle damage assessment
BDI	Belief, desire, intention
BLOS	Beyond line of sight
C2	Command and Control
C3	Command, Control, and Communications
C4ISR	Command, Control, Communications, Computers, Intelligence, Surveillance, and Reconnaissance
CA	Civil Affairs
CASOS	Center for Computational Analysis of Social and Organizational Systems, located at Carnegie Mellon University
CCM	Cultural consensus model
CIC	Combat Information Center
CL	Computational laboratory
CMO	Civil Military Operations
CMYK	Cyan, magenta, yellow, black
CNO	Computer Network Operations
COA	Course of action
Cog-Aff	Cognitive-affective
COP	Constraint Optimization Problem
COTS	Commercial off the shelf
CPE	Commander's Predictive Environment
CPM	Critical Path Method
DARPA	Defense Advanced Research Projects Agency
D-COG	Distributed Cognition
DFT	Decision Field Theory
DIME	Diplomatic, information, military, and economic
DMO/MR	Distributed Mission Operations/Mission Rehearsal
DMSO	Defense Modeling and Simulation Office
DoD	Department of Defense
DSM	Design Structure Matrix
EBO/EBP	Effects-Based Operations/Effects-Based Planning
EM	A model of appraisal using OCC (see below) theory
ERGM	Exponential Random Graph Model
ES	Expert system
EW	Electronic warfare

FCS	Future combat system
FHA	Federal Housing Administration
FOB	Forward operating base
FPS	First person shooter (game)
GIG	Global information grid
GOTS	Government off the shelf
GUI	Graphical user interface
HBR	Human behavior representation
HIL	Human in the loop
ICCS	International Conference on Complex Systems
IDE	Integrated development environment
IED	Improvised explosive device
INTEL	Intelligence
IO	Influence operations; also information operations
IOS	Individual, organizational, and societal
JDEP	Joint Distributed Engineering Plant
JFCOM	Joint Forces Command
JSF	Joint Strike Fighter
JUO	Joint Urban Operations
LISP	List Processing (programming language)
LOS	Line of sight
M&S	Modeling and simulation
MAD	Mutually assured destruction
MAS	Multiagent systems
MC2C	Multisensor Command and Control Constellation
MCO	Major combat operations
MI	Military intelligence
MMOG	Massively multiplayer online game
MOOTW	Military operations other than war
MOUT	Military operations on urban terrain
MP	Military police
MSIAC	Modeling and Simulation Information Analysis Center
MSRR	Modeling and Simulation Resource Repository (at DMSO)
MUO	Major urban operations
NGO	Nongovernmental organization

NLOS	Nonline of sight
NMSO	Navy Modeling and Simulation Office
NRC	National Research Council
ONR	Office of Naval Research
OPS	Operations
OPSEC	Operations Security
OR	Operations Research
OSD	Office of the Secretary of Defense
PERT	Program Evaluation and Review Technique
PMESII	Political, military, economic, social, information, and infrastructure
PSYOP	Psychological Operations
QDR	Quadrennial Defense Review
R&D	Research and development
R&S	Reconnaissance and surveillance
RBA	Revolution in business affairs
RGB	Red, green, blue
RMA	Revolution in Military Affairs
RPD	Recognition-primed Decision Making
RPG	Rocket-propelled grenade
S&T	Science and Technology
SAB	(U.S. Air Force) Scientific Advisory Board, also USAF SAB
SAFIA	Expert system for controlling blast furnace operations
SASO	Stability and Support Operations
SCCS	Standard Cross-Cultural Survey
SDB	Soar DeBugger
SNA	Social network analysis
SOA	State of the art
SSC	(U.S. Army) Soldier Systems Center
STEAM	Shell for TEAMwork
UA	Unit of action
USECT	Understand, shape, engage, consolidate, and transition
VV&A	Verification. validation, and accreditation

WMD   Weapons of mass destruction

XML   Extensible Markup Language
XSLT   XSL Transformations

Appendix B

Exemplary Scenarios and Vignettes to Illustrate Potential Model Uses

To support the analysis effort and focus subsequent discussions of potential model utility, we present here a detailed scenario describing key operational aspects of a real-life scenario containing many of the Quadrennial Defense Review (U.S. Department of Defense, 2006) and Joint Urban Operations (JUO) considerations posed earlier. Researchers and model developers might believe that there are any number of scenarios available on which one might build one's analyses, but this is not the case. It is very difficult to find one that embraces all of the likely future combat conditions, since official publications state that realistic scenarios must include

- modernized industrial age forces with high-tech systems and more primitive paramilitary and insurgent forces;
- complex terrain and urban environments;
- failed states (the norm) with the internal society fractured and crime rampant;
- international interest/involvement in the region with nongovernmental organizations or information operations (IOs) engaged;
- national will at issue;
- use of IOs including media-mediated psychological operations (PSYOPS) and computer network operations;
- soft influences ongoing in parallel, including diplomatic, infrastructure, military, and economic activities;
- time criticality; and
- potential for inclusion of diverse missions.

GENERAL SETTING AND
FRIENDLY FORCE ORGANIZATIONAL STRUCTURE

The scenario elements included here are derivative of the one detailed in TRADOC PAM 525-3-90 O&O 22 JUL 2002 (U.S. Army, 2002) and include all these aspects. For purposes of this study, three vignettes have been extracted and distilled. The three vignettes provide a construct for the purpose of addressing potential of behavioral models supporting a brigade combat team (BCT) as part of a joint campaign. As stated in the TRADOC pamphlet: "They are presented for illustrative purposes only and are cast incidentally in the trans-Caucasus region to account for the realistic, tough range of variables and conditions, as well as the difficulty of the tactical dilemmas presented" (U.S. Army, 2002, p. F-1). The pamphlet, in its seven sections, provides a very detailed mission operational setting in the trans-Caucasus region (see Figure B-1). It includes three relevant vignettes:

FIGURE B-1 Trans-Caucasus region for TRADOC PAM 525-3-90 scenario. SOURCE: U.S. Army (2002, p. F-1).

1. tactical operations in entry operation (Entry),
2. operational maneuver by air, combined arms operation for urban warfare (Transition), and
3. secure portion of a major urban area (JUO).

The design purpose of these vignettes is to develop requirements, seek new tactical concepts, and seek new organizational design principles. The pamphlet emphasizes joint operations, and it explicitly describes new tactical principles based on development of the situation in and out of contact with the enemy. In addition, the trans-Caucasus region includes long-standing fault lines of bitter ethnic rivalry dating back millennia and thus supports strong components of scenario design for purposes of assessing particular behavioral model applications with religious, political, social, economic, and cultural impacts.

The nature of these "soft" regional factors emphasizes the need to appreciate and leverage political and informational domains to advantage.

The BCT will be the basic building block of future combat forces (U.S. Army, 2002). It will have the capability to command and control up to six maneuver battalions, will be able to employ a range of supporting capabilities, and will be able to perform a variety of missions, including reinforcing fires, engineers, military police air defense, PSYOPS, civil affairs, etc. The BCT will not be a fixed organization but "must be absolutely superior in complex situations where sophisticated political and informational skills are required in small unit leadership. Adversaries will leverage information, the media, and ethnic and religious fractures to maximum advantage" (U.S. Army, 2002, p. 21).

The BCT must have the ability to see, understand and act first, then finish decisively. Mid-grade and junior leaders must effectively recognize and solve problems in complex situations with political and informational dimensions. In the past, uncertainty about enemy and friendly conditions on the battlefield often dictated cautious movements, expenditure of time and resources to develop the situation, followed by initiation of decisive action at times and places not necessarily of the commander's choosing. The BCT will not be constrained in this way. Future commanders will develop the situation before making contact, maneuver to positions of advantage largely out of contact, and, when ready, initiate decisive action with initiative, speed, and agility.

The supporting 81-person military intelligence (MI) unit, organized as illustrated in Figure B-2, is an important component of the BCT. It is the primary focal point for management and analysis of information pulled from the full spectrum of intelligence, surveillance, and reconnaissance (ISR) resources. The MI company provides all of the brigade's timely, relevant, accurate, and synchronized intelligence, emitter mapping, electronic

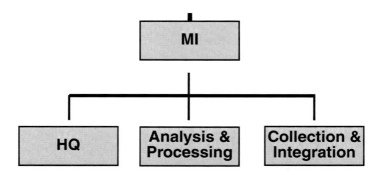

FIGURE B-2 Military intelligence unit organization.
SOURCE: U.S. Army (2002, p. 32).

attack, targeting information, and battle damage assessment support during the planning and execution of multiple, simultaneous decisive actions by means of information and intelligence collection, analysis, processing, integration, and dissemination. The purpose of this organization is analysis, fusion, and integration of ISR from external sources, organic UA R&S, combat battalion reconnaissance detachments, and troops in contact.

The MI unit has available to it ISR assets that are either organic (effectively owned and operated by the unit) or nonorganic (loaned to them for temporary use by sister or higher echelon units). The reliance on these two classes of assets changes over the course of an engagement, as illustrated in Figure B-3.

THREE PHASES OF THE SCENARIO

This scenario develops vignettes occurring during three phases of the scenario:

1. Entry: Combat forces enter the area of operations, Azerbaijan, and establish Forward Operating Base (FOB) Alpha.
2. Transition: Combat forces depart FOB and maneuver to Baku.
3. Major urban operations: Combat forces attack to seize Baku city center to facilitate its return to the host nation's control.

These vignettes are scaled back to depict only one BCT employed in combat operations. In this scenario, the BCT will conduct tactical operations in three distinct phases.

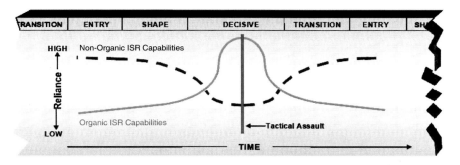

FIGURE B-3 Reliance on organic and nonorganic ISR assets over time.
SOURCE: U.S. Army (2002, p. 89).

Entry operations: The BCT uses military and commercial strategic lift to arrive in FOB Alpha ready to fight, fully synchronized with other elements of the joint force. For example, the BCT will have access to networked fires or "NetFires"[1] as soon as it touches down in the FOB.

This is a fundamental change in current approaches to deploying forces to theaters of operation. The future intent is also for intelligence already available from national and theater assets, as well as information on friendly forces, weather, and geospatial products provided through the global information grid, routed through the combat information centers, to be pushed directly to the BCT, allowing the commanders to do planning and rehearsals en route. When the FOB is secure, the BCT will enter the transition phase, a movement to contact, prior to entering their objective area, Baku.

Transition operations: Until recently, the operational significance of transition operations was underestimated. This attitude has changed: "Transitions—going from offense to defense and back again, projecting power through airheads and beachheads, transitioning from peacekeeping to warfighting and back again—sap operational momentum. Mastering transitions is key to winning decisively. Forces that can do so provide strategic flexibility to the National Command Authorities, who need as many options as possible in a crisis" (U.S. Army white paper, *Concepts for the Objective Force*, [quoted in U.S. Army, 2002, p. 61]). Operational

---

[1] "NetFires will enable the dynamic application of lethal and nonlethal destructive and suppressive effects. It will be integrated fully from the theater level to the tactical platform level, allowing the commander to establish, alter and terminate linkages between sensors and line-of-sight (LOS), beyond-line-of-sight (BLOS), non-line-of-sight (NLOS) division/corps and joint systems to achieve a wide set of lethal and nonlethal effects" (Haithcock, 2006, p. 25).

transitions are required as the force shifts from deployment operations, to smaller scale contingencies, to major combat operations. The transition from securing the FOB to the movement to contact at Baku will provide the enemy 304th Brigade with time and space to recover and attempt to exploit BCT vulnerabilities.

The BCT will plan and rehearse carefully to eliminate these dangerous transition areas. Because of its ability to keep situational understanding during a tactical operation, the BCT can transition immediately and aggressively to movement to contact. The BCT will initiate a series of deliberate attacks against a moving enemy under hasty conditions. Such an operation is graphically depicted in Figure B-4.

The enemy 304th Brigade will marshal all the resources available in the locale and use every means possible to disrupt, attrite, and destroy elements of the BCT. Hasty and deliberate attacks resembling cold war maneuvers, crowds laced with suicide bombers, attacks by fire, mines, and improvised explosive devices will be used by the enemy at every possible opportunity.

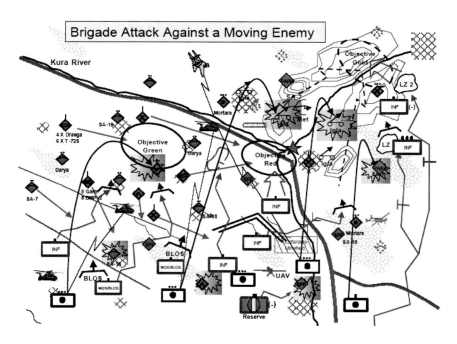

FIGURE B-4 BCT attack against a moving enemy.
SOURCE: U.S. Army (2002, p. 63).

During this phase the BCT will use three primary tenets—speed, precision, and knowledge—to successfully complete the transition in preparation for major urban operations (MUOs).

Major urban operations: The brigade's mission is to seize Baku city center in order to facilitate its return to host nation control. It will have made some preparation for MUO during the movement to contact and transition phases, but the less built-up areas encountered en route to Baku will bear very little resemblance to Baku itself.

Baku is a third-world city of 2 million composed of massed and heavy-clad framed buildings, which are dispersed in circular street patterns. Currently, the enemy is occupying company strong point defenses within the city, and they have activated terrorist cells and other paramilitary units to control critical areas. The Baku city center with BCT objectives is shown in Figure B-5.

Insurgent clans and terrorists will move to reinforce elements of the enemy 304th Brigade. The clans will "pile on" to join in the attrition of

FIGURE B-5 Baku city center.
SOURCE: U.S. Army (2002, p. F-19).

the BCT. In accordance with joint doctrine, "Close assault is a central aspect of urban engagements, both due to the nature of the terrain and enemy as well as the need to minimize collateral damage and preserve critical infrastructure. Small unit effectiveness and empowered leadership are critical to the success of these operations. Close urban assault has a significant dismounted character, requiring a robust infantry capability to engage and sustain the urban fight. . . . These units will exploit handheld and unmanned ISR tools and the common operational picture (COP). Target acquisition and engagement is difficult in the close confines of the urban environment. Fleeting targets can be acquired and killed using the BCT ISR capabilities and advanced weapons systems . . . The BCT must be able to sustain operational momentum through multiple battles by cycling forces in and out of contact" (U.S. Army, 2002, pp. E-2–E-3).

REFERENCES

Haithcock, J., Jr. (2006). Networked fires. *Field Artillery, Jan–Feb*, 22–27.

U.S. Army. (2002). *U.S. Army Training and Doctrine Command (TRADOC) pamphlet (Pam) #525-3-90, The United States Army objective force operational and organizational plan for maneuver unit of action*. Washington, DC: U.S. Army Training and Doctrine Command.

U.S. Department of Defense. (2006). *Quadrennial defense review report*. Washington, DC: Author.

Appendix C

Candidate DIME/PMESII Modeling Paradigms

A variety of modeling formalisms could be considered for DIME/ PMESII modeling efforts. We review some of them here.

Table C-1 compares selected modeling techniques by tabulating them against key characteristics which ultimately determine modeling utility. In the remainder of this appendix, we define the characteristics, and provide a very brief overview of each modeling formalism.

Expressivity of a modeling paradigm refers to its ability to capture and express an analyst's knowledge in terms of the constructs the paradigm offers. The expressivity of a concept graph is very high as it keeps the phrases used by the analysts intact in the model. In contrast, a neural network model is only able to keep the input-output relationships in the model. More expressive models are better able to capture the richness of PMESII domains and are typically easier to build, use, and understand by the modeler.

The *executable* feature of a modeling technique refers to whether some useful information that is implicit in a model (e.g., degree of influence of one variable onto another) can be derived from the model via some kind of inferencing algorithm. A causal graph, for example, is an executable paradigm as it offers propagation algorithms, and so also is a trained neural network. In contrast, the concept-mapping model does not have such an algorithm. Nonexecutable modeling techniques are useful for visualizing complex models for human understanding and analysis; executable models are useful for providing automated analysis of the models.

Reasoning of a modeling paradigm refers to the paradigm's ability to detect the direction of influence (not just connection) of one variable to

TABLE C-1 PMESII Modeling Paradigms and Their Characteristics

Characteristics: Modeling Paradigms:	Expressivity	Executable	Reasoning	Adaptability	Tools	Exemplary Products
Concept map	High	No	Forward Backward	Medium	Free (Research)	CmapTools (cmap.ihmc.us)
					COTS	Decision Explorer (www.banxia.com/demain.html)
Concept graph	Medium	No (Graphviz)	Forward Backward	Low	Free (Limited)	Graphviz (graphviz.org)
		Yes (OCCAM)			GOTS	OCCAM (http://www.cra.com/contract-r-d/cognitive-systems-occam.asp)
Social networks	Medium	Yes	Forward Backward	Medium	GOTS	OCCAM (cra.com/contract-r-d/cognitive-systems-occam.asp)
Causal graph	Medium	Yes	Forward Backward	Medium	COTS	BNet (cra.com/bnet)
					Free (Limited)	C4.5 (http://www.rulequest.com/Personal/)
System dynamics model	Medium	Yes	Forward Backward	Low	Free (Research)	Ptolemy (http://ptolemy.berkeley.edu)
Neural network	Low	Yes	Forward	High	COTS	NeuroSolutions (www.neurosolutions.com)
					Free (Research)	Xerion (www.cs.toronto.edu/~xerion/)
Situation theory	Medium	Yes	Forward	Low	In-House	PRISM (www.cra.com)

another. A belief network propagation algorithm, for example, incorporates both deductive and abductive reasoning, and thus is able to detect both forward and backward influences. On the other hand, the standard back propagation neural-network modeling paradigm is limited only to forward reasoning. Different modeling tasks require different kinds of reasoning. It is sometimes useful to be able to look at a state and reason about likely future outcomes (forward reasoning). For instance, one might want to attempt to predict the likelihood of social unrest by evaluating the current social, political, and economic state of affairs. Other times it is useful to look at externally available information and diagnose the likely underlying causes (backward reasoning). For instance, one might want to reason from observed social unrest back to the likely underlying political, economic, and social causes in order to properly address the causes of the unrest. For these reasons, it is important to support both forms of reasoning with the modeling tools we provide.

Adaptability of a modeling paradigm refers to automatic adjustments by models, which are necessary to take into account new observations. It is hard to adjust structures of graphical models as they are built in consultation with subject matter experts. But the strength of relationships among a set of variables within a model (e.g., probabilities in a belief network model or activation levels within a neural model) can be adjusted based on observations without changing their structure. Having models that can easily be adapted to represent new concepts and incorporate new data are generally preferable.

Tools of a modeling paradigm refers to the currently available software tools implementing the paradigm, that is, whether such a tool is commercial off the shelf (COTS), government off the shelf (GOTS), open source, or freely available for research/commercial purposes.

We now briefly describe the different modeling techniques shown in the table.

CONCEPT MAPS

Concept maps are a result of research into human learning and knowledge construction (Novak, 1998). In concept maps, the primary elements of knowledge are concepts, and relationships between concepts are propositions. Concept maps are a graphical two-dimensional display of concepts, connected by directed arcs encoding brief relationships (e.g., linking phrases) between pairs of concepts forming propositions. Each concept node is labeled with a noun, adjective, or short phrase, and each edge is labeled with verbs or verb phrases describing the relation between the connected concepts. Concepts maps are highly effective in quickly capturing domain knowledge along DIME/PMESII dimensions.

A popular tool for concept mapping is the CmapTools (Canas et al., 2004) package developed at the Institute for Human and Machine Cognition (see http://www.ihmc.us). The package is freely available for both commercial and noncommercial use, and has many advantages over using sticky notes or a more general diagramming tool (e.g., it can record the entire mapmaking process). There are also COTS tools that can be used, such as Banxia's Knowledge Explorer.

CONCEPT GRAPHS

Concept graphs are a formal system of logic based on the existential graphs of C.S. Peirce and semantic networks. Concept graphs explicitly represent entities/concepts and relationships between entities as nodes in a directed graph. They are mathematically precise and computationally tractable structures, which have a graphic representation that is humanly readable. For this reason, concept graphs have been used in a variety of applications for computer linguistics, knowledge representation, information retrieval, and database design. Their ease of use and generality make them immediately useful for modeling a wide variety of domains, including PMESII domains.

Figure C-1 is an example concept graph encoding a generic behavioral model of a terrorist leader.

SOCIAL NETWORKS

Social networks are similar to concept graphs, but they represent social structures. The nodes of the social network typically represent individuals

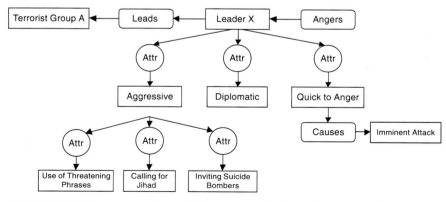

FIGURE C-1 Concept graph model for terrorist leader behavior.

and the links between them represent social relationships. Social network analysis (SNA) provides tools for reasoning about social networks, their strengths and weaknesses, the structural roles played by particular individuals, and their dynamics over time. Because of the focus on the analysis of social structures, SNA is directly applicable to a range of PMESII modeling tasks.

SNA tools can be extended in a number of directions. For example, one can build on traditional SNA functionality by providing additional representational and analytic power by having nodes representing not only individuals, but also arbitrary entities, especially including groups. Links can be similarly extended to represent not only individual-to-individual relationships, but also individual-to-group relationships (e.g., member-of) and group-to-group relationships (e.g., rival political party). By providing built-in Bayesian and rule-based reasoning capabilities, one could enable automated analysis of the graph. For instance, a Bayesian network might represent that members of a group might have a high probability of holding views that are promoted by that group, where the group, the individual, and the ideology are all represented in the network as nodes with appropriate links between them. In this case, an enhanced SNA tool could automatically create a new *believes* link between the individual and the ideology and annotate it with a particular probability.

CAUSAL GRAPHS

A *causal graph* (e.g., a belief network) (Jensen, 1996) is a graphical, probabilistic knowledge representation of a collection of variables describing some domain. The strength of causal graphs are their ability to represent both the causal structure of a domain and the probabilistic elements of those causal relationships (X causes Y with some probability), thus enabling the modeling of both qualitative and quantitative details of the model. In addition, the ability of causal graphs to handle both forward (causal) reasoning and backward (diagnostic or abductive) reasoning makes them especially well suited to domains with many sources of data, some of which are uncertain, unreliable, or potentially missing. Many PMESII modeling problems fall within such a scope.

Influence diagrams are a specialization of causal networks, augmented with decision variables and utility functions to solve decision problems. *Decision trees* are specialized influence diagrams that help to choose between options by projecting likely outcomes as utilities. Such extensions to causal graphs make it possible to also reason about the costs and benefits of possible decisions. This functionality can be used to both support intelligent decision making and to model likely decisions on the part of the entities being modeled.

Bayesian reasoning tools, such as those provided by Microsoft (MSBN; see http://research.microsoft.com/research/dtg/#bayesian), Norsys (Netica; see http://www.norsys.com/index.html), and Charles River Analytics (BNet; see http://www.cra.com), can support construction and reasoning with causal graphs. There are also other existing COTS solutions to modeling influence diagrams and decision trees, such as C4.5 (see http://www.rulequest.com/Personal/).

SYSTEM DYNAMICS MODELS

As described in Chapter 4, system dynamics models, such as the Stabilization and Reconstruction Operations Model (SROM) (Robbins et al., 2005) can be used to analyze the organizational hierarchy, dependencies, interdependencies, exogenous drivers, strengths, and weaknesses of a country's PMESII systems to enable more efficient resource expenditure. SROM models PMESII systems at the national and regional levels, including the interactions between regions. They also take into account demographic data, insurgent and coalition military, critical infrastructure, law enforcement, indigenous security institutions, and public opinion.

The SROM models developed by the AFRL/IF NO'EM group were built using the Ptolemy heterogeneous modeling software (see http://ptolemy.berkeley.edu), which is developed and supported by the Electrical Engineering and Computer Science department of the University of California, Berkeley. While developed primarily for modeling of real-time embedded systems, its heterogeneous processing model makes it an effective tool for integrating a variety of data processing algorithms.

NEURAL NETWORKS

A *neural network* is a nonlinear information-processing paradigm that models complex systems with a large number of highly interconnected processing elements (*a.k.a.* neurons or nodes), arranged in multiple layers, working in unison to solve specific problems. Neural networks offer some of the most versatile ways of mapping or classifying a nonlinear process or relationship. Neural networks have been successfully used in diverse paradigms, such as recognition of speakers in communications, diagnosis of hepatitis, recovery of telecommunications from faulty software, interpretation of multimeaning Chinese words, undersea mine detection, texture analysis, three-dimensional object recognition, hand-written word recognition, and facial recognition. Neural networks would be useful in building PMESII models for those domains that have highly complex nonlinear relationships between input and output variables.

A large number of neural network construction kits and runtime engines exist, including the Xerion tool from the University of Toronto (see http://www.cs.toronto.edu/~xerion/) and the NeuroSolutions tools from NeuroSolutions (see http://www.nd.com/products/nsv3.htm).

SITUATION THEORY

Situation theory models information processing and flow, that is, how an agent extracts information from the world and how it is subsequently transferred between agents. Situation theory provides a paradigm for describing the world, an ontology for representing it, and a suite of inferences for reasoning about it. Situation theory is unique in that it places situations alongside individuals, relations, and locations as first-class members of its ontology. Situations provide partial descriptions of the world in terms of the features individuated by some agent. They are defined in terms of the relationships they support; that is, they represent relationships between relationships. Situations provide a powerful representation of complex events spread over both space and time and, therefore, serve as a natural representation of a variety of PMESII models. Situation theory has been applied to a variety of fields including natural language understanding (Barwise and Perry, 1983), information visualization (Lewis, 1991), cooperative social interaction (Devlin and Rosenberg, 1991), and both Level 2 (Steinberg and Bowman, 2004) and Level 3 (Steinberg, 2005) data fusion.

REFERENCES

Barwise, J., and Perry, J. (1983). *Situations and attitudes.* Cambridge, MA: MIT Press.

Canas, A., Hill, G., Carff, R., Suri, N., Lott, J., Eskridge, T.C., Gómez, Arroyo, M., and Carvajal, R. (2004). CmapTools: A knowledge modeling and sharing environment. In *Proceedings of First International Conference on Concept Mapping* (pp. 125–133). Pamplona, Spain: Universidad Pública de Navarra.

Devlin, K., and Rosenberg, D. (1991). Situation theory and cooperative action. In *Proceedings of the Third Conference on Situation Theory and Its Applications,* Oiso, Japan.

Jensen, F.V. (1996). Bayesian networks basics. *AISB Quarterly, 94,* 9–22.

Lewis, C.M. (1991). Visualization and situations. In J. Barwise, J.M. Gawron, G. Plotkin, and S. Tutiya (Eds.), *Situation theory and its applications II* (pp. 553–580). Stanford, CA: CSLI.

Novak, J.D. (1998). *Learning, creating, and using knowledge: Concept maps as facilitative tools in schools and corporations.* Mahwah, NJ: Lawrence Erlbaum Associates.

Robbins, M., Deckro, R.F., and Wiley, V.D. (2005). *Stabilization and reconstruction operations model (SROM).* Presented at the Center for Multisource Information Fusion Fourth Workshop on Critical Issues in Information Fusion: The Role of Higher Level Information Fusion Systems Across the Services, University of Buffalo. Available: http://www.infofusion.buffalo.edu/ [accessed Feb. 2008].

Steinberg, A. (2005). An approach to threat assessment. In *Proceedings of the Eighth International Conference on Information Fusion, Volume 2*, Philadelphia, PA. Available: http://ieeexplore.ieee.org/iel5/10604/33513/01592001.pdf?isNumber= [accessed Feb. 2008].

Steinberg, A., and Bowman, C. (2004). Rethinking the JDL data fusion levels. In *Proceedings of National Symposium on Sensor and Data Fusion*. Available: http://www.infofusion.buffalo.edu/tm/Dr.Llinas'stuff/Rethinking%20JDL%20Data%20Fusion%20Levels_BowmanSteinberg.pdf [accessed Feb. 2008].

Appendix D

Biographical Sketches of Committee Members and Staff

Greg L. Zacharias *(Cochair)* is principal scientist at Charles River Analytics, Inc. He guides research in cognitive systems engineering and computational intelligence to support the development of intelligent agents for a broad range of systems applications, including decision support systems. Before cofounding Charles River, he was a senior scientist at BBN Technologies, a research engineer at C.S. Draper Labs, and an Air Force research attaché for NASA's space shuttle program at the Johnson Space Center. At the National Research Council, he was a member of the Committee on Human Factors from 1995 to 2007 and served on the Panel on Modeling Human Behavior and Command Decision Making. He has served as a technical reviewer for the Dayton Area Graduate Studies Institute, the Human Factors & Ergonomics Society Ely Award Committee for "annual best paper," and the Department of Defense's Human Systems Technology Area Review and Assessment Panel. He serves on the Air Force's Scientific Advisory Board (SAB) as a technical program reviewer for the Air Force Research Laboratory, in human effectiveness and information systems, and was a member of the Air Combat Command Advisory Group, and chair of the Human System Wing Advisory Group at Brooks Air Force Base. He has also chaired SAB studies in UAV mission management, and human systems integration in USAF Acquisition. He is on the board of the Embry-Riddle Research Advisory Board and the Small Business Technology Coalition. He has a Ph.D. from the Department of Aeronautics and Astronautics of the Massachusetts Institute of Technology.

Jean MacMillan *(Cochair)* is chief scientist at Aptima, Inc. Her 25-year research career has spanned a broad range of topics in human-machine interaction and user-centered system design, including adaptive instructional design, team decision making, command center design, and human performance measurement in simulation environments. At Aptima, she has served as the principal investigator for projects focusing on the development of reliable and valid performance measures for teams of F-16 pilots training in a distributed simulation facility, the design of synthetic entities that function as team members for simulation-based training of teamwork skills, the assessment of team performance and development of team performance measures for AWACS weapons director teams, and the development of optimized model-based manning reduction strategies that will allow the operation of Navy ships with a reduced number of personnel. She also codesigned a computer-based adaptive tutor for improving the reading comprehension of adult readers. She is a member of the Human Factors and Ergonomics Society and one of the founding members of its Cognitive Engineering and Decision-Making Technical Group. She has served on the editorial boards of the *Human Factors* and *Cognitive Engineering and Decision Making* journals. She has a B.A. from Antioch College, an M.C.P. from Harvard University, and a Ph.D. in cognitive psychology from Harvard University.

Holly Arrow is a member of both the Department of Psychology and the Institute for Cognitive and Decision Sciences at the University of Oregon. She studies the emergence and transformation of structure—including norms, influence hierarchies, and the cognitive networks of members—in small groups. She is coauthor (with Joseph McGrath and Jennifer Berdahl) of *Small Groups as Complex Systems: Formation, Coordination, Development, and Adaptation.* She is currently serving a second term as president of the Society for Chaos Theory in Psychology and Life Sciences. Current research projects include a computational model exploring the evolution of war and in-group altruism. She has a Ph.D. from the University of Illinois and did her initial training in complexity theory at the Santa Fe Institute Complex Systems Summer School (1995).

Stephen P. Borgatti is the Chellgren endowed chair of management at the Gatton School of Business at the University of Kentucky. During the writing of the report, he was an associate professor of organizational studies in the Carroll School of Management at Boston College. Previously he worked as a management scientist for a Boston-area consulting firm, specializing in modeling adoption of new consumer products. His research interests focus on how social networks constrain and enable the acquisition of individual beliefs, particularly the subset called knowledge. He is the past director of the National Science Foundation's Summer Institute for Ethnographic

Research Methods and the developer of ANTHROPAC, a computer program for cultural domain analysis. He is also the principal author of UCINET, a leading software package for social network analysis. He is a past president of the International Network for Social Networks Analysis (INSNA), the professional association for social network researchers, and is currently senior editor of *Organization Science*, as well as member of the editorial boards of several other journals. He has a B.A. in cultural anthropology (1977) from Cornell University and a Ph.D. in mathematical anthropology (1989) from the University of California, Irvine.

Richard M. Burton is professor of business administration in the Fuqua School of Business at Duke University. He is also professor of management at the European Institute for Advanced Studies in Management in Brussels and honorary professor at the University of Southern Denmark. His research focuses on organizational design, particularly its relationship to strategy for the firm. He has authored numerous articles and books on strategy, organization, and management science. *Strategic Organizational Diagnosis and Design: The Dynamics of Fit* (coauthored with Borge Obel) is in its third edition. With the associated software, OrgCon, the book provides an integrated theoretical and practical approach to organizational design for strategy implementation. He teaches executive M.B.A. courses in organizational design and international management. In the Ph.D. program, he teaches the theory course in organization theory and an advanced course in computational organization theory applications. He is active on a number of editorial boards and has been department editor for strategy, organizational design, and performance for *Management Science*. Currently, he is senior editor for *Organization Science*. He has B.S., M.B.A., and Ph.D. degrees, the latter in business administration, from the University of Illinois.

Kathleen M. Carley is professor in the Department of Social and Decision Sciences at Carnegie Mellon University. She specializes in research on organization theory, dynamic network analysis, social networks, multiagent systems, and computational social science. In her work, she examines how cognitive, social, and institutional factors affect individual, team, social, and policy outcomes. She is the author or coauthor of numerous books and articles in the areas of computational social and organizational science and dynamic network analysis. At the National Research Council, she served on the organizing committee for the Workshop on Statistical Analysis of Networks and the Panel on Modeling Human Behavior and Command Decision Making: Representations for Military Simulations. She is a member of the Academy of Management, Informs, the INSNA, the American Sociological Society, the American Association for the Advancement of

Science, and Sigma XI. In 2001 she received the lifetime achievement award from the sociology and computers section of the American Sociological Association. She is a founding and the current editor of *Computational and Mathematical Organization Theory*. She has a Ph.D. in sociology from Harvard University.

Catherine Dibble is assistant professor in the Department of Geography at the University of Maryland. She uses agent-based computational laboratories to model such processes as long-run regional development and land use changes, epidemics among highly mobile populations, and the effects of spatial technology networks on the evolution of inequality. She has served on the International Steering Committee for the GeoComputation Conference Series since its inception in 1996 and is an experienced designer of spatial evolutionary algorithms and relevance filters. Her academic background includes graduate work in formal microeconomic theory and international trade theory, public finance (which includes theory and practice relating to overuse of shared resources, externalities, and public goods), and game theory (especially axiomatic bargaining theory, a formal structure for evaluating fairness in resource allocation problems) in the Department of Economics at the University of Rochester. Her professional experience includes computer science and many years as a professional software designer and developer (simulations, executive information systems, and national and international software patents). She has an M.A. in economics from the University of Rochester and a Ph.D. in geography from the University of California, Santa Barbara.

Eva Hudlicka is a principal scientist and president of Psychometrix Associates, Inc. Prior to founding Psychometrix Associates in 1995, she was a senior scientist at Bolt, Beranek, and Newman. Her research focuses on cognitive modeling, with emphasis on the development of computational models of individual differences and emotion in cognitive architectures. Key features of these models are the explicit representation of affect appraisal and the effects of emotion and personality on decision making and performance. Her prior research included user interface design, decision support system design, and knowledge elicitation. She has authored several book chapters, organized a number of workshops focusing on emotion modeling, and served on conference program committees. She was a guest editor for a special issue of the *International Journal of Human-Computer Studies*, focusing on affective computing. She is a member of the editorial board of *PRESENCE*. She has a B.S. in biochemistry from Virginia Polytechnic and State University (1977), an M.S. in computer science from the Ohio State University (1979), and a Ph.D. in computer science from the University of Massachusetts, Amherst (1986).

Jeffrey C. Johnson is a senior scientist at the Institute for Coastal and Marine Resources and professor in the Department of Sociology and adjunct in the departments of Anthropology, Biology, and Biostatistics at East Carolina University. He has been active in research projects funded by Sea Grant and the National Oceanic and Atmospheric Administration for more than two decades. He has conducted an extensive long-term research project supported by the National Science Foundation comparing group dynamics of the overwintering crews at the American South Pole Station with those at the Polish, Russian, Chinese, and Indian Antarctic Stations. In addition, he is interested in network models of complex biological systems and is currently working with several ecologists on the examination of problems associated with trophic dynamics in food webs. His recent work involves the development and testing of cognitive models of Inupiaq understandings of the Kotzebue Sound ecosystem in the Arctic. He has published extensively in anthropological, sociological, and marine science journals. The founder and former editor-in-chief of the *Journal of Quantitative Anthropology* and the current coeditor of the journal *Human Organization*, he is also the author of *Selecting Ethnographic Informants*. He has a Ph.D. in social science from the University of California, Irvine.

Scott E. Page is professor of complex systems, political science, and economics at the University of Michigan and an external faculty member of the Santa Fe Institute. He is the principal investigator on a study of computational modeling in the social sciences. He is the author of numerous papers on the applications of mathematical and computational modeling to questions in the social sciences. His most recent work, which appeared in the *Proceedings of the National Academy of Sciences*, is on the aggregative properties of individual-level diversity. He has an undergraduate degree from the University of Michigan, an M.S. in mathematics from the University of Wisconsin, and a Ph.D. in managerial economics and decision sciences from Northwestern University.

Andrew P. Sage is founding dean emeritus, university professor, and First American Bank professor in the Department of Systems Engineering and Operations Research at George Mason University. His interests include systems engineering and management efforts in a variety of application areas, including systems integration and architecting, reengineering, engineering economic systems, and sustainable development. He is an elected fellow of the Institute of Electrical and Electronics Engineers (IEEE), the American Association for the Advancement of Science, and the International Council on Systems Engineering. He is editor of the John Wiley textbook series on systems engineering and management, the INCOSE Wiley journal *Systems Engineering,* and is coeditor of *Information, Knowledge, and Systems*

Management. In 1994 he received the Donald G. Fink Prize from the IEEE and a public service award for his service on the CNA Corporation Board of Trustees from the U.S. Secretary of the Navy. In 2000, he received the Simon Ramo Medal from the IEEE in recognition of his contributions to systems engineering and an IEEE third millennium medal. He was elected to the National Academy of Engineering in 2004 for contributions to the theory and practice of systems engineering and systems management. He has a B.S.E.E. degree from the Citadel, an S.M.E.E. degree from the Massachusetts Institute of Technology, and a Ph.D. (1960) from Purdue University.

Leigh S. Tesfatsion is professor of economics and mathematics in the Department of Economics at Iowa State University. Her current research focuses on agent-based computational economics (ACE), the computational study of economic processes modeled as dynamic systems of interacting agents. Her particular interest is the development of empirically based ACE frameworks for the study of restructured electricity markets. She currently serves as an editorial board member for Edward Elgar's *New Dimensions in Networks* book series. She is also an associate editor (or editorial board member) for the *Journal of Economic Dynamics and Control*, the *Journal of Economic Interaction and Coordination*, the *Journal of Energy Markets*, and *Applied Mathematics and Computation*. She is active in several IEEE working groups and task forces, the Society for Computational Economics (SCE), the SCE special interest group on ACE. She is a participating faculty member in the Graduate Program on Human-Computer Interaction and an affiliate faculty member of the Center for Computational Intelligence, Learning and Discovery. She has a Ph.D. in economics, with a minor in mathematics, from the University of Minnesota, Minneapolis (1975).

Susan B. Van Hemel *(Study Director)* is a senior program officer in the Division of Behavioral and Social Sciences and Education at the National Research Council. She currently manages a study of developmental outcomes and assessments for young children for the Administration for Children and Families of the U.S. Department of Health and Human Services. Previous projects at the National Research Council include a study of staffing standards for aviation safety inspectors at the Federal Aviation Administration, studies of Social Security disability determination for individuals with visual and hearing impairments, and workshops on technology for adaptive aging and on decision making in older adults. She has also done work for a previous employer on vision requirements for commercial drivers and on commercial driver fatigue. For over 25 years she has managed and performed studies on a variety of topics related to human performance and training. She is a member of the Human Factors and Ergonomics Society and its technical groups on perception and performance and aging. She has a Ph.D. in experimental psychology from the Johns Hopkins University.

Michael J. Zyda is director of the GamePipe Laboratory, a professor of engineering practice in the Department of Computer Science, and a staff member of the Information Sciences Institute at the University of Southern California. He was the founding director of the MOVES Institute, located at the Naval Postgraduate School in Monterey, California, and a professor in the Department of Computer Science there as well. His research interests include computer graphics, large-scale, networked 3-D virtual environments, agent-based simulation, modeling human and organizational behavior, interactive computer-generated stories, computer-generated characters, video production, entertainment/defense collaboration, modeling and simulation, and serious and entertainment games. He has a lifetime appointment as a national associate of the National Academies, an appointment made by the Council of the National Academy of Sciences in November 2003, awarded in recognition of "extraordinary service" to the National Academies. He is a member of the Academy of Interactive Arts and Sciences. He served as the principal investigator and development director of the PC game America's Army, which transformed Army recruiting. He has a B.A. in bioengineering from the University of California, San Diego, an M.S. in computer science from the University of Massachusetts, Amherst, and a D.Sc. in computer science (1984) from Washington University, St. Louis.